D1293341

HEPATITIS VIRUSES

edited by

J.-H. James Ou
University of Southern California

KLUWER ACADEMIC PUBLISHERS
Boston / Dordrecht / London

Distributors for North, Central and South America:
Kluwer Academic Publishers
101 Philip Drive
Assinippi Park
Norwell, Massachusetts 02061 USA
Telephone (781) 871-6600
Fax (781) 681-9045
E-Mail <kluwer@wkap.com>

Distributors for all other countries:
Kluwer Academic Publishers Group
Distribution Centre
Post Office Box 322
3300 AH Dordrecht, THE NETHERLANDS
Telephone 31 78 6392 392
Fax 31 78 6546 474
E-Mail <services@wkap.nl>

Electronic Services <http://www.wkap.nl>

Library of Congress Cataloging-in-Publication Data

A C.I.P. Catalogue record for this book is available
from the Library of Congress.

Printed on acid-free paper.
Printed in the United States of America

The Publisher offers discounts on this book for course use and bulk purchases. For further information, send email to <joanne.tracy@wkap.com> .

This book is dedicated to hepatitis patients and their loved ones.

TABLE OF CONTENTS

LIST OF CONTRIBUTORS

Keril J. Blight, Ph.D.
Center for the Study of Hepatitis C, The Rockefeller University, Box 64, 1230 York Avenue, New York, NY 10021

Jinah Choi, Ph.D.
Department of Molecular Microbiology and Immunology, University of Southern California, Keck School of Medicine, 2011 Zonal Avenue, Los Angeles, CA 90033

Michael P. Curry, M.B., MRCPI.
Department of Medicine, Beth Israel Deaconess Medical Center and Harvard Medical School, One Deaconess Rd., Boston, MA 02215

Arash Grakoui, Ph.D.
Center for the Study of Hepatitis C, The Rockefeller University, Box 64, 1230 York Avenue, New York, NY 10021

Robert K. Hamatake, Ph.D.
ICN Pharmaceuticals, 3300 Hyland Avenue, Costa Mesa, CA 92626

Holly L. Hanson, Ph.D.
Center for the Study of Hepatitis C, The Rockefeller University, Box 64, 1230 York Avenue, New York, NY 10021

Zhi Hong, Ph.D.
ICN Pharmaceuticals, 3300 Hyland Avenue, Costa Mesa, CA 92626

Margaret James Koziel, M.D.
Department of Infectious Diseases, Beth Israel Deaconess Medical Center, Associate Professor of Medicine, Harvard Medical School, One Deaconess Rd., Boston, MA 02215.

Michael M. C. Lai, M.D., Ph.D.
Department of Molecular Microbiology and Immunology, and Howard Hughes Medical Institute, University of Southern California Keck School of Medicine, 2011 Zonal Avenue, Los Angeles, CA 90033

Johnson Y. N. Lau, M.D.
ICN Pharmaceuticals, 3300 Hyland Avenue, Costa Mesa, CA 92626

Thomas P. Leary, Ph.D.
Virus Discovery Group, Abbott Diagnostic Division, Dept. 90D, Bldg. L3, North Chicago, IL 60064-6269

Stanley M. Lemon, M.D.
University of Texas Medical Branch, Galveston, TX 77555

Annette Martin, Ph.D.
Institut Pasteur, Paris, France

Thomas B. Macnaughton, Ph.D.
Department of Molecular Microbiology and Immunology, University of Southern California Keck School of Medicine, 2011 Zonal Avenue, Los Angeles, CA 90033

Isa K. Mushahwar, Ph.D.
Distinguished Research Fellow, Abbott Laboratories, North Chicago, IL 60064-4000

Jing-hsiung James Ou, Ph.D.
Department of Molecular Microbiology and Immunology, University of Southern California, Keck School of Medicine, 2011 Zonal Avenue, Los Angeles, CA 90033

Gregory R. Reyes, M.D., Ph.D.
Infectious Diseases and Tumor Biology, Schering-Plough Research Institute, 2015 Galloping Hill Road, Kenilworth, NJ 07033

Charles M. Rice, Ph.D
Center for the Study of Hepatitis C, The Rockefeller University, Box 64, 1230 York Avenue, New York, NY 10021

Jack R. Wands, M.D.
Liver Research Center, Rhode Island Hospital, Department of Medicine, Brown University School of Medicine, 55 Claverick Street, 4th Floor, Providence, RI 02903

Marcus W. Wiedmann, M.D.
Liver Research Center, Rhode Island Hospital, Department of Medicine, Brown University School of Medicine, 55 Claverick Street, 4th Floor, Providence, RI 02903

T. S. Benedict Yen, M.D., Ph.D.
Pathology Service 113B, San Francisco Veterans Affairs Medical Center, 4150 Clement Street, San Francisco, CA 94121, and Department of Pathology, University of California, San Francisco, CA 94143-0506

Weidong Zhong, Ph.D.
ICN Pharmaceuticals, 3300 Hyland Avenue, Costa Mesa, CA 92626

FOREWORD

The hepatitis viruses are a major cause of human illness worldwide. They belong to several different families and differ in their mode of replication and transmission, but they have in common a pronounced tropism for the liver, thereby causing a similar disease, hepatitis. In fact, the disease symptomology is so similar that they cannot be differentiated on this basis alone, and they have been named hepatitis A virus, hepatitis B virus, etc., more or less in the order of their discovery. In textbooks, these viruses are usually described together with other members of the families to which they belong, which may cause quite different diseases. The current book deals only with the various hepatitis viruses, which puts emphasis on the disease process rather than the evolutionary history of the viruses.

Hepatitis A virus (a picornavirus, the family that also includes, among others, the rhinoviruses, causative agents of the common cold, and poliovirus, causative agent of poliomyelitis) and hepatitis E virus (distantly related to the caliciviruses, some of which cause colds or gastroenteritis in humans, and at one time classified in the calicivirus family) cause acute hepatitis with, usually, low mortality, from which recovery is complete. In contrast, hepatitis B virus (a hepadnavirus, distantly related to the retroviruses, which include human immunodeficiency virus) and hepatitis C virus (belonging to the flavivirus family and distantly related to yellow fever virus and the dengue viruses) often result in lifelong chronic infection that may lead to cirrhosis or hepatocellular carcinoma after many years. An estimated 8-10% of the world's population is chronically infected by one or the other of these viruses, and these infections are responsible for the great majority of human liver cancers. Hepatitis D virus is not technically a true virus, but a subviral agent related to the viroids responsible for much plant disease. It is a parasite of hepatitis B virus, undergoing a full replication cycle only in cells infected with this virus. Chronic infection with both B and D results in more severe hepatitis. In addition to these now well characterized viruses, several other viruses have been described recently that may cause hepatitis in humans. The possible importance in human disease of such viruses, which include a virus often called hepatitis G virus, is not yet clear.

It is interesting that the hepatitis viruses include no true DNA virus. All have RNA genomes except for hepatitis B virus, whose DNA genome replicates through an RNA intermediate. Although some DNA viruses, as well as some RNA viruses not classified as hepatitis viruses, can and do replicate in the liver, the primary symptomology of disease caused by these viruses is not hepatitis. It is the pronounced tropism of the hepatitis viruses for the liver to the virtual exclusion of replication in other tissues that sets them apart.

Because of the great importance of the hepatitis viruses in causing serious human illness, this volume is a welcome addition to the literature of viruses and human disease. I take some pride in its appearance because two of the principals, James Ou and Charles Rice, are former students of mine, and, thus, I can take some credit, albeit quite distant credit, for it. James joined my laboratory at Caltech as a graduate student in the late 70s, where he studied the molecular biology of alphaviruses. He went on to become deeply involved in the hepatitis viruses and has been an important contributer in this area. Charlie was a graduate student at about the same time, and studied not only alphaviruses but also began the study of flaviviruses in my laboratory. He has gone on to become a leader in the study of hepatitis C virus and has recently founded the Center for the Study of Hepatitis C Virus at Rockefeller University. I am also well acquainted with several of the other principals involved and have followed their contributions to the literature of the hepatitis viruses for years. All of the authors have been well chosen for their knowledge and expertise in the area of the hepatitis viruses, ensuring that this volume is an accurate and up to date treatment of what is known about this important group of pathogens. The book will be particularly useful for those interested in hepatitis and its long term sequellae, and to those interested in the disease potential of viruses and mechanisms of viral pathogenesis.

James H. Strauss, Ph.D.
Ethel Wilson and Robert Bowles
Professor of Biology
California Institute of Technology
Pasadena, CA

FOREWORD

The advances in molecular biology over the last two decades have provided unprecedented opportunities in biomedical research and revolutionized the way we diagnose, manage and treat diseases. This "molecular revolution" has invariably emboldened the scientific progress in the field of viral hepatitis. The cloning of hepatitis B virus genome in the late 1970s is one of the first few human pathogens whose complete genome was characterized molecularly. The cloning and characterization of hepatitis A and D viruses soon followed. Over the next decade, the molecular technology has evolved to the extent that a pathogen can be identified and characterized without ever being detected and visualized by conventional techniques. This change in paradigm has culminated in the cloning of hepatitis C and E viruses that have eluded the scientific community for many years.

The approaches to study the various forms of viral hepatitis have also benefited from the rapid advances in molecular biology. For years, scientists working on viral hepatitis have faced a daunting challenge in advancing the field, that is, how can one study a virus that does not replicate easily in tissue culture and that has a rather restricted tissue and species tropism? Again the advances in molecular techniques have spurred the establishment of alternative model systems. Using these model systems, great strides have been made in understanding the viral life cycle and mechanisms of viral pathogenesis. The molecular technology also has its immeasurable imprint on the development of vaccine and therapy for viral hepatitis. Hepatitis B vaccine is the first vaccine that was developed using recombinant DNA techniques. This widely successful vaccine bespeaks the power of molecular biology in improving the quality of people's lives. Similarly, various therapeutic modalities against the hepatitis viruses are being pursued based on what we have learned about the molecular pathways of viral replication.

In this book, the triumphs of molecular biology in viral hepatitis are amply illustrated. The virology and pathogenesis of each hepatitis virus is covered in a series of comprehensive chapters with exhaustive review of the literature. The strength of this book lies in the impact of molecular biology on the current state of knowledge in viral hepatitis. This cumulative knowledge about each hepatitis virus as a result of advances in molecular biology will undoubtedly pave the path for our continual success in controlling this global public health problem.

T. Jake Liang, M.D.
Liver Diseases Section
National Institute of Diabetes,
Digestive and kidney Diseases
National Institutes of Health
Bethesda, MD

FOREWORD

Since the description of the so-called "serum" and "infectious" forms of viral hepatitis in the 1950's and 1960's, unrelenting progress has been made in our understanding of the etiological agents involved in this very common , global disease of mankind. Today, we now appreciate the existence of at least five very different infectious agents whose primary target organ is the liver - the hepatitis viruses A, B, C, D & E. While they all cause similar clinical symptoms, they each represent a very different type of viral agent with varying transmission modes and outcomes of infection. Hepatitis B, C & D are caused by blood-borne agents that can result in both acute and chronic hepatitis. Persistence of these viruses contributes greatly to the global burden of chronic liver diseases including chronic hepatitis, liver cirrhosis and hepatocellular carcinoma. Hepatitis A and E on the other hand, are usually caused by ingestion of contaminated food or water supplies and typically result in acute, self-limiting infections. Hepatitis A and E usually occur therefore in developing countries where sanitation is poor. A wide spectrum of disease is associated with these infectious agents varying from asymptomatic through fatal infections. Hepatitis B & C are associated with most of the global mortality rates associated with viral hepatitis although hepatitis E can often be fatal in pregnant women.

Our acquired knowledge of these infectious agents has led in turn to considerable success in controlling them within the human population. They can now be accurately serodiagnosed and the blood supply is now very safe in developed countries. Very effective vaccines are available for hepatitis A & B and since the hepatitis D agent is a defective viral agent that requires the hepatitis B virus as a helper virus, effective control over HDV can also be exerted through the use of the hepatitis B vaccine. In addition, animal studies suggest that a vaccine for HEV may not be that far away and although considered historically to be a very difficult vaccine target, we now have real hope that a vaccine may be possible for HCV. On the therapeutic front, partially effective but nonetheless very valuable drugs are available for hepatitis B & C and much work is in progress on new and better treatments. We can therefore be hopeful that in the first half of the 21st century, we will have very effective control of this historic scourge of mankind. This comprehensive volume brings together our current knowledge of the molecular virology of all of the viral hepatitis agents including information on their natural histories, host interactions ,immunology, detection and treatments. In addition, animal viruses related to the hepatitis C virus are also discussed because of their value as models for their human relative. As such, this volume will be invaluable to viral and medical practitioners as well as to the student and general community.

Michael Houghton, Ph.D.
Chiron Corporation
Emeryville, CA

PREFACE

The research with hepatitis viruses started more than fifty years ago. The names of hepatitis A and hepatitis B were introduced in 1947 when it became clear that there were two types of hepatitis that were transmitted either enterically or parenterally. It became apparent in the 1970's that there were additional hepatitis viruses distinct from hepatitis A and hepatitis B, and hence the term non-A, non-B hepatitis was introduced. The non-A, non-B hepatitis was further divided into post-transfusion non-A, non-B hepatitis and enterically-transmitted non-A, non-B hepatitis in the 1980's. By the end of the 1980's, both post-transfusion non-A, non-B virus and enterically-transmitted non-A, non-B virus had been identified and renamed hepatitis C virus and hepatitis E virus, respectively. Hepatitis delta antigen was first recognized as an antigen associated with hepatitis B virus infection in the 1970's. In the early 1980's, a virus was isolated and named hepatitis delta virus. These five different hepatitis viruses have distinct replication pathways and are major heath concerns. The infections by hepatitis A virus and hepatitis E virus are prevalent in underdeveloped countries, and at least 400 million people in the world are chronic carriers of hepatitis B virus, hepatitis C virus and hepatitis delta virus. Many of the chronic hepatitis patients are infected by two or more hepatitis viruses.

The goal of this book is to provide a comprehensive up-to-date review of these hepatitis viruses to the readers. The first chapter of this book provides an overview that describes the epidemiology and the natural history of infection of these viruses. This chapter is aimed at readers of diverse backgrounds. Chapters 2-6 review the molecular biology of individual hepatitis viruses. Chapter 7 reviews the molecular biology of GB-viruses. GB virus-C was also named hepatitis G virus. Although GB viruses are no longer considered to be major causative factors of viral hepatitis, they are distantly related to hepatitis C virus. In particular, GB virus-B, which can cause hepatitis in non-human primates, has become a surrogate model for studying hepatitis C virus. While the focus of these chapters is on the molecular biology of the virus, some other topics are also discussed. For example, in the hepatitis E virus chapter, extensive discussions on viral pathogenesis and diagnostics are included. Chapter 8 examines the host-virus interactions and the immunobiology of hepatitis viruses and Chapter 9 discusses the hepatocellular oncogenesis induced by hepatitis viruses. The final chapter discusses the antiviral agents for hepatitis viruses and their current states of research.

A tremendous amount of research information about hepatitis viruses has been generated during the past two decades. Although there are many books that address individual hepatitis viruses and the clinical aspects of viral hepatitis, books that discuss all the hepatitis viruses are scarce. This book is to fill that vacuum.

The completion of this book required a tremendous amount of work. It would not have been possible without the help of my colleagues who contributed individual chapters. All of them are experts and well-respected scientists in their respective research areas. The completion of this book would also not have been possible without the clerical help of Dr. Jinah Choi and Ms. Anne Strohecker and the moral support of my beloved family members: my wife Susan Ker-hwa, my daughter Elaine and my son Evan.

Jing-hsiung James Ou, Ph.D.
Professor of Molecular Microbiology and Immunology
University of Southern California
Los Angeles, CA

Chapter 1

HEPATITIS VIRUSES: THE NATURAL HISTORY OF INFECTION

Jinah Choi and Jing-hsiung Ou*

Department of Molecular Microbiology & Immunology, University of Southern California Keck School of Medicine, 2011 Zonal Avenue, Los Angeles, California 90033

1. INTRODUCTION

Hepatitis viruses are important human pathogens that affect millions of people worldwide. These viruses, as their names imply, cause inflammation of the liver and therefore, were mainly identified for their shared abilities to cause transmissible hepatitis in human. During the last several decades, studies have revealed the diverse molecular nature of these viruses that, in turn, have greatly aided their control by various preventative and therapeutic measures. However, despite our concerted efforts to control and eradicate them, hepatitis viruses remain as one of the major challenges to scientific and health communities worldwide.

Transmissible hepatitis has long been recognized before the actual identification of the hepatitis viruses (1-4). With time, two types of viral hepatitis were recognized: the "infective hepatitis" and the "serum hepatitis," also known as hepatitis A and B (5-7). In 1960's and 70's, hepatitis A virus (HAV) was discovered as an agent that caused the infective (i.e., fecal-oral transmitted) hepatitis (8). HAV was later shown to be a single, positive-stranded RNA virus that belongs to the *Picornaviridae* family (Table 1) (9). In 1960's, Baruch Blumberg and his colleagues discovered the Australia antigen (10, 11) which eventually led to the identification of hepatitis B virus (HBV), also referred to as the "Dane particle" (12). The Australia antigen was none other than the HBV envelope glycoprotein, now known as the HBV surface antigen (HBsAg). HBV was determined to be a member of *Hepadnaviridae* and a major cause of the serum (i.e., parenterally transmitted) hepatitis. In the 1980's, several HBV-related virus genomes were cloned, including those of woodchuck hepatitis

B virus (15). These viruses have been used widely as model systems to study HBV.

A "new" antigen, found in the liver specimens from chronic HBV patients, led to the identification of another hepatitis virus in 1977 (16). This antigen came to be known as the delta antigen and thus, the virus was named hepatitis delta virus or HDV. Interestingly, HDV was found only in a fraction of HBV patients and showed the same parenteral mode of transmission as HBV. Today, HDV is known to be a defective virus that requires HBV to support its own replication (17). Thus, HDV closely resembles plant viroids that depend on helper viruses to propagate (18).

Soon after the discovery of these viral agents, diagnostic assays were developed for their detection. However, it also became increasingly apparent that some viral hepatitis could not be attributed to either HAV or HBV infection. These viral hepatitis were thus named "non-A, non-B (NANB) hepatitis" for their lack of association with HAV or HBV. Indeed, a viral agent responsible for the enterically transmitted, NANB hepatitis was later characterized and named hepatitis E virus for its enteric nature of transmission (19-21). Subsequently, HCV was demonstrated to be the major etiologic agent of the parenterally transmitted NANB hepatitis. The discovery of HCV may be considered the hallmark of the recent advances in molecular biology, as its complete viral genome was cloned in 1989 even before the isolation or the visualization of the virus (22, 23).

Yet, non-A, -B, -C, -D, and -E (nonA-E) hepatitis have also been described. The causative agents for nonA-E hepatitis continue to be sought after, and the candidates include hepatitis F virus (HFV), hepatitis G virus (HGV), TT virus (TTV), and SEN-V, a DNA virus (24). HFV was initially identified as a putative fecal-oral transmitted hepatitis virus. However, there is insufficient evidence to support this claim and its identity is in doubt. HGV and TTV were isolated from hepatitis patients. Recent research indicates that these two viruses may not be associated with hepatitis (25, 26). Note that the name of TTV was derived from the initial of a Japanese patient from which this virus was isolated (27). However, TTV has also been referred to as the "transfusion transmitted virus" (28). Sen-V is a small DNA virus that is closely related to TTV (29). Its possible role in viral hepatitis is still being debated (30).

HGV is also known as GB virus C. GB is the initial of the patient from which the virus was isolated. There are three related GB viruses, named GB virus A (GBV-A), GB virus B (GBV-B) and GB virus C (GBV-C). While these viruses may not cause hepatitis in human, GBV-B has been shown to cause hepatitis in macaques and tamarins (31, 32). GBV-B is closely related to HCV and has become a surrogated model system for studying HCV (33). Chapter 7 of this book summarizes recent advances in the molecular characterization of GB viruses. The molecular characteristics of hepatitis viruses A-E are compared in Table 1. Molecular biology of these viruses is also the main subject of the subsequent chapters. The following sections of this chapter summarize our current knowledge on hepatitis viruses A-E and their epidemiology and pathogenesis.

2. PREVALENCE AND TRANSMISSION ROUTES

Hepatitis viruses show characteristic geographical and demographical distribution that is often associated with their mode of transmission. Important information can be gained by analyzing the prevalence, common age of infection, and the primary mode of transmission of individual hepatitis viruses, especially for the identification of risk factors and prevention strategies to help control them. The primary source of each virus and their route of transmission are summarized in Table 1.

2.1 Enterically Transmitted Hepatitis Viruses

HAV is a non-enveloped virus that is transmitted by the fecal-oral route (34). Indeed, its stability against bile as well as other extreme environmental conditions has been suggested to allow its transmission by this pathway (35). HAV infection tends to be directly correlated with poor sanitation and commonly occurs early in life in the economically developing countries (34, 36). Studies have shown that the seroprevalence rate of HAV in the highly endemic regions can approach 100%, as compared to ~33% in the U.S (Figure 1) (36). Thus, traveling to the developing countries poses a risk for HAV infection (37). Person-to-person contact has been identified as a major pathway for HAV infection (37). Specifically, close contact with HAV infected persons in household settings (37), mixing with young children at day-care centers (38), and male homosexual activity (37) are some of the important risk factors for contracting HAV. HAV outbreaks are uncommon in the U.S., but occasional HAV outbreaks have been associated with contaminated food (39), such as when food is handled by an HAV infected personnel (36) or imported from the developing countries (40, 41). Waterborne HAV outbreaks are rare in the U.S. (42). Working in sewage plants may be considered a risk factor in some countries (43) but not in the U.S. (44). HAV transmission has also been reported in hospital settings (45). In one report, an immune-deficient patient who was tested negative for HAV antibody was the source of infection (46). Nevertheless, according

Table 1. Characteristics of hepatitis viruses A-E

Viruses	**A**	**B**	**C**	**D**	**E**
Classification	*Picornaviridae*[#]	*Hepadnaviridae*[#]	*Flaviviridae*[#]	*Deltavirinae*[*]	*unclassified*
Genome	RNA	DNA	RNA	RNA	RNA
Envelope	No	Yes	Yes	Yes	No
Primary source	feces; food/water-borne	blood and blood products	blood and blood products	blood and blood products	feces; water-borne
Primary mode of transmission	fecal-oral	parenteral	parenteral	parenteral	fecal-oral
Type of hepatitis	acute hepatitis	acute and chronic hepatitis	acute and chronic hepatitis	acute and chronic hepatitis	acute hepatitis

[#] Family; * Genus

to the Center for Disease Control (CDC)'s guidelines, healthcare workers are not considered a risk group for HAV infection.

HEV is another nonenveloped and enterically-transmitted virus that is associated with an inadequate sanitation (47-49). Fecally contaminated drinking water is by far the most well-defined route of HEV transmission (50). Person-to-person transmission of HEV is considered a relatively uncommon event (51). HEV infection appears to be largely confined to the developing countries, particularly around the tropics (Figure 2) (48). The highest rates of HEV-related disease are found in young to middle-age adults (51). Most of HEV infections in the industrialized countries have been correlated with travel to the HEV endemic areas (52-54).

Sporadic cases of HEV infection do seem to occur, however, even in the developed countries (47, 51, 55-58). Although sporadic cases may help maintain the virus between the epidemics, it has been speculated that HEV, unlike HAV, may be zoonotic, using some non-human mammalian hosts as a reservoir (48). Domestic livestocks, such as swine, have been suggested as possible reservoirs for HEV transmission to human (59, 60).

Interestingly, HEV incidence is high among hemophiliacs, hemodialysis patients, and post-transfusion patients, suggesting possible parenteral mode of transmission (47) and in this regard, HEV resembles the blood-borne hepatitis viruses, B, C, and D. HAV transmission by blood or blood-products have also been documented (61-63). It should be noted that

Figure 1. Geographic distribution of HAV infection (Center for Disease Control; www.cdc.gov./ncidod/diseases/hepatitis/slideset/hep_a/slide_11.htm.).

both hepatitis A and hepatitis E incidence rates are high among illicit drug users (64, 65). Once again, this indicates possible parenteral mode of HAV and HEV transmission, although other risk factors may also play a role.

2.2 Parenterally Transmitted Hepatitis Viruses

Today, there are about 350 million chronic HBV carriers in the world (66), and this includes about 1.25 million in the U.S. (Figure 3) (67). Despite the availability of vaccines, about 140 – 320,000 new HBV infections are estimated by CDC to occur per year in the U.S. alone. The chronically infected individuals are the major source of HBV infection (68, 69). A majority of these chronic HBV carriers are located in Asia and Africa (70).

HBV is blood-borne and is vertically transmitted from mother to baby or horizontally transmitted between individuals by percutaneous or mucous membrane exposure to infective blood or other body fluids, such as seminal fluid (71). The principal route of transmission, however, tends to vary with the prevalence of HBV. For example, in highly endemic areas, the perinatal transmission is the main route of HBV transmission (71). Infections usually occur early in life, such as at birth to early childhood. In contrast, in low endemic areas such as the U.S., the horizontal transmission is the primary mode of HBV transmission (67, 71). Sexual activity (70), contaminated needle sharing, and occupational exposure to infected blood or

Figure 2. Geographic distribution of hepatitis E (Center for Disease Control; www.cdc.gov./ncidod/diseases/hepatitis/slideset/hep_e/slide_5.htm.). Shaded areas represent outbreaks or confirmed infection in greater than 25% of sporadic non-ABC hepatitis.

other body fluids (67) constitute important risk factors. Household contact, hemodialysis, organ transplantation, and receiving blood products are also implicated in the HBV transmission, although to a lesser degree (67). Importantly, HBV is more prevalent among male than female (72).

Approximately 5% of the chronic HBV carriers in the world are also infected with HDV (73). As HDV requires HBV for replication, its transmission pathway closely resembles that of HBV (74). Percutaneous exposure to blood, such as that occurs in the illicit drug users, appears to be one of the most efficient routes of HDV transmission (75, 76). Hemophiliacs receiving clotting factors are also at an elevated risk (77). However, sexual transmission of HDV is not as efficient as that of HBV (69, 74), and vertical transmission is relatively rare (74). HDV has been reported to cause outbreaks, such as observed in the Amazon River Basin (78, 79). These outbreaks are often associated with a more severe disease outcome.

Interestingly, however, the geographical distribution of HDV does not always correlate with that of HBV (Figure 4) (74). In regions where HBV is less prevalent, HDV prevalence is generally also low. However, in regions where HBV is highly prevalent, HDV prevalence can range anywhere from high to very low. Nevertheless, when the HBV-HDV dual infection does occur, it may have serious clinical consequences (see below).

HCV has emerged as an important global health problem. Currently, it is estimated that more than 170 million people in the world are infected with HCV (Figure 5) (80). The prevalence rate of HCV varies from region to region and can be as high as 30% in highly endemic areas (81-83). The Third National Health and Nutrition Examination Survey (NHANES III) estimated that about 3.9 million Americans in the U.S. have been infected with HCV (84). This represents about 2% of the U.S. population (85).

As HCV is blood-borne, one of the leading causes of HCV infection used to be blood transfusion and receipt of clotting factor concentrates (86). The incidence of HCV infection has dropped substantially since the early 1990's after the development of HCV diagnostic assays for blood screening and the procedures to inactivate the virus in all immune globulin products (86-88). Today, blood transfusion is responsible for only about 6% of HCV infections in the U.S. (86). However, it is estimated that there are still about 36,000 new infections per year in the U.S. (86). About two-thirds of HCV incidence in the U.S. can be attributed to drug use (85). Organ transplantation (89), blood transfusion (90), receiving blood clotting factors (91), sexual exposure (92, 93), and perinatal transmission (94) are also identified as potential risk factors. Furthermore, nosocomial and occupational HCV transmission can occur if the workers are exposed to HCV-contaminated blood (95).

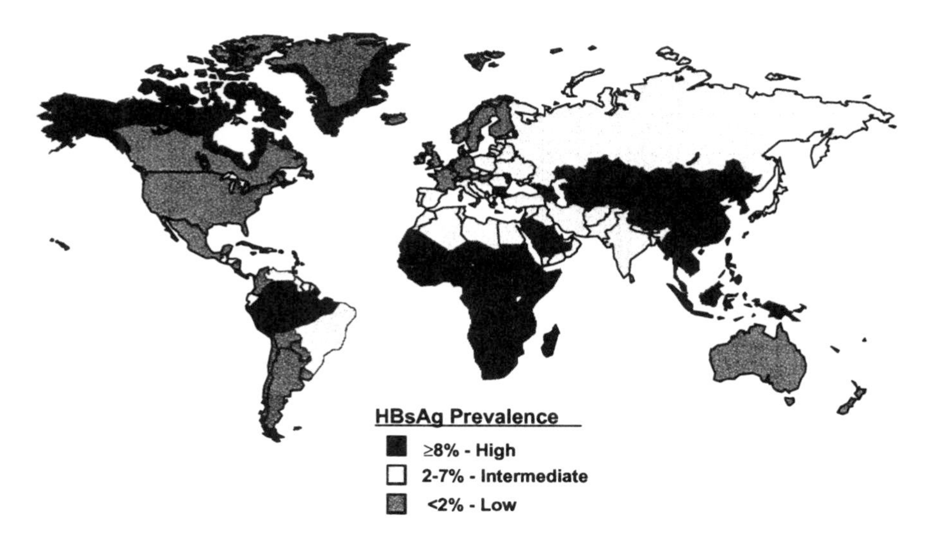

Figure 3. Geographic distribution of chronic HBV infection (Center for Disease Control; www.cdc.gov./ncidod/diseases/hepatitis/slideset/hep_b/slide_9.htm)

Figure 4. Geographic distribution of HDV infection (Center for Disease Control; www.cdc.gov./ncidod/diseases/hepatitis/slideset/hep_d/slide_6.htm)

The age-specific HCV prevalence data indicate that in low endemic areas like the U.S., most infections are found in young adults and people in their middle ages (96). As HCV-related liver complications tend to manifest later in life, it may be predicted that the HCV will become a greater burden to health communities as this group ages (97). The clustering of HCV incidence in the young adult age group also suggests that the virus was contracted relatively recently, such as by illicit drug use (96). In regions with higher prevalence, HCV infections are commonly found in the older population, suggesting that the risk of HCV infection was the greatest in the more distant past. Unsafe medical practices, such as using contaminated needles, may explain this trend (98). In highly endemic areas such as Egypt, however, HCV infection may be prevalent in all age groups, indicating some ongoing risk factor (96).

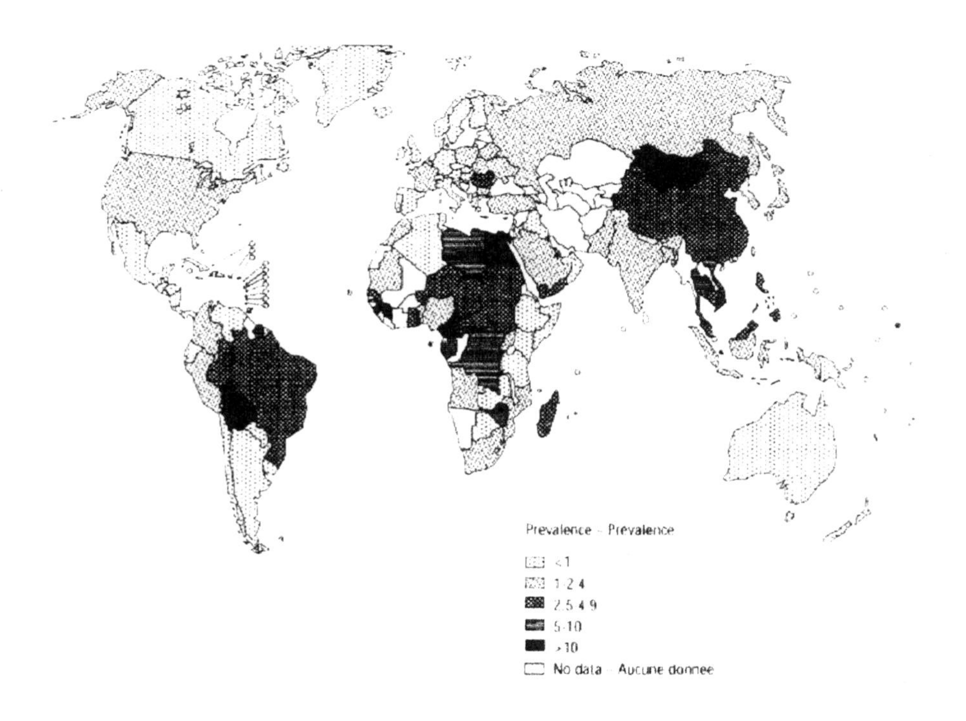

Figure 5. Global prevalence of hepatitis C (World Health Organization; www.who.int/emc/images/hepacmap.jpg)

3. PATHOGENESIS AND THE NATURAL COURSE OF INFECTIONS

3.1 Enterically Transmitted Hepatitis Viruses

Both HAV and HEV induce acute hepatitis. The incubation period for HAV is on the average of about 30 days while that of HEV is about 40 days (39, 51). Many HAV infections are asymptomatic. In the symptomatic patients, hepatitis A symptoms occur abruptly, and the patients typically complain of excessive fatigue, nausea, vomiting, malaise, fever, headache, loss of appetite, abdominal pain, and diarrhea, the latter of which tends to occur especially in children (35). Dark urine (conjugated bilirubinuria), light clay-colored stool, and jaundice may be present. Acute yellow atrophy of the liver is a well-known pathological change associated with hepatitis A (3). Persons infected with HEV show similar clinical signs and symptoms as those infected with HAV, including abdominal pain, malaise, nausea, fever, and dark urine (51, 99, 100). In terms of serum chemistry, serum aminotransferase activities rise with both HEV and HAV infections, indicating liver injury (35, 51). Serum bilirubin and alkaline phosphatase activities usually also increase (35, 101). Other abnormalities may also be detected, such as the presence of rheumatoid factor (IgM anti-IgG) in the serum (35). HAV and HEV coinfection is often associated with a more pronounced disease (102).

HAV and HEV infections are usually self-limiting. The strong cellular immune response seems to be effective at both clearing the virus and conferring lifelong immunity to HAV (36). The IgG, generated in response to HEV, confers at least temporary immunity to HEV (101, 103). However, both HAV and HEV can cause more severe disease in older patients (48). In fact, HAV infection can be fatal although the mortality rate tends to be low (39). Fulminant hepatitis, acute liver failure, meningoencephalitis, pancreatitis, cholestatic hepatitis A, and acute renal failure, though rare, have all been associated with HAV infection (36, 39). Hemolysis may be observed in certain patients who are predisposed for developing such problems (104). The overall mortality rate due to HEV infection is about 0.5 to 3% (51). Interestingly, HEV causes significantly higher degree of mortality in pregnant women (20, 105), and the rate may be as high as 20% among pregnant women in their trimester. However, neither HAV nor HEV is known to cause chronic infection (36, 106). Although some reports claim that some HAV patients relapse, they eventually recover completely from the viral infection.

As HAV does not seem to have direct cytopathic effect on the hepatocytes (107), the host immune system, such as CD8+ T lymphocytes and natural killer cells, has been implicated in the liver injury (108). Such involvement of host immune systems in the hepatocyte injury is implicated not only in HAV infections but appears to be a common mechanism of pathogenesis of viral hepatitis (67, 101).

3.2 Parenterally Transmitted Hepatitis Viruses

In contrast, parenterally transmitted hepatitis viruses, in addition to causing acute hepatitis, can cause chronic infection. They are also associated with severe liver diseases, such as cirrhosis and hepatocellular carcinoma (HCC) and more frequently associated with liver failure and death than HAV and HEV. Chronic viral hepatitis is indeed the leading cause for liver transplantation in the U.S. and is responsible for thousands of deaths per year in the U.S. alone (86, 109).

3.2.1 Hepatitis B and D viruses

About 1% of acute hepatitis B patients develop acute liver failure (110). A significant proportion of acute hepatitis B patients also become chronic HBV carriers (71). Interestingly, patients who develop strong immune response to HBV tend to clear the virus better than those who do not and progress to chronic infection (67). In addition, as the host immune response plays a pivotal role in HBV-associated liver injury, those with stronger immune response develop more serious liver injury. An interesting correlation is found between the age at infection and the risk of chronic infection (67). For example, about 95% of persons who are infected with HBV before age 5 become chronically infected because of their immature immune systems versus only 3 - 5%, if infected as adults.

Chronic HBV infection drastically increases the risk for progressive liver diseases. The relative risk of death due to cirrhosis and HCC can be as high as 79 and 148, respectively, in hepatitis B patients as compared to the normal individuals (67, 111). It is estimated that about 15 – 25% of chronic HBV carriers will die prematurely due to various liver complications (112). The rate of progression to cirrhosis and HCC depends on multiple factors that include the patient's age, male gender, immune status, the serology of HBV infection, and the environmental factors (67, 113-115). Aflatoxin, for instance, is an important environmental factor that promotes the development of HCC among the HBV patients (116). A more detailed discussion on the HCC is found in chapter 8.

A small fraction of hepatitis B patients are infected with mutant viruses and do not show an HBV antigen, the e antigen (HBeAg), in their serum (67, 117). The mutations apparently disrupt or suppress the synthesis of this viral protein without affecting the viral replication. It is debatable whether the presence of HBeAg is associated with more serious hepatitis symptoms. However, infants born to HBeAg-positive mothers often become chronically infected, while those from the HBeAg-negative mothers do not (118). Such correlation suggests a role of HBeAg in the establishment of chronic HBV infections.

Furthermore, patients who are infected with both HBV and HDV may acquire more serious liver diseases than those infected with HBV alone. For example, more cases of fulminant hepatitis are reported with HDV coinfection (i.e., simultaneous HBV and HDV infection) and superinfection (i.e., HDV infection of chronic HBV carriers) (119-122). The chronic HBV carriers who are superinfected by HDV often become chronic HDV carriers with an accelerated risk for developing chronic liver diseases (123, 124). However, chronic HDV infection does not increase the risk of HBV patients for HCC (125, 126). Coinfection of HBV with human immunodeficiency virus (HIV) has also been reported. Despite HIV's detrimental effect on the host immune system, however, the presence of this virus is suggested not to affect the natural course of HBV infection, and vise versa (127).

3.2.2 Hepatitis C virus

Approximately 75-85% of patients infected by HCV become chronically infected (86, 128-130). These HCV carriers often remain asymptomatic for one or two decades before the onset of liver diseases. Therefore, the diagnosis for HCV infection often occurs during routine physical checkups and blood donor screening (86). HCV, like HBV, can cause liver cirrhosis and HCC. It is estimated that about 10 to 20% of chronic hepatitis C patients develop liver cirrhosis in 20 – 30 years (86, 131). About 1 to 5% of chronic hepatitis C patients develop HCC (86, 132, 133). Currently, chronic HCV infection is also the leading cause for liver transplantation in the U.S. (86). Chronic HCV infection is also believed to induce various immune dysfunctions such as cryoglobulinemia (134).

Alcohol intake, existence of alcoholic liver disease, and the age at the time of infection appear to serve as important risk factors for progression to liver diseases (83, 135). Specifically, being older than 40 years at the time of infection is correlated with a more serious clinical outcome (86). The risk of developing HCC also increases with the onset of liver cirrhosis (134). In addition, for reasons yet unknown, male gender is also associated with a more severe disease (86, 135).

The natural course of HCV infection also varies with the co-existence of other viral infections. It has been suggested, for example, that

HCV and HAV coinfections can lead to a greater rate of fulminant hepatitis (136). Similarly, HBV and HCV coinfections are correlated with a more serious disease than either one alone (137). HCV and HIV coinfections are reported, such as in the intravenous (IV) drug users, and the disruption of the host immune system by HIV is likely to affect the natural course of HCV infection as well as complicate the antiviral therapy. Perinatal transmission of HCV from mother to baby, for instance, is enhanced in women who are also infected with HIV (94, 138).

HCV shows significant genetic variations that have led to the classification of this virus into six major genotypes (139). This genetic diversity of HCV and its quasispecies nature have become a major obstacle for the development of an effective vaccine. Furthermore, different genotypes have been suggested to influence the clinical outcome of HCV infections. For example, HCV genotype 1b has been reported to cause chronic infection in 92% of hepatitis C patients whereas other genotypes showed much lower (i.e., 33 to 50%) rate of chronic infection (140). However, whether different genotypes indeed show different level of pathogenicity is controversial (139). A clearer difference between HCV genotypes has been identified in regards to the efficacy of interferon therapy. In short, HCV genotypes 1a and 1b appear to be more resistant to interferon therapy than the other genotypes (141). Indeed, the HCV genotypes have become an important factor in considering the antiviral therapy for patients.

4. CONCLUDING REMARKS

The research on hepatitis viruses has progressed rapidly during the past three decades. We have also witnessed the technology breakthrough that led to the cloning of the HCV genome prior to the isolation of the virus. Apparently, some hepatitis viruses have been characterized more extensively than the others due to their involvement in more serious liver diseases. These studies have significantly reduced the incidence of viral hepatitis in different parts of the world by enabling the development of both therapeutic and preventative measures. A detailed outline of such therapeutic strategies is presented in chapter 10 of this book. Regrettably, however, no vaccine is yet available for HCV, and more studies will be needed to positively delineate the nonA-E hepatitis viruses. Finally, it is important to consider that despite the availability of vaccines, new HAV and HBV infections continue to occur. To this date, many countries have not implemented the public health policy to vaccinate every child against HBV

(71). More effective therapy and preventative health measures will be needed to control the hepatitis virus infections around the world.

5. ACKNOWLEDGMENTS

This work was supported by a research grant from the National Institutes of Health and the Research Scholar Grant #PF-01-037-01-MBC from the American Cancer Society.

REFERENCES

1. Cockayne E.A. Catarrhal jaundice, sporadic and epidemic, and its relation to acute yellow atrophy of the liver, Q J Med 1912; 6:1-29.
2. Findlay G.M., Dunlop J.L., Brown H.C. Observations on epidemic catarrhal jaundice, Trans R Soc Trop Med Hyg 1931; 25:7-24,
3. McDonald S. Acute yellow atrophy of the liver, Edin Med J 1908; 1:83-88.
4. Lurman A. Eine icterusepidemie. Berl Klin Wochenschr 1885; 22:20.
5. Anonymous. Homologous serum hepatitis. Lancet 1947; ii: 691-692.
6. Krugman S., Giles J.P., and Hammond J. Infectious hepatitis, Evidence for two distinctive clinical, epidemiological, and immunological types of infection. JAMA 1967; 200:365-373.
7. MacCallum F.O. Homologous serum jaundice. Lancet 1947; 2:691.
8. Purcell R.H. The discovery of the hepatitis viruses. Gastroenterology 1993; 104:955-963.
9. Francki R.I.B., Fauquet C.M., Knudson D.L., Brown F. Classification of nomenclature of viruses. Fifth report of the International Committee on Taxonomy of Viruses. Arch Virol 1991; 2(Suppl):320-326.
10. Blumberg B.S. The nature of Australia antigen: infectious and genetic characteristics. Prog Liver Dis 1972; 4:367-379.
11. Blumberg B.S., Gerstley B.J., Hungerford D.A., London W.T., Sutnick A.I. A serum antigen (Australia antigen) in Down's syndrome, leukemia, and hepatitis. Ann Intern Med 1967; 66:924-931.
12. Dane D.S., Cameron C.H., Briggs M. Virus-like particles in serum of patients with Australia-antigen-associated hepatitis, Lancet 1970; 1: 695-698.
13. Galibert F., Chen T. N., Mandart E. Nucleotide sequence of a cloned woodchuck hepatitis virus genome: comparison with the hepatitis B virus sequence. J Virol 1982; 41:51-65.
14. Seeger C., Ganem D., Varmus H.E. Nucleotide sequence of an infectious molecularly cloned genome of ground squirrel hepatitis virus. J Virol 1984; 51:367-375.
15. Mason W.S., Seal G., Summers J. Virus of Pekin ducks with structural and biological relatedness to human hepatitis B virus. J Virol 1980; 36:829-836.
16. Rizzetto M., Canese M.G., Arico S., Crivelli O., Trepo C., Bonino F., Verme G. Immunofluorescence detection of new antigen-antibody system (delta/anti-delta)

associated to hepatitis B virus in liver and in serum of HBsAg carriers. Gut 1977; 18:997-1003.

17. Lai, M.M. Molecular biologic and pathogenetic analysis of hepatitis delta virus. J Hepatol 1995; 22:127-131.
18. Diener T.O. Viroid replication. New York: Plenum, 1987; pp 117-166.
19. Balayan M.S., Andjaparidze A.G., Savinskaya S.S., Ketiladze E.S., Braginsky D.M., Savinov A.P., Poleschuk V.F. Evidence for a virus in non-A, non-B hepatitis transmitted via the fecal-oral route. Intervirology 1983; 20:23-31.
20. Kane M.A., Bradley D.W., Shrestha S.M., Maynard J.E., Cook E.H., Mishra R.P., Joshi D.D. Epidemic non-A, non-B hepatitis in Nepal. Recovery of a possible etiologic agent and transmission studies in marmosets. JAMA 1984; 252:3140-3145.
21. Reyes G.R., Purdy M.A., Kim J.P., Luk K.C., Young L.M., Fry K.E., Bradley D.W. Isolation of a cDNA from the virus responsible for enterically transmitted non-A, non-B hepatitis. Science 1990; 247:1335-1339.
22. Choo Q.L., Kuo G., Weiner A.J., Overby L.R., Bradley D.W., Houghton M. Isolation of a cDNA clone derived from a blood-borne non-A, non-B viral hepatitis genome. Science 1989; 244:359-362.
23. Choo Q.L., Richman K.H., Han J.H., Berger K., Lee C., Dong C., Gallegos C., Coit D., Medina-Selby R., Barr P.J., et al. Genetic organization and diversity of the hepatitis C virus. Proc Natl Acad Sci USA 1991; 88:2451-2455.
24. Bowden S. New hepatitis viruses: contenders and pretenders, J Gastroenterol Hepatol 2001; 16:124-31.
25. Gimenez-Barcons M., Forns X., Ampurdanes S., Guilera M., Soler M., Soguero C., Sanchez-Fueyo A., Mas A., Bruix J., Sanchez-Tapias J. M., Rodes J., Saiz J.C. Infection with a novel human DNA virus (TTV) has no pathogenic significance in patients with liver diseases. J Hepatol 1999; 30:1028-1034.
26. Karayiannis P. Thomas H.C. Hepatitis G virus: identification and prevalence. Br J Hosp Med 1996; 56:238-240.
27. Nishizawa T., Okamoto H., Konishi K., Yoshizawa H., Miyakawa Y., Mayumi M. A novel DNA virus (TTV) associated with elevated transaminase levels in posttransfusion hepatitis of unknown etiology, Biochem Biophys Res Commun 1997; 241:92-97.
28. Zuckerman A.J. The acronym TTV. Transfusion transmitted virus. Lancet 1999; 353:932.
29. Tanaka Y., Primi D., Wang R.Y., Umemura T., Yeo A.E., Mizokami M., Alter H.J., Shih J.W. Genomic and molecular evolutionary analysis of a newly identified infectious agent (SEN virus) and its relationship to the TT virus family, J Infect Dis 2001; 183:359-367.
30. Umemura T., Yeo A. E., Sottini A., Moratto D., Tanaka Y., Wang R. Y., Shih J. W., Donahue P., Primi D., Alter H.J. SEN virus infection and its relationship to transfusion-associated hepatitis. Hepatology 2001; 33:1303-1311.
31. Cheng Y., Zhang W., Li J., Li B., Zhao J., Gao R., Xin S., Mao P., Cao Y. Serological and histological findings in infection and transmission of GBV-C/HGV to macaques. J Med Virol 2000; 60:28-33.
32. Schaluder G.G., Dawson G.J., Simons J.N., Pilot-Matias T.J., Gutierrez R.A., Heynen C.A., Knigge M.F., Kurpiewski G.S., Buijk S.L., Leary T.P., et al. Molecular and serologic analysis in the transmission of the GB hepatitis agents. J Med Virol 1995; 46:81-90.
33. Beames B., Chavez D., Guerra B., Notvall L., Brasky K.M., Lanford R.E. Development of a primary tamarin hepatocyte culture system for GB virus- B: a surrogate model for hepatitis C virus. J Virol 2000; 74:11764-11772.

34. Havens W.P.J., Ward R., Drill V.A., Paul J.R. Experimental production of hepatitis by feeding icterogenic materials. Proc Soc Exp Biol Med 1944; 57:206-208.
35. Lemon S.M. Type A viral hepatitis: epidemiology, diagnosis, and prevention. Clin Chem 1997; 43:1494-499.
36. Koff R.S. Hepatitis A. Lancet 1998; 351:1643-1649.
37. Francis D.P., Hadler S.C., Prendergast T.J., Peterson E., Ginsberg M.M., Lookabaugh C., Holmes J.R., Maynard J.E. Occurrence of hepatitis A, B, and non-A/non-B in the United States. CDC sentinel county hepatitis study I. Am J Med 1984; 76:69-74.
38. Hadler S.C., Webster H.M., Erben J.J., Swanson J.E., Maynard J. E. Hepatitis A in day-care centers. A community-wide assessment. N Engl J Med 1980; 302:1222-1227.
39. Cuthbert J.A. Hepatitis A: old and new. Clin Microbiol Rev 2001; 14:38-58.
40. Hutin Y.J., Pool V., Cramer E.H., Nainan O.V., Weth J., Williams I.T., Goldstein S.T., Gensheimer K.F., Bell B.P., Shapiro C.N., Alter M.J., Margolis H.S. A multistate, foodborne outbreak of hepatitis A. National Hepatitis A Investigation Team. N Engl J Med 1999; 340:595-602.
41. Henkel J. Food firm gets huge fine for tainted strawberry harvest, FDA Consum 1999; 33:37-38.
42. Bloch A.B., Stramer S.L., Smith J.D., Margolis H.S., Fields H.A., McKinley T.W., Gerba C.P., Maynard J.E., Sikes R.K. Recovery of hepatitis A virus from a water supply responsible for a common source outbreak of hepatitis A. Am J Public Health 1990; 80:428-430.
43. De Serres G. Laliberte D. Hepatitis A among workers from a waste water treatment plant during a small community outbreak. Occup Environ Med 1997; 54:60-62.
44. Trout D., Mueller C., Venczel L., Krake A. Evaluation of occupational transmission of hepatitis A virus among wastewater workers, J Occup Environ Med 2000; 42:83-87.
45. Goodman R.A. Nosocomial hepatitis A, Ann Intern Med 1985; 103:452-454.
46. Burkholder B.T., Coronado V.G., Brown J., Hutto J.H., Shapiro C.N., Robertson B., Woodruff B.A. Nosocomial transmission of hepatitis A in a pediatric hospital traced to an anti-hepatitis A virus-negative patient with immunodeficiency. Pediatr Infect Dis J 1995; 14:261-266.
47. Irshad M. Hepatitis E virus: an update on its molecular, clinical and epidemiological characteristics. Intervirology 1999; 42:252-262.
48. Harrison T.J. Hepatitis E virus -- an update. Liver 1999; 19:171-176.
49. Corwin A., Jarot K., Lubis I., Nasution K., Suparmawo S., Sumardiati A., Widodo S., Nazir S., Orndorff G., Choi Y., et al. Two years' investigation of epidemic hepatitis E virus transmission in West Kalimantan (Borneo), Indonesia. Trans R Soc Trop Med Hyg 1995; 89:262-265.
50. Drabick J.J., Gambel J.M., Gouvea V.S., Caudill J.D., Sun W., Hoke C.H., Jr., Innis B.L. A cluster of acute hepatitis E infection in United Nations Bangladeshi peacekeepers in Haiti, Am J Trop Med Hyg 1997; 57:449-454.
51. Mast E.E. Krawczynski K. Hepatitis E: an overview. Annu Rev Med 1996; 47:257-266.
52. De Cock K.M., Bradley D.W., Sandford N.L., Govindarajan S., Maynard J.E., Redeker A.G. Epidemic non-A, non-B hepatitis in patients from Pakistan, Ann Intern Med 1987; 106:227-230.
53. Center for Disease Control and Prevention. Hepatitis E among U.S. travelers, 1989-1992. Morb Mortal Wkly Rep 1993; 42:1-4.
54. Wu J.C., Sheen I.J., Chiang T.Y., Sheng W.Y., Wang Y.J., Chan C.Y., Lee S.D. The impact of traveling to endemic areas on the spread of hepatitis E virus

infection: epidemiological and molecular analyses. Hepatology 1998; 27:1415-1420.

55. Dawson G.J., Chau K.H., Cabal C.M., Yarbough P.O., Reyes G.R., Mushahwar I.K. Solid-phase enzyme-linked immunosorbent assay for hepatitis E virus IgG and IgM antibodies utilizing recombinant antigens and synthetic peptides. J Virol Methods 1992; 38:175-186.
56. Kwo P.Y., Schlauder G.G., Carpenter H.A., Murphy P.J., Rosenblatt J.E., Dawson G.J., Mast E.E., Krawczynski K., Balan V. Acute hepatitis E by a new isolate acquired in the United States, Mayo Clin Proc 1997; 72:1133-1136.
57. Mast E.E., Kuramoto I.K., Favorov M.O., Schoening V.R., Burkholder B.T., Shapiro C.N., Holland P.V. Prevalence of and risk factors for antibody to hepatitis E virus seroreactivity among blood donors in Northern California. J Infect Dis 1997; 176:34-40.
58. Thomas D.L., Yarbough P.O., Vlahov D., Tsarev S.A., Nelson K.E., Saah A.J., Purcell R.H. Seroreactivity to hepatitis E virus in areas where the disease is not endemic. J Clin Microbiol 1997; 35:1244-1247.
59. Meng X.J., Purcell R.H., Halbur P.G., Lehman J.R., Webb D.M., Tsareva T.S., Haynes J.S., Thacker B.J., Emerson S.U. A novel virus in swine is closely related to the human hepatitis E virus. Proc Natl Acad Sci USA 1997; 94:9860-9865.
60. Meng X.J., Halbur P.G., Shapiro M.S., Govindarajan S., Bruna J.D., Mushahwar I.K., Purcell R.H., Emerson S.U. Genetic and experimental evidence for cross-species infection by swine hepatitis E virus. J Virol 1998; 72:9714-9721.
61. Noble R.C., Kane M.A., Reeves S.A., Roeckel I. Posttransfusion hepatitis A in a neonatal intensive care unit, JAMA 1984; 252:2711-2715.
62. Meyers J.D., Huff J.C., Holmes K.K., Thomas E.D., Bryan J.A. Parenterally transmitted hepatitis A associated with platelet transfusions. Epidemiologic study of an outbreak in a marrow transplantation center. Ann Intern Med 1974; 81:145-151.
63. Weisfuse I.B., Graham D.J., Will M., Parkinson D., Snydman D.R., Atkins M., Karron R.A., Feinstone S., Rayner A.A., Fisher R.I., et al. An outbreak of hepatitis A among cancer patients treated with interleukin-2 and lymphokine-activated killer cells, J Infect Dis 1990; 161: 647-652.
64. Zanetti A.R. Dawson G.J. Hepatitis type E in Italy: a seroepidemiological survey. Study Group of Hepatitis E. J Med Virol 1994; 42:318-320.
65. Grinde B., Stene-Johansen K., Sharma B., Hoel T., Jensenius M., Skaug K. Characterisation of an epidemic of hepatitis A virus involving intravenous drug abusers--infection by needle sharing? J Med Virol 1997; 53:69-75.
66. Zuckerman A.J. More than third of world's population has been infected with hepatitis B virus. BMJ 1999; 318:1213.
67. Lee W.M. Hepatitis B virus infection. N Engl J Med 1997; 337:1733-1745.
68. Boag F. Hepatitis B: heterosexual transmission and vaccination strategies. Int J STD AIDS 1991; 2:318-324.
69. Margolis H.S., Alter M.J., Hadler S.C. Hepatitis B: evolving epidemiology and implications for control. Semin Liver Dis 1991; 11:84-92.
70. Gust I.D. Epidemiology of hepatitis B infection in the Western Pacific and South East Asia. Gut 1996; 38:S18-23.
71. Maddrey W.C. Hepatitis B: an important public health issue. J Med Virol 2000; 61:362-366.
72. McQuilla, G.M., Coleman P.J., Kruszon-Moran D., Moyer L.A., Lambert S.B., Margolis H.S. Prevalence of hepatitis B virus infection in the United States: the National Health and Nutrition Examination Surveys, 1976 through 1994. Am J Public Health 1999; 89:14-18.

73. Rizzetto M., Ponzetto A., Forzani I. Epidemiology of hepatitis delta virus: overview, Prog Clin Biol Res 1991; 364:1-20.

74. Polish L.B., Gallagher M., Fields H.A., Hadler S.C. Delta hepatitis: molecular biology and clinical and epidemiological features. Clin Microbiol Rev 1993; 6:211-229.

75. Lange W.R., Cone E.J., Snyder F.R. The association of hepatitis delta virus and hepatitis B virus in parenteral drug abusers. 1971 to 1972 and 1986 to 1987. Arch Intern Med 1990; 150:365-368.

76. Mandelli C., Cesana M., Ferroni P., Lorini G.P., Aimo G.P., Tagger A., Bianchi P. A., Conte D. HBV, HDV and HIV infections in 242 drug addicts: two-year follow-up. Eur J Epidemiol 1988; 4:318-321.

77. Becherer P.R., Wang J.G., White G.C., Lesesne H., Janco R.L., Hanna W.T., Davis C., Johnson C.A., Poon M.C., Andes A., Lemon S.M. "Hepatitis delta virus infection in hemophiliacs." In *Viral hepatitis and liver disease,* Hollinger F.B., Lemon S.M., Margolis H.S., eds., Baltimore: The Williams & Wilkins Co., pp 492-494.

78. Hadler S.C., De Monzon M., Ponzetto A., Anzola E., Rivero D., Mondolfi A., Bracho A., Francis D.P., Gerber M.A., Thung S., et al. Delta virus infection and severe hepatitis. An epidemic in the Yucpa Indians of Venezuela. 1984; Ann Intern Med 100:339-344.

79. Bensabath G., Hadler S.C., Soares M.C., Fields H., Dias L.B., Popper H., Maynard J.E. Hepatitis delta virus infection and Labrea hepatitis. Prevalence and role in fulminant hepatitis in the Amazon Basin. JAMA 1987; 258: 479-483.

80. World Health Organization. Hepatitis C-global prevalence (update). Weekly Epidemiological Record 2000; 75:18-19.

81. Abdel-Wahab M.F., Zakaria S., Kamel M., Abdel-Khaliq M.K., Mabrouk M.A., Salama H., Esmat G., Thomas D.L., Strickland G.T. High seroprevalence of hepatitis C infection among risk groups in Egypt. Am J Trop Med Hyg 1994; 51:563-567.

82. Arthur R.R., Hassan N.F., Abdallah M.Y., el-Sharkawy M.S., Saad M.D., Hackbart B.G., Imam I.Z. Hepatitis C antibody prevalence in blood donors in different governorates in Egypt. Trans R Soc Trop Med Hyg 1997; 91:271-274.

83. Thomas D.L., Astemborski J., Rai R.M., Anania F.A., Schaeffer M., Galai N., Nolt K., Nelson K.E., Strathdee S.A., Johnson L., Laeyendecker O., Boitnott J., Wilson L.E., Vlahov D. The natural history of hepatitis C virus infection: host, viral, and environmental factors. JAMA 2000; 284:450-456.

84. McQuillan G.M., Alter M.J., Moyer L.A., Lambert S.B., Margolis H.S. "A population based serologic study of hepatitis C virus infection in the United States." In *Viral hepatitis and liver disease,* Rizzetto M., Purcell R.H., Gerin J.L., Verme G., eds., Turin: Edizioni Minerva Medica, 1997; pp 267-270.

85. Alter M.J. Epidemiology of hepatitis C. Hepatology 1997; 26:62S-65S.

86. Center for Disease Control and Prevention. Recommendations for prevention and control of hepatitis C virus (HCV) infection and HCV-related chronic disease. Morb Mortal Wkly Rep 1998; 47(RR19):1-39.

87. Blajchman M.A., Bull S.B., Feinman S.V. Post-transfusion hepatitis: impact of non-A, non-B hepatitis surrogate tests. Canadian Post-Transfusion Hepatitis Prevention Study Group. Lancet 1995; 345:21-25.

88. Donahue J.G., Munoz A., Ness P.M., Brown D.E., Jr., Yawn D.H., McAllister H.A., Jr., Reitz B.A., Nelson K.E. The declining risk of post-transfusion hepatitis C virus infection. N Engl J Med 1992; 327:369-373.

89. Pereira B.J., Milford E.L., Kirkman R.L., Quan S., Sayre K.R., Johnson P.J., Wilber J.C., Levey A.S. Prevalence of hepatitis C virus RNA in organ donors

positive for hepatitis C antibody and in the recipients of their organs. N Engl J Med 1992; 327:910-915.
90. Schreiber G.B., Busch M.P., Kleinman S.H., Korelitz J.J. The risk of transfusion-transmitted viral infections. The Retrovirus Epidemiology Donor Study. N Engl J Med 1996; 334:1685-1690.
91. Bresee J.S., Mast E.E., Coleman P.J., Baron M.J., Schonberger L.B., Alter M.J., Jonas M.M., Yu M.Y., Renzi P.M., Schneider L.C. Hepatitis C virus infection associated with administration of intravenous immune globulin. A cohort study. JAMA 1996; 276:1563-1567.
92. Capelli C., Prati D., Bosoni P., Zanuso F., Pappalettera M., Mozzi F., De Mattei C., Zanella A., Sirchia G. Sexual transmission of hepatitis C virus to a repeat blood donor. Transfusion. 1997; 37:436-440.
93. Thomas D.L., Zenilman J.M., Alter H.J., Shih J.W., Galai N., Carella A.V., Quinn T.C. Sexual transmission of hepatitis C virus among patients attending sexually transmitted diseases clinics in Baltimore--an analysis of 309 sex partnerships. J Infect Dis 1995; 171:768-775.
94. Zanetti A.R., Tanzi E., Paccagnini S., Principi N., Pizzocolo G., Caccamo M.L., D'Amico E., Cambie G., Vecchi L. Mother-to-infant transmission of hepatitis C virus. Lombardy Study Group on Vertical HCV Transmission. Lancet 1995; 345:289-291.
95. Mitsui T., Iwano K., Masuko K., Yamazaki C., Okamoto H., Tsuda F., Tanaka T., Mishiro S. Hepatitis C virus infection in medical personnel after needlestick accident. Hepatology 1992; 16:1109-1114.
96. Wasley A. Alter M.J. Epidemiology of hepatitis C: geographic differences and temporal trends. Semin Liver Dis 2000; 20:1-16.
97. Wong J.B., McQuillan G.M., McHutchison J.G., Poynard T. Estimating future hepatitis C morbidity, mortality, and costs in the United States. Am J Public Health 2000; 90:1562-1569.
98. Guadagnino V., Stroffolini T., Rapicetta M., Costantino A., Kondili L.A., Menniti-Ippolito F., Caroleo B., Costa C., Griffo G., Loiacono L., Pisani V., Foca A., Piazza M. Prevalence, risk factors, and genotype distribution of hepatitis C virus infection in the general population: a community-based survey in southern Italy. Hepatology 1997; 26:1006-1011.
99. Morrow R.H., Jr., Smetana H.F., Sai F.T., Edgcomb J.H. Unusual features of viral hepatitis in Accra, Ghana. Ann Intern Med 1968; 68:1250-1264.
100. Khuroo M.S. Study of an epidemic of non-A, non-B hepatitis. Possibility of another human hepatitis virus distinct from post-transfusion non-A, non-B type. Am J Med 1980; 68:818-824.
101. Labrique A.B., Thomas D.L., Stoszek S.K., Nelson K.E. Hepatitis E: an emerging infectious disease. Epidemiol Rev 1999; 21:162-179.
102. Arora N.K., Nanda S.K., Gulati S., Ansari I.H., Chawla M.K., Gupta S.D., Panda S.K. Acute viral hepatitis types E, A, and B singly and in combination in acute liver failure in children in north India. J Med Virol 1996; 48:215-221.
103. Bryan J.P., Tsarev S.A., Iqbal M., Ticehurst J., Emerson S., Ahmed A., Duncan J., Rafiqui A.R., Malik I.A., Purcell R.H., et al. Epidemic hepatitis E in Pakistan: patterns of serologic response and evidence that antibody to hepatitis E virus protects against disease. J Infect Dis 1994; 170:517-521.
104. Huo T.I., Wu J.C., Chiu C.F., Lee S.D. Severe hyperbilirubinemia due to acute hepatitis A superimposed on a chronic hepatitis B carrier with glucose-6-phosphate dehydrogenase deficiency. Am J Gastroenterol 1996; 91:158-159.

105. Tsega E., Hansson B.G., Krawczynski K., Nordenfelt E. Acute sporadic viral hepatitis in Ethiopia: causes, risk factors, and effects on pregnancy. Clin Infect Dis 1992; 14:961-965.
106. Khuroo M.S., Saleem M., Teli M.R., Sofi M.A. Failure to detect chronic liver disease after epidemic non-A, non-B hepatitis. Lancet 1980; 2:97-98.
107. Provost P.J. Hilleman M.R. Propagation of human hepatitis A virus in cell culture in vitro. Proc Soc Exp Biol Med 1979; 160:213-221.
108. Fleischer B., Fleischer S., Maier K., Wiedmann K.H., Sacher M., Thaler H., Vallbracht A. Clonal analysis of infiltrating T lymphocytes in liver tissue in viral hepatitis A. Immunology 1990; 69:14-19.
109. National Institutes of Health. National Institutes of Health Consensus Development Conference Panel statement: management of hepatitis C. Hepatology 1997; 26:2S-10S.
110. Lee W.M. Acute liver failure. N Engl J Med 1993; 329:1862-1872.
111. McMahon B.J., Alberts S.R., Wainwright R.B., Bulkow L., Lanier A.P. Hepatitis B-related sequelae. Prospective study in 1400 hepatitis B surface antigen-positive Alaska native carriers. Arch Intern Med 1990; 150:1051-1054.
112. Alter M.J. Epidemiology of hepatitis C. Eur J Gastroenterol Hepatol 1996; 8:319-323.
113. Fattovich G., Giustina G., Schalm S.W., Hadziyannis S., Sanchez-Tapias J., Almasio P., Christensen E., Krogsgaard K., Degos F., Carneiro de Moura M., et al. Occurrence of hepatocellular carcinoma and decompensation in western European patients with cirrhosis type B. The EUROHEP Study Group on Hepatitis B Virus and Cirrhosis. Hepatology 1995; 21:77-82.
114. Liaw Y.F., Tai D.I., Chu C.M., Chen T.J. The development of cirrhosis in patients with chronic type B hepatitis: a prospective study. Hepatology 1988; 8:493-496.
115. Chen C.J., Yu M.W., Liaw Y.F. Epidemiological characteristics and risk factors of hepatocellular carcinoma. J Gastroenterol Hepatol 1997; 12:S294-308.
116. Rajagopalan M.S., Busch M.P., Blum H.E., Vyas G.N. Interaction of aflatoxin and hepatitis B virus in the pathogenesis of hepatocellular carcinoma. Life Sci 1986; 39:1287-1290.
117. Buckwold V.E., Ou J.-H. Hepatitis B virus C-gene expression and function: the lessons learned from viral mutants. Current Topics in Virology 1999; 1:71-81.
118. Okada K., Kamiyama I., Inomata M., Imai M., Miyakawa Y. e antigen and anti-e in the serum of asymptomatic carrier mothers as indicators of positive and negative transmission of hepatitis B virus to their infants. N Engl J Med 1976; 294:746-749.
119. Smedile A., Farci P., Verme G., Caredda F., Cargnel A., Caporaso N., Dentico P., Trepo C., Opolon P., Gimson A., Vergani D., Williams R., Rizzetto M. Influence of delta infection on severity of hepatitis B. Lancet 1982; 2:945-947.
120. Govindarajan S., Chin K.P., Redeker A.G., Peters R.L. Fulminant B viral hepatitis: role of delta agent. Gastroenterology 1984; 86:1417-1420.
121. Tassopoulos N.C., Koutelou M.G., Macagno S., Zorbas P., Rizzetto M. Diagnostic significance of IgM antibody to hepatitis delta virus in fulminant hepatitis B. J Med Virol 1990; 30:174-177.
122. Saracco G., Macagno S., Rosina F., Rizzetto M. Serologic markers with fulminant hepatitis in persons positive for hepatitis B surface antigen. A worldwide epidemiologic and clinical survey. Ann Intern Med 1988; 108:380-383.
123. De Cock K.M., Govindarajan S., Chin K.P., Redeker A.G. Delta hepatitis in the Los Angeles area: a report of 126 cases. Ann Intern Med 1986; 105:108-114.
124. Farci P., Smedile A., Lavarini C., Piantino P., Crivelli O., Caporaso N., Toti M., Bonino F., Rizzetto M. Delta hepatitis in inapparent carriers of hepatitis B surface

antigen. A disease simulating acute hepatitis B progressive to chronicity. Gastroenterology 1983; 85:669-673.

125. Govindarajan S., Hevia F.J. Peters R.L. Prevalence of delta antigen/antibody in B-viral-associated hepatocellular carcinoma. Cancer 1984; 53:1692-1694.

126. Kew M.C., Dusheiko G.M., Hadziyannis S.J., Patterson A. Does delta infection play a part in the pathogenesis of hepatitis B virus related hepatocellular carcinoma? Br Med J (Clin Res Ed) 1984; 288:1727.

127. Scharschmidt B.F., Held M.J., Hollander H.H., Read A.E., Lavine J.E., Veereman G., McGuire R.F., Thaler M.M. Hepatitis B in patients with HIV infection: relationship to AIDS and patient survival. Ann Intern Med 1992; 117: 837-838.

128. Shakil A.O., Conry-Cantilena C., Alter H.J., Hayashi P., Kleiner D.E., Tedeschi V., Krawczynski K., Conjeevaram H.S., Sallie R., Di Bisceglie A.M. Volunteer blood donors with antibody to hepatitis C virus: clinical, biochemical, virologic, and histologic features. The Hepatitis C Study Group. Ann Intern Med 1995; 123:330-337.

129. Esteban J.I., Lopez-Talavera J.C., Genesca J., Madoz P., Viladomiu L., Muniz E., Martin-Vega C., Rosell M., Allende H., Vidal X., et al. High rate of infectivity and liver disease in blood donors with antibodies to hepatitis C virus. Ann Intern Med 1991; 115:443-449.

130. Seeff L.B., Buskell-Bales Z., Wright E.C., Durako S.J., Alter H.J., Iber F.L., Hollinger F. ., Gitnick G., Knodell R.G., Perrillo R.P., et al. Long-term mortality after transfusion-associated non-A, non-B hepatitis. The National Heart, Lung, and Blood Institute Study Group. N Engl J Med 1992; 327:1906-1911.

131. Fattovich G., Giustina G., Degos F., Tremolada F., Diodati G., Almasio P., Nevens F., Solinas A., Mura D., Brouwer J.T., Thomas H., Njapoum C., Casarin C., Bonetti P., Fuschi P., Basho J., Tocco A., Bhalla A., Galassini R., Noventa F., Schalm S.W., Realdi G. Morbidity and mortality in compensated cirrhosis type C: a retrospective follow-up study of 384 patients. Gastroenterology 1997; 112:463-472.

132. Kiyosawa K., Sodeyama T., Tanaka E., Gibo Y., Yoshizawa K., Nakano Y., Furuta S., Akahane Y., Nishioka K., Purcell R.H., et al. Interrelationship of blood transfusion, non-A, non-B hepatitis and hepatocellular carcinoma: analysis by detection of antibody to hepatitis C virus. Hepatology 1990; 12:671-675.

133. Di Bisceglie A.M., Order S.E., Klein J.L., Waggoner J.., Sjogren M.H., Kuo G., Houghton M., Choo Q.L., Hoofnagle J.H. The role of chronic viral hepatitis in hepatocellular carcinoma in the United States. Am J Gastroenterol 1991; 86:335-338.

134. Koff R.S. Dienstag J.L. Extrahepatic manifestations of hepatitis C and the association with alcoholic liver disease. Semin Liver Dis 1995; 15:101-109.

135. Poynard T., Bedossa P., Opolon P. Natural history of liver fibrosis progression in patients with chronic hepatitis C. The OBSVIRC, METAVIR, CLINIVIR, and DOSVIRC groups. Lancet 1997; 349: 825-832.

136. Vento S., Garofano T., Renzini C., Cainelli F., Casali F., Ghironzi G., Ferraro T., Concia E. Fulminant hepatitis associated with hepatitis A virus superinfection in patients with chronic hepatitis C. N Engl J Med 1998; 338:286-290.

137. Chu C.M., Yeh C.T., Liaw Y.F. Fulminant hepatic failure in acute hepatitis C: increased risk in chronic carriers of hepatitis B virus. Gut 1999; 45:613-617.

138. Weintrub P.S., Veereman-Wauters G., Cowan M.J., Thaler M.M. Hepatitis C virus infection in infants whose mothers took street drugs intravenously. J Pediatr 1991; 119:869-874.

139. Zein N.N. Clinical significance of hepatitis C virus genotypes. Clin Microbiol Rev 2000; 13:223-235.

140. Amoroso P., Rapicetta M., Tosti M.E., Mele A., Spada E., Buonocore S., Lettieri G., Pierri P., Chionne P., Ciccaglione A.R., Sagliocca L. Correlation between virus genotype and chronicity rate in acute hepatitis C. J Hepatol 1998; 28:939-944.
141. Zein N.N., Rakela J., Krawitt E.L., Reddy K.R., Tominaga T., Persing D.H. Hepatitis C virus genotypes in the United States: epidemiology, pathogenicity, and response to interferon therapy. Collaborative Study Group. Ann Intern Med 1996; 125:634-639.

Chapter 2

THE MOLECULAR BIOLOGY OF HEPATITIS A VIRUS

Annette Martin[1] and Stanley M. Lemon[2]

[1]Institut Pasteur, Paris, France and [2]University of Texas Medical Branch, Galveston, TX 77555

1. INTRODUCTION

Hepatitis A virus (HAV) is a positive-strand RNA virus classified within the family *Picornaviridae*. The mature HAV virion is a small 27 nm, generally spherical, non-enveloped particle (1), with three major capsid proteins encapsidating a single-stranded, positive-sense RNA genome about 7.5 kb in length. Although there are many attributes of the virus that distinguish it from other picornaviruses, HAV shares a number of features in common with other members of the family *Picornaviridae*, particularly those in the genus *Aphthovirus* (foot and mouth disease virus, FMDV) and genus *Cardiovirus* (e.g., encephalomyocarditis virus, EMCV). Its genomic organization is also similar to that of all other picornaviruses, including members of the *Enterovirus* (e.g., poliovirus) and *Rhinovirus* genera. In recognition of its distinctive features, however, HAV is classified as the type species of a separate genus, the genus *Hepatovirus*. An important human pathogen, HAV is a common cause of acute viral hepatitis and one of five human viruses typically associated with this disease.

Alone among the human hepatitis viruses, HAV may be propagated in cell cultures with moderate efficiency. A variety of primate cell lines are permissive for HAV, but primary isolation of wild-type virus is difficult and frequently entails a period of several weeks (or longer) between inoculation of cell cultures and the first detection of viral antigen (2, 3). Moreover, most cell-culture adapted strains of HAV do not produce a cytopathic effect, and there is no shutdown in host cell protein synthesis such as occurs with other prototypical picornaviral infections (4, 5). The slow and generally noncytolytic growth of the virus, coupled with virus yields that are much lower than those routinely achieved with many other picornaviruses, including poliovirus, combine to make HAV a difficult virus to study. These features of the virus have both limited the number of investigators in the field, and restricted the amount of information available concerning the molecular aspects of the life cycle of HAV. However, several cell culture-adapted HAV variants replicate with sufficient efficiency to allow their use in the production of formalin-inactivated vaccines. Such viruses also have been shown to be highly attenuated in their ability to cause disease in otherwise susceptible primates, and have been evaluated as candidate attenuated vaccines.

2. TYPE A VIRAL HEPATITIS: THE PATHOBIOLOGIC CONTEXT OF HAV REPLICATION

HAV is present in a world-wide distribution, with the highest prevalence of infection occurring in regions where low standards of sanitation promote transmission of the virus (6). Infection with HAV usually occurs by the fecal-oral route of transmission, and is associated with extensive shedding of the virus in feces during the 3-6 week incubation period and extending into the early days of the illness (7). Although the virus appears capable of infecting intestinal epithelial cells (8), the primary target cell is the hepatocyte and most virus shed in the feces is derived from the liver. The passage of the virus through the biliary tract is facilitated by the absence of a lipid containing envelope, and a very stable naked capsid structure (9). Maximal fecal shedding generally precedes the onset of liver injury, indicating that the replication of virus occurs without cell death. Thus, the virus appears to be secreted by a specific mechanism across the apical membrane of hepatocytes, into the biliary canaliculi, from which it reaches the intestines. A viremia persists throughout the course of the acute infection, but becomes greatly reduced in magnitude (as does the fecal shedding of virus) with the onset of acute hepatitis and the appearance of antibodies to HAV (10).

Acute liver injury, marked by a rise in hepatocellular-derived enzymes in the blood, jaundice, and acute inflammation within the liver, occurs as a result of a robust immunologic response to the infection. This response involves both cellular and humoral arms of the immune system. Virus-neutralizing antibodies appear early in the course of the infection (11, 12), and HLA-restricted, virus-specific, cytotoxic, CD8+ T cells have been recovered from the liver of patients developing acute hepatitis A (13-15). Virus-specific cytotoxic T cells have been shown to secrete γ-interferon, which likely stimulates the recruitment of additional, nonspecific inflammatory cells to the site of virus replication within the liver, and may act as a key factor in T-cell promoted clearance of HAV-infected hepatocytes (16, 17). This immune response leads to the elimination of the infection over a period of several weeks. Recovery is complete, although a very small proportion of infected individuals develop a fatal, fulminant hepatitis in association with the infection. Although fecal shedding of the virus may be prolonged in premature infants, chronic hepatitis and true persistent infection have not been associated with HAV in humans (18-20).

Chimpanzees, as well as several species of tamarins and New World owl monkeys are susceptible to HAV and may be infected by either oral or parenteral administration of virus (21, 22). Although disease in these primates is usually mild compared with symptomatic infections in adult humans, the course of the infection is otherwise very similar. The absence of a non-primate animal model for hepatitis A, however, has significantly impeded research in this field.

Control of the infection is facilitated by the fact that there are no significant antigenic differences among strains collected from humans in widely separated geographic regions, despite significant genetic heterogeneity (23, 24). With only a single serotype of HAV, antibodies directed against any strain are highly protective against all other HAV strains. Pooled human immunoglobulin contains neutralizing antibody to HAV, and is highly protective against disease when administered within 2 weeks of exposure to the virus. Formalin-killed whole-virus vaccines are also very effective in prevention of clinical hepatitis A, and several different vaccines have been licensed during the past decade (25).

3. HEPATITIS A VIRUS: CHARACTERISTICS OF THE VIRION

HAV particles were first identified by immune electron microscopy in the feces of an experimentally-infected chimpanzee. The mature virion is a small 27 nm, generally spherical, non-enveloped particle (1), with a predominant buoyant density of 1.325 and a sedimentation value of 155S. Three capsid proteins (VP1, VP2 and VP3, ranging from ~220 to ~275 amino acid residues in length) have been demonstrated in purified virus preparations (26, 27). Each is likely to be present at 60 copies per particle, based on the known structures of other picornaviruses. Unfortunately, however, although the atomic structures of representative viruses from each of the other major picornaviral genera have been solved by X-ray crystallography (28, 29), multiple attempts to crystallize HAV particles have not succeeded. The existence of a much smaller, fourth polypeptide (VP4) is inferred from its presence in other picornaviruses and the fact that the polyprotein of HAV (see below) contains a probable VP4 polypeptide of 21-23 amino acids in length at its amino terminus, but it has never been demonstrated directly in purified virus preparations.

As mentioned above, only a single HAV serotype has been identified. Substantial evidence suggests that the critical antigen(s) of HAV are conformationally determined and are thus comprised of "assembled" rather than "linear" epitopes. Analysis of HAV using neutralizing murine monoclonal antibodies indicates the presence of an immunodominant antigenic site on the virus capsid that is involved in antibody-mediated neutralization. Genomic sequencing of viral escape mutants selected during growth of the virus in the presence of neutralizing monoclonal antibodies indicates that such resistance can be conferred by substitutions of a number of different amino acid residues within VP3 and VP1 which therefore contribute to this site (30). The location of these residues in alignments of the HAV capsid protein sequences with the capsid protein sequences of other picornaviruses for which the atomic structure has been solved suggests that the HAV particle shares a general structural organization with these other viruses, but with some significant differences in the proximity of VP1 and VP3 residues. The HAV particle has recently been visualized by cryoelectron microscopy (31), and it is likely that this approach will provide additional information on the structure of the particle.

Like all picornaviruses, the virion contains a single copy of a single-stranded, positive sense, and non-segmented RNA genome. In the case of HAV, this RNA is approximately 7.5 kb in length.

4. ORGANIZATION OF THE HAV GENOME

The genomic RNA contains a 5' nontranslated segment that is ~730 bases in length, followed by a single long open reading frame encoding a polyprotein of ~2227 amino acids (32,33) (Figure 1). This polyprotein undergoes processing directed by both viral and cellular proteinases, resulting in 11 different polypeptides having various functions in viral assembly and replication. The polyprotein-coding segment of the genome is followed by a short 3' noncoding region which terminates in a 3' poly(adenylic acid) tract (Figure 1). A small genome-encoded protein (3B, also called VPg) is attached covalently to the 5' end of virion RNA (34), and is likely to play a role in priming of RNA synthesis, as has been shown with poliovirus (35). As is the case with other picornaviruses, RNA transcribed from genome-length, cloned HAV cDNA is infectious when transfected into permissive cells (36).

Protein synthesis can initiate from either of two AUG codons located two amino acids apart from each other at the 5’ end of the long open reading frame (37), and gives rise to a polyprotein with the (L)-P1-P2-P3 organization typical of other picornaviruses. The amino terminal third of the polyprotein (P1 segment) contains the structural proteins that comprise the viral capsid: in order from the amino terminus, VP4 (otherwise known as 1A, and possibly containing a very short N-terminal leader, L, polypeptide, as discussed below), VP2 (or 1B), VP3 (1C), and VP1 (1D) (Figure 1). Downstream of VP1 is the 2A polypeptide, which is now known to function not in RNA replication but in particle assembly as a carboxy-terminal extension of VP1 (that is, as VP1-2A, otherwise known as the PX polypeptide, see below). Thus, the VP4-2A segment comprises proteins involved in the structure and assembly of the HAV particle.

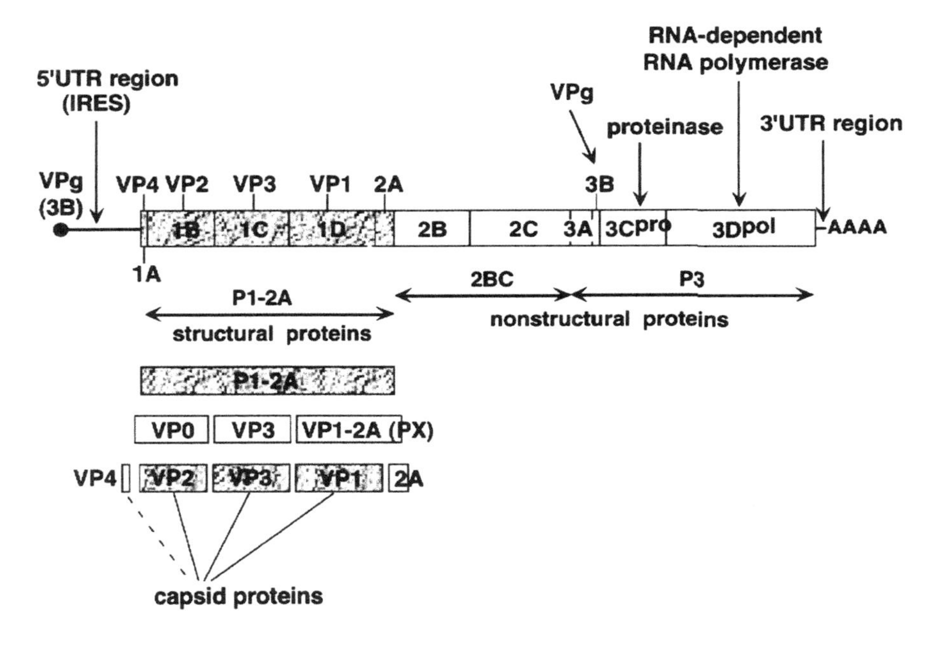

Figure 1. Schematic representation of the HAV genome. The polyprotein-coding region (long open reading frame) appears as a box, with the structural polypeptides and capsid protein precursors highlighted.

All of the polypeptides located downstream of the 2A sequence in the polyprotein play essential, albeit incompletely defined, roles in RNA replication, as described in the following sections (Figure 1). These include 2B, a large, hydrophobic protein of unknown function within which mutations have been shown to confer important adaptive properties related to the replication of the virus in cultured cells (38-40), and 2C, a putative RNA helicase. Further downstream is 3A, which by analogy with poliovirus is likely to play a role in anchoring replication complexes to cellular membranes via hydrophobic sequences located near its amino terminus, and 3B (VPg), the small genome linked protein which likely functions to prime viral RNA synthesis. The $3C^{pro}$ polypeptide is a cysteine proteinase that is responsible for directing most cleavage events in the viral polyprotein, while $3D^{pol}$, located at the carboxyl end of the polyprotein, is the putative RNA-dependent, RNA polymerase of HAV. Many if not all of these polypeptides are likely to have multiple functions in viral replication. Moreover, by analogy with other picornaviruses, some polypeptides may have unique functions as partially uncleaved processing intermediates (such as 3CD, in the case of poliovirus) that may differ from their functions as fully cleaved products of the processing cascade ($3C^{pro}$ and $3D^{pol}$). In the case of HAV, the 3ABC polypeptide appears to be a relatively stable precursor. It is detected in HAV-infected cells, and may have specific functions in the virus life cycle, particularly in the processing and assembly of capsid proteins (41, Martin et al., unpublished results).

5. EXPERIMENTAL SYSTEMS IN WHICH HAV REPLICATION HAS BEEN STUDIED

The fact that the hepatocyte represents the primary site of replication of the virus greatly complicates the study of HAV replication mechanisms, since fully differentiated hepatocytes exist *in vivo* in a unique environment and cannot be continuously cultured *ex vivo* without significant loss of differentiated functions. Thus, almost all experimental studies of HAV replication mechanisms have either depended on the use of cell culture-adapted viruses replicating in stable primate fibroblast-like cell lines (such as BSC-1 cells derived from African green monkey kidney or FRhK-4 cells derived from rhesus monkey kidney), or involved the functional characterization of viral polypeptides expressed from recombinant cDNA. Unfortunately, neither approach can provide a complete picture of the interaction of the virus with human hepatocytes *in vivo*. Although recent studies have begun to explore virus-host interactions in cell lines derived from human hepatocytes (42) or intestinal crypt cells (43), even these approaches fall short of providing a truly native environment for defining HAV-host cell interactions.

Furthermore, the noncytolytic and relatively inefficient replication cycle of most HAV strains, as well as the general absence of any associated shut-off in host cell synthetic processes, has limited the kind of studies that can be accomplished with HAV in infected cell cultures. As a result, much of our present understanding of the molecular aspects of HAV replication is extrapolated from knowledge of the replication mechanisms of other picornaviruses (such as poliovirus).

Virus can be rescued from synthetic full-length RNA following its transfection into permissive cell cultures (36). This has allowed the application of "reverse molecular genetics" to studies of HAV replication. Such studies remain tedious and labor intensive, however, due to difficulties in assaying for infectious HAV in cell culture. The recent development of subgenomic RNA replicons, expressing luciferase in lieu of the P1 capsid protein segment (42), greatly simplifies such experiments, and is likely to facilitate studies exploring the viral genetics of RNA replication.

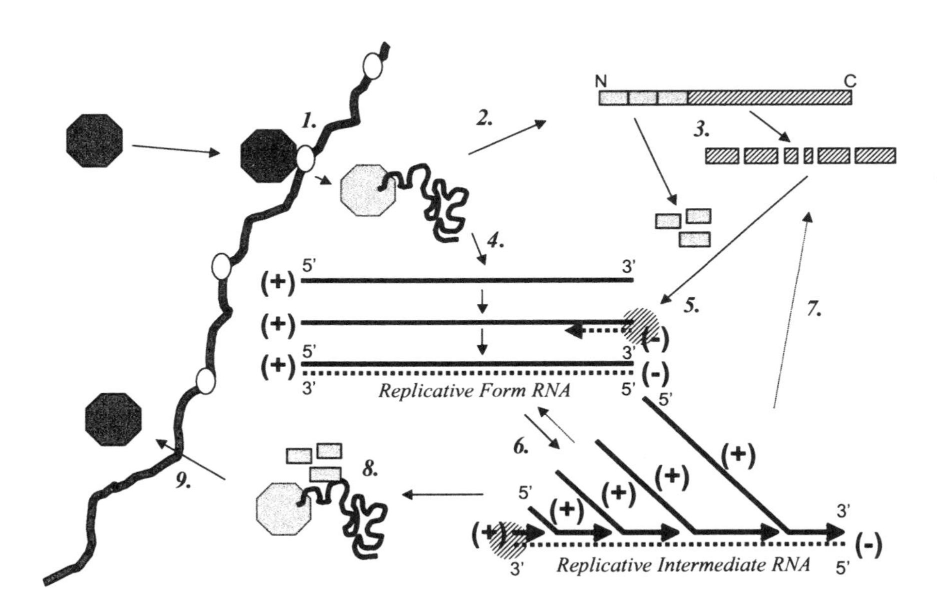

Figure 2. Schematic showing the major events that are thought to occur during the replication of HAV: (*1*) viral attachment and entry occur through a cellular receptor, and result in release of the positive-strand genome into the cytoplasm of the cell; (*2*) virion RNA is translated into a large polyprotein; (*3*) the polyprotein undergoes processing directed by $3C^{pro}$; (*4*) virion RNA is also used as template for synthesis of negative-strand RNA under direction of a (*5*) viral replicase complex containing $3D^{pol}$ and other viral proteins; (*6*) new positive-strand RNA molecules are synthesized from the negative-strand template, and (*7*) direct the translation of additional copies of the polyprotein, serve as template for additional negative-strand molecules, or (*8*) are packaged into virions that are then (*9*) transported out of the cell. All replication events occur in close association with cellular membranes.

All available evidence suggests that the replication scheme of HAV resembles that of other picornaviruses (Figure 2). Virion RNA acts directly as messenger RNA, programming the translation of a large polyprotein which is cotranslationally processed by the virus-encoded $3C^{pro}$ proteinase into both structural and nonstructural proteins (44). Virion RNA also serves as template for negative-strand RNA synthesis, which acts in turn as template for synthesis of new positive-strand genomic RNA. The newly synthesized positive-strand RNA is either used for further rounds of replication, or packaged within the capsid proteins for export from the hepatocyte. As is the case with other picornaviruses, available evidence suggests that HAV replication occurs in the cytoplasm of the infected cell and that RNA transcription proceeds asymmetrically, with an excess of positive-strand molecules synthesized under direction of the virus-encoded RNA-dependent RNA polymerase, $3D^{pol}$. Despite these similarities with other picornaviruses, a number of recent studies have begun to define characteristics of viral translation, RNA replication, and particle assembly that clearly distinguish HAV from other picornaviruses. The following sections explore what is known and still unknown of these processes as related to infection of the human hepatocyte by HAV.

6. VIRUS ENTRY AND THE INITIAL STEPS IN VIRAL REPLICATION

The infecting HAV particle presumably arrives at the apical membrane surface of the hepatocyte, within the hepatic sinusoid, via the portal circulation from the intestines. Although there is evidence that the virus may undergo primary replication within crypt cells of the small intestine (8), studies with infected Caco-2 cells suggest that the release of virus from such cells occurs predominantly across the apical, not the basolateral, cell surface, and thus back into the lumen of the intestine (43). Given this observation, it seems probable that the virus (whether ingested or replicated in crypt cells) may be directly transported across the intestinal epithelium via specialized transport ("M") cells in the distal ileum, as has been proposed for poliovirus (45).

Based on studies with multiple other picornaviruses, the initial step in invasion of the hepatocyte by the virus is likely to be the interaction of the virus with a specific cellular receptor. In the case of poliovirus, the cellular receptor may play a major role in determining the host range of the virus (46). This does not appear to be the case with HAV, as a cell culture-adapted variant of HAV has been shown to be capable of replicating in GPE, SP 1K, and IB-RS-2 D10 cells of guinea pig, dolphin, and pig origin, respectively (47, 48). This suggests that the cell surface receptor(s) and other host factor(s) required for HAV replication are present in many different types of cells, including cells of nonhepatic origin from species that are not considered to be susceptible to HAV infection.

Kaplan and coworkers have identified a surface glycoprotein belonging to the mucin family that appears to be involved in the binding and entry of HAV into cultured African green monkey kidney cells (49). Antibodies reactive to this putative cellular receptor protein (HAVcr-1) block infection of these cells, and its expression in murine cells was considered to confer a moderate increase in permissiveness for viral replication. Therefore, HAVcr-1 may not be the only molecule involved in HAV binding and entry. In addition, an immunoadhesin containing the HAV-binding domain of HAVcr-1 only partially neutralizes HAV infectivity, although, as noted by the authors, the extent of viral neutralization was of an order of magnitude similar to that achieved with neutralizing monoclonal antibodies (50). A human analog of this putative primate HAV receptor molecule (huHAVcr-1) has also been identified and found to be expressed in many human tissues (51). Expression of huHAVcr-1 confers HAV-binding activity and a limited ability to support HAV infection in normally nonpermissive cells of canine origin. While these results suggest that huHAV-cr1 may function as an HAV-binding protein, and that it may facilitate entry of HAV into cells, the involvement of this mucin-like molecule in infection of hepatocytes *in vivo* is yet to be clearly established.

Interestingly, HAV-specific immunoglobulin A (IgA) has been shown to mediate infection of hepatocytes with HAV by the binding and internalizing of IgA molecules via the hepatocyte asialoglycoprotein receptor (52). Although these observations were made in a mouse hepatocyte model system, human hepatocytes also were shown to ingest complexes of HAV and anti-HAV IgA, resulting in infection with the virus. Thus, the entry of virus into IgA receptor-positive hepatocytes by HAV may occur via a surrogate receptor mechanism, in which IgA produced by the gastrointestinal mucosa-associated lymphoid tissue may serve as a carrier molecule, in the form of virus-antibody complexes transporting virus from the intestinal tract in a receptor-targeted fashion to the liver. Once again, however, the relevance of these findings to the situation in acute hepatitis A is not entirely clear. Despite the fact that HAV-specific IgA antibodies appear early in the course of the infection, viral antigens have been identified within the hepatocytes of HAV-permissive New World owl monkeys as early as 4 days after oral challenge (8), many days prior to the appearance of any antibody to the virus. Thus, IgA-mediated transport of virus into the hepatocyte is not the only means by which the virus gains entry to its primary cell target. If it plays a significant role during natural infection, it is likely to occur later in the course of the infection. It may possibly contribute to rare episodes of recurrent hepatitis A (52).

Virtually nothing is known of the mechanisms surrounding the entry of HAV into hepatocytes, or the process by which the viral capsid disassembles to release the viral RNA into the cytoplasm once the virus has entered the cell. However, the uncoating of HAV particles is known to occur in a protracted and asynchronous fashion in cultured cells. This may be related to the presence of high proportions of less infectious, immature particles (provirions) in populations of virus produced *in vitro*. These provirions need to develop into mature virions by cleavage of the VP4-VP2 (VP0) capsid protein precursor (see below), prior to uncoating. This particular feature of HAV particles, a slower rate of virion maturation compared to other picornaviruses, may account for the asynchronous nature of HAV infections in cell culture (53, 54).

7. VIRAL TRANSLATION

As is the case for other picornaviral RNAs, the 5'UTR of HAV is much longer (~730 nts) than that of most cellular mRNAs, and it contains numerous AUG codons that do not serve as translational start sites (Figure 3). In addition, the positive-strand RNA molecule lacks the typical 5'm7GpppG cap structure that plays a critical role in the initiation of translation on most eucaryotic mRNAs. In contrast, the 5'UTRs of picornaviral RNAs efficiently initiate translation in the absence of a 5'm7GpppG cap by directing the 40S ribosomal subunit to bind to an internal site located hundreds of nucleotides downstream of the 5' end of the molecule (55). Although there are significant differences in how this is accomplished by different picornaviruses, in each the process is dependent upon a highly ordered RNA structural element (the internal ribosome entry site, or IRES) that extends throughout much of the 5'UTR sequence. While details of the mechanism underlying internal ribosomal entry are not fully known, translation initiation occurs via the interaction of a group of secondary and/or tertiary structural elements within the IRES with the 40S ribosomal subunit and a set of cellular proteins that include both canonical and noncanonical translation factors (56, 57).

Support for the concept of internal ribosome entry comes from translation studies carried out both *in vivo* and *in vitro* in which the insertion of 5'UTR sequences within the intercistronic space of bicistronic RNAs promoted translation of the second cistron in a manner that was independent of translation of the upstream cistron. The existence of an IRES between nts 161-734 of the 5'UTR of HAV (Figure 3) has been demonstrated by analyzing translation in rabbit reticulocyte lysates programmed with bicistronic RNAs containing the 5'UTR of HAV in the intercistronic space (58, 59). The structure of the IRES sequence was deduced by a combination of phylogenetic comparative sequence analysis, thermodynamic modeling, and nuclease probing of the secondary structure of synthetic RNAs. Although the results of these studies indicate that the secondary structure of this segment of the HAV 5'UTR is unique, it shares a number of features in common with the IRES of the *Cardiovirus* and *Aphthovirus* genera of the family *Picornaviridae* (60, 61). These structural similarities exist despite a lack of significant nucleotide sequence identity between the viruses.

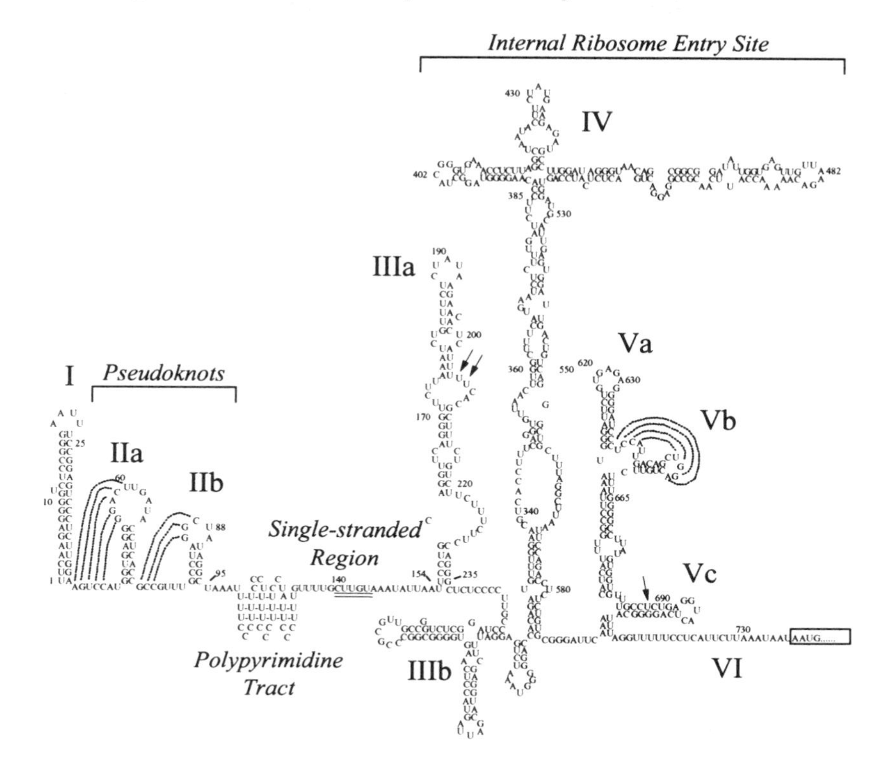

Figure 3. Proposed RNA structure of the 5'UTR of HAV. Dotted lines represent potential base-pair interactions in putative RNA pseudoknots; the box indicates the 5' end of the open reading frame. Stem-loops IIIa-VI (nts 155-735) comprise the IRES. Arrows indicate the site of mutations in cell culture-adapted HAV that facilitate translation. Deletion of the underlined sequence (nts 140-144) leads to a temperature-sensitive phenotype.

The IRES-directed initiation of translation of the HAV polyprotein, however, appears to differ in at least two important aspects. First, other picornaviral IRES elements do not require the presence of intact eucaryotic translation initiation factor eIF-4G, the p220 component of the cellular cap-binding complex, for efficient translation initiation. These IRES elements function in the presence of a C-terminal cleaved form of the eIF-4G molecule. Since intact eIF-4G is essential for the normal, cap-dependent translation of cellular mRNAs, this has allowed several other picornaviruses (notably poliovirus and FMDV) to evolve viral proteinases that efficiently cleave the eIF-4G molecule, thereby abrogating cellular protein translation in the face of continued, active viral protein translation initiation (62). In the case of poliovirus, this activity is carried out by the viral 2A proteinase, while the Lb protein of FMDV is responsible for this cleavage event. HAV translation initiation is quite different in that it is effectively inhibited by the expression of the poliovirus 2A protein (63), and has been shown to require an intact eIF-4G molecule (64). A second important difference is the level of efficiency with which the HAV IRES is capable of directing the internal initiation of translation on the downstream RNA. This appears to be many-fold lower than the efficiency of the IRES elements of EMCV or poliovirus, or for that matter the unrelated IRES element of hepatitis C virus (HCV), both *in vitro* (58, 65) and *in vivo*, in various cell lines including a cell line derived from human liver (63, 66). The reason for this difference is not understood, nor is it clear that this is the case in the infected hepatocyte *in situ*, but it is possible that it contributes significantly to the slow and inefficient replication of HAV in cultured cells.

A number of studies have emphasized the importance of various cellular proteins in the internal initiation of picornaviral translation, both *in vivo* and in cell-free systems *in vitro*. The cellular polypyrimidine tract-binding protein (PTB) has been shown to facilitate translation directed by picornaviral IRES elements in rabbit reticulocyte lysates *in vitro* (55, 67). This protein is normally present primarily in the nucleus of the cell, but we have recently demonstrated that the overexpression of PTB stimulates translation directed by the IRES elements of HAV as well as poliovirus in several different types of cells (68). PTB expression also stimulates translation directed by the HCV IRES. PTB has been shown to be bound directly by these viral RNAs. Since these IRES elements share little if any common secondary or tertiary RNA structure, these results suggest that PTB may act to stabilize RNA structure in a nonspecific fashion and that it may function as a general RNA folding chaperone (55).

In addition to PTB, the glycolytic enzyme, glyceraldehyde 3'-dehydrogenase (GAPDH), is also bound efficiently (through its NADH binding groove) to U-rich segments of the HAV 5'UTR, including stem-loop IIIa within the HAV IRES (Figure 3) (69). Although GAPDH and PTB bind to overlapping segments of the IRES, these two cellular proteins appear to have opposing effects on both the higher ordered RNA structure and internal initiation of translation. GAPDH suppresses translation by destabilizing essential secondary RNA structures within the IRES (70), while PTB may reverse this effect by stabilizing structure, or by effectively competing with GAPDH for the overlapping sites on the IRES to which these proteins bind (68, 69).

The cellular protein poly(rC) binding protein 2 (PCBP2) also has been reported to bind to the 5'UTR of HAV. In this case, binding of the protein has been mapped to the terminal 157 nucleotides of the 5'UTR of HAV (which contains a pyrimidine-rich tract and also binds PTB and

GAPDH) (Figure 3) (71). *In vitro* depletion and replenishment experiments suggest that PCBP2 may be required for HAV IRES activity. However, it is possible that the interaction of PCBP2 with the viral RNA may play a greater role in RNA replication than in translation, since the RNA sequence to which it binds appears to function primarily in new RNA synthesis (see below), not protein translation.

In the case of poliovirus, host factors specific to permissive cells appear to facilitate virus translation or correct aberrant patterns of translation initiation that occur in rabbit reticulocyte lysates (72). These observations suggest that the interaction of picornavirus 5'UTRs with cellular proteins may be an important determinant of the ability of the virus to replicate, and that such interactions may profoundly influence host-range and possibly virulence (73, 74). Similarly, interactions between the HAV IRES and specific cellular proteins appear to be important during the adaptation of HAV to replication in cultured cells (75). Certain mutations within the IRES of cell culture-adapted HAV variants promote the growth of the virus in a cell-specific fashion (76-79). These mutations appear to alter the interaction of the 5'UTR with cellular proteins (PTB, GAPDH) in a manner that favorably influences viral translation, a step in replication which appears to be rate-limiting step in some cell types (70, 78, 80). However, although mutations in the IRES influence the ability of the virus to replicate in specific cell types, mutations within the 2B and 2C proteins are the major genetic determinants of adaptation of the virus to growth in cultured cells (38-40, 81). These latter mutations have been shown to result in more efficient replication of subgenomic RNA replicons lacking the P1, capsid-coding sequence of the virus (42).

Although it is certain that translation of the HAV polyprotein is initiated via IRES-mediated, internal entry of ribosomes on the viral RNA, it is not at all clear what cellular proteins (other than the canonical eucaryotic translation initiation factors) may be important to this process *in vivo*. One relatively early report suggests that an hepatocyte-specific factor may be capable of enhancing HAV translation *in vitro* (82), but this report has never been confirmed. Recently, however, we have found that translation directed by the HAV IRES appears to be many-fold more efficient in Huh7 cells, which are derived from human hepatocellular carcinoma cells, that in BSC-1 or FRhK-4 cells which are derived from the kidneys of nonhuman primates (42).

8. PROCESSING OF THE POLYPROTEIN

The single polyprotein encoded by the picornaviral genome undergoes both co- and post-translational proteolytic cleavage under the direction of one or more virus-encoded proteinases. This occurs according to well-characterized schemes that differ in their details among the four major picornaviral genera, as reviewed by Palmenberg (83) (Figure 4). In the case of enteroviruses and rhinoviruses, the primary cleavage of the polyprotein occurs at the VP1/2A junction, allowing the release of the P1 precursor containing the capsid proteins from the non-structural protein precursor (P2-P3). This occurs via a *cis*-action of the 2A proteinase at its own N-terminus. In the case of cardioviruses and aphthoviruses, the primary cleavage event occurs at the 2A/2B junction, releasing the L-P1-2A (cardioviruses) or P1-2A (aphthovirus) precursor for capsid proteins from non-structural proteins (2BC-P3), with the proteolytic cleavage event dependent on the carboxy-terminal sequence of the 2A protein, which contains a highly conserved tripeptide sequence at its extreme terminus (Asn-Pro-Gly), and the N-terminal residue (Pro) of the 2B protein. In each of these four picornaviral genera, all other cleavages within the polyprotein are carried out by the major $3C^{pro}$ proteinase, with the exception of the L/P1 cleavage directed by the Lb proteinase in the case of aphthoviruses, and of the maturation cleavage between the VP4 and VP2 capsid proteins (see below).

The processing of the HAV polyprotein appears to follow the general scheme that is evident in the cardioviruses, but with some striking differences. Most importantly, unlike the polyproteins of each of the other major picornaviral genera that contain at least two, and in some cases, three distinct proteinase activities as described above, the polyprotein of HAV contains only a single proteinase, $3C^{pro}$, that acts both *in cis* and *in trans* to effect cleavage of the polyprotein. This cysteine proteinase was the first picornaviral proteinase for which an atomic-level resolution structure was deduced by X-ray crystallography (84, 85). Although HAV processing events have generally proven to be difficult to analyze in infected cells due to the protracted replication cycle of the virus and the failure of the virus to inhibit host cell protein synthesis, recent studies have shown that the primary cleavage event within the HAV polyprotein takes place at the 2A/2B junction. This cleavage has been precisely mapped by N-terminal sequencing of the 2B polypeptide, and is carried out by the $3C^{pro}$ proteinase (86, 87). The P1-2A capsid protein precursor is probably released from the non-structural protein precursor (2BC-P3) as soon as $3C^{pro}$ is synthesized, as no full-length polyprotein has been observed. The P1-2A fragment is subsequently readily cleaved by $3C^{pro}$ to generate VP0 (VP4-VP2), VP3 and VP1-2A (also termed PX) in experiments in which purified bacterially- expressed $3C^{pro}$ is incubated with P1-2A generated by cell-free translation of synthetic RNA (44). These events have also been confirmed with HAV polypeptides expressed from recombinant vaccinia viruses (88, 89). Thus, HAV appears to be unique among the picornaviruses in terms of the absence of proteinase activity associated with the 2A polypeptide, and in having the primary cleavage ofthe polyprotein directed by the same viral proteinase, $3C^{pro}$, that directs most other cleavage events in the polyprotein (Figure 4).

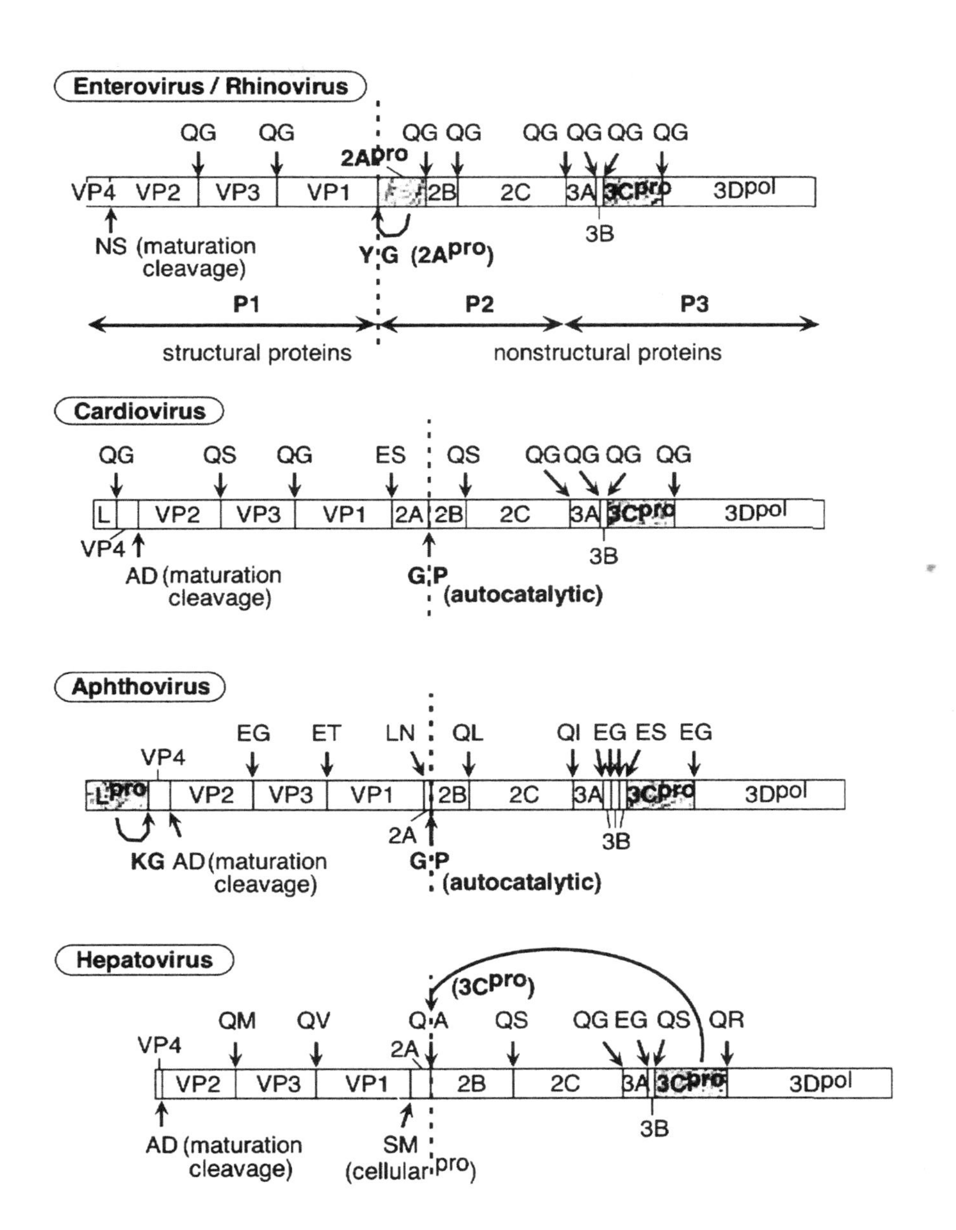

Figure 4. Proteolytic processing of picornavirus polyproteins. Proteinases are highlighted in the polyproteins of each genus. The locations of primary cleavage events are indicated by a dashed line. Dipeptides cleaved by $3C^{pro}$ and dipeptides cleaved by other proteolytic activities (2A, Lb, or unknown) are indicated above and below each polyprotein, respectively.

A second unique feature of HAV processing concerns the VP1-2A (PX) polypeptide that is found, together with VP0 and VP3, in pentamer assemblies that represent the earliest intermediate in the morphogenesis process (see below). The mature VP1 capsid protein is generated from the VP1-2A precursor at a later point in the assembly process, by an as yet incompletely defined mechanism. Several lines of evidence, however, argue strongly against the involvement of HAV $3C^{pro}$ proteinase in this maturation process. Using a heterologous expression system based on recombinant vaccinia viruses directing the expression of full-length or C-terminally-truncated capsid protein precursors and the VP1 protein isolated from HAV-infected cells, we mapped the approximate location of the C-terminus of the mature VP1 capsid protein (88). We found that mutagenesis of the only possible dipeptide sequence that could serve to target the $3C^{pro}$ proteinase to this region of the polyprotein neither abolished the infectivity of RNA transcripts, nor prevented normal maturation of VP1 in the virus rescued from the mutated RNA. This result argues strongly against the involvement of the $3C^{pro}$ proteinase in the VP1/2A cleavage (88). Independent studies mapped the carboxy-terminus of VP1 from purified virions by mass spectrometry. These results also show that the mature capsid protein is not produced by a $3C^{pro}$-mediated cleavage event (90). Thus, at present it is considered likely that the VP1/2A cleavage occurs under the direction of an as yet unidentified cellular proteinase (Figure 4). This aspect of polyprotein processing is unique to HAV among-picornaviruses, and appears to lead to some degree of heterogeneity at the carboxy terminus of VP1 (90). The only possible parallel to this observation among other picornaviruses is found in the cardiovirus genus, where the VP1 capsid protein undergoes a post-assembly, carboxy-terminal trimming of three amino acid residues directed by an apparent cellular proteinase (91). However, this occurs only after $3C^{pro}$-mediated cleavage at the VP1/2A junction (92).

Processing of the non-structural protein precursor (2BC-P3) occurs through both *cis*- and *trans*-proteinase activities of $3C^{pro}$. These cleavage events are not well defined in terms of the temporal order in which they occur, but the most efficient cleavages appear to be those at the 2C/3A and 3C/3D junctions, probably releasing 2BC, 3ABC, and 3D polypeptides (93). Each of the proteins that arise by processing of the nonstructural precursor(s) appear to play a role in the synthesis of new viral RNA molecules. This includes the proteinase $3C^{pro}$ (or possibly a 3ABC precursor) which contains a prominent RNA-binding site on the face of the molecule opposite the active site of the proteinase, and which, based on studies with poliovirus, is likely to play a key role in RNA synthesis (85).

9. VIRAL RNA SYNTHESIS

The replication of HAV, and of subgenomic HAV RNA replicons, proceeds in cultured cells at a much slower rate, and with a much lower overall efficiency, than similar events in other, well studied picornaviruses such as poliovirus. However, it appears that similar general replication mechanisms are at work. Nonetheless, it is important to note that relatively few studies have examined the events involved in the replication of HAV in any kind of detail. One important feature that is common to the replication of all positive-strand RNA viruses is that only the viral RNA, and not unrelated cellular mRNA, undergoes amplification. This occurs despite the fact that the virion RNAs of picornaviruses and cellular mRNAs both share in common a 3' poly(adenylic acid) tail. This specificity of RNA amplification may be accounted for either by compartmentalization of the replication machinery and its co-sequestration with template RNA, and/or the presence of specific replicase-recognition signals in the viral RNA. The evidence suggests that both of these factors may contribute to the specificity of viral RNA amplification in picornaviral infections.

In the case of poliovirus, the most extensively studied of the picornaviruses, the synthesis of new viral RNA molecules takes place in cytoplasmic, membranous structures that arise by a reorganization of intracellular membranes directed by the nonstructural proteins 2C and/or 2BC (94). A similar reorganization of intracellular membranes appears to occur during infection with HAV, albeit to a lesser extent consistent with the lower level of replication of this virus (95). Ultrastructural studies of HAV-infected cells show a tubular-vesicular network in close proximity to the rough endoplasmic reticulum. These membranous structures stain positively with antibodies to the 2B and 2C proteins of the HAV, can be induced by the sole expression of 2B protein, and are likely to be the site of ongoing viral RNA synthesis (96). Thus, 2B and 2C are essential for RNA replication.

In contrast, the deletion of the C-terminal half of the 2A sequence has no effect on the kinetics of virus replication (Cohen, Bénichou, and Martin, manuscript in preparation). Moreover, all of the 2A sequence can be removed from subgenomic HAV RNA replicons lacking the P1 capsid protein region without loss of their capacity to undergo self-amplification in transfected cells (our unpublished results). Consistent with the fact that 2A is not involved in RNA replication, infectious virus can be recovered from recombinant HAV genomes containing exogenous protein-coding sequences inserted in-frame at the 2A/2B junction and flanked by consensus $3C^{pro}$ cleavage sites. These observations demonstrate that the 2A polypeptide does not function *in cis* with the remaining nonstructural proteins (97).

Based on studies with poliovirus and other picornaviruses, replication of HAV RNA is also likely to involve the P3 proteins, 3A, 3B (the genome-linked protein, VPg), $3C^{pro}$, and $3D^{pol}$, as well as one or more processing intermediates. However, the events leading up to the initiation of either minus-strand synthesis or the production of new copies of positive-strand RNA have not been studied in detail in the case of HAV. It is also important to note that the mechanisms involved in poliovirus RNA replication remain incompletely defined and are still the object of intense study.

By analogy, however, it is likely that the HAV proteins described above assemble as a replicase complex at or near the 3' end of the positive-strand, virion RNA. Synthesis of the minus-strand intermediate most probably occurs via a primed reaction in which 3B (VPg), modified by a cellular enzyme so that it is covalently linked to a U-U dinucleotide sequence (VPg-pU-pU) through a conserved tyrosine residue, serves as primer on the poly(adenylic acid) sequence located at the 3' end. The uridylation of the VPg protein of other picornaviruses appears to be templated by a small stem-loop RNA structure (the *cis*-acting replication element, or *cre*) that is located within the polyprotein coding sequences of these viruses (98-100), and that is essential for minus-strand synthesis (101, 102). A similar element is likely to exist in the HAV genome, but it has yet to be specifically identified.

The 3A protein may play a role in this process, by helping to anchor the 3AB (or larger) processing intermediate to the 2B and 2C-containing membranous structures described above. The HAV 3AB precursor, expressed in bacteria, has been shown to bind to membranes and to HAV RNA (103). The $3D^{pol}$ polypeptide contains the classic sequence motifs (*e.g.*, Glu-Asp-Asp) for an RNA-dependent, RNA polymerase. There is no question that it is responsible for the subsequent synthesis of the new minus-strand RNA, but it is peculiar that attempts to demonstrate polymerase activity in purified preparations of this protein have consistently met with failure, despite positive results obtained with the analogous poliovirus $3D^{pol}$ protein (104).

Other than the 3' terminal poly(adenylic acid) tail, the role of the untranslated RNA sequence at the 3' end of the genome is uncertain in these events. However, conserved secondary RNA structures exist within the 3'UTR, just upstream of the poly(adenylic acid) sequence in the HAV genome. They may assist the replicase in positioning itself near the initiation site. Further evidence of a specific interaction between the picornavirus 3'UTR and $3D^{pol}$ has been provided by the demonstration of second-site mutations in the active-site cleft of $3D^{pol}$ that occurred in response to replication-limiting mutations in the 3'UTR (105). However, infectious poliovirus has been rescued from RNA following the deletion of the entire 3'UTR, showing that this sequence is not absolutely required for viral replication (106).

Relatively little minus-strand RNA is produced during HAV infections (107), and it is typically exceedingly difficult to demonstrate its presence in infected cells. This intermediate does, however, serve as template for the production of significantly larger quantities of new positive-strand RNA molecules. Consistent with this asymmetric amplification of the viral

RNA, the initiation events in positive-strand synthesis appear to differ from those in minus-strand synthesis. The most appealing hypothesis, based almost exclusively on studies with poliovirus, is that the template for new positive-strand synthesis is the duplex RNA molecule that results from the synthesis of a complete minus-strand copy on the input virion RNA (so-called "replicative form" (RF) RNA species) (Figure 2). In the case of poliovirus, a complex formed by cellular poly(rC)-binding protein as well as viral (3AB, and 3CD) proteins forms at the extreme 5' end of the positive-strand molecule which contains a conserved cloverleaf-like RNA structure (108,109). Presumably, this structure serves as a site of assembly of the replicase complex, which initiates synthesis of a new positive-strand RNA molecule by a primer-dependent mechanism in which VPg-pU-pU again serves as the primer (108). Multiple copies of nascent positive-strand RNA may initiate off a single duplex molecule, as the replication forks move in a 5' direction along the negative-strand template, resulting in the "replicative intermediate" (RI) RNA species (Figure 2).

It is not known how much of this model is relevant to the replication of HAV. HAV also contains highly conserved RNA secondary and tertiary structure within its 5' 98 nucleotides (a large stem-loop followed by two probable RNA pseudoknots) (60) (Figure 3). These structures could serve functions similar to the cloverleaf of poliovirus in positive-strand RNA replication, but there are few data that address this issue. Downstream of these sequences lies a loosely conserved polypyrimidine tract of ~40 nucleotides in length (Figure 3). Using a mobility shift assay and UV cross-linking, it has been shown that HAV $3C^{pro}$ binds to RNA transcripts representing the first 150 nucleotides of the 5'NTR (110). These observations are reminiscent of the binding of poliovirus 3CD to the 5' cloverleaf of poliovirus. In addition, the deletion of two nucleotides from the second of the putative pseudoknots is sufficient to knock out the capacity of synthetic HAV RNA to replicate (111). Deletion of the entire polypyrimidine tract does not affect virus replication in cell culture, but extending the deletion by 5 nucleotides into an immediately downstream single-stranded region of the RNA results in a thermosensitive replication phenotype for the virus (111). Since this temperature-sensitive phenotype appears to be due to a defect in RNA synthesis (112), this region (which is immediately upstream of the IRES, Figure 3) may be recognized by proteins involved in RNA synthesis.

Much of the foregoing remains speculative, however. Significant questions persist concerning the mechanisms of replication of picornaviral RNAs in general. Unfortunately, even less is known about the specific mechanisms underlying the replication of HAV RNA.

10. VIRAL ASSEMBLY AND RELEASE

RNA packaging probably occurs as the capsid assembles, although the process by which this happens is not understood at present. Empty capsids are present in most virus preparations, at times approaching a significant proportion of the particles. These empty capsids contain the VP4-VP2 precursor (VP0), in contrast to the presence of VP2 in the mature, RNA-containing virion. As in all picornaviruses, the "maturation cleavage" between the VP4 and VP2 capsid proteins occurs by an as yet unexplained mechanism that is dependent on the presence of viral RNA within the capsid (54).

Thus, as indicated above, the mature hepatitis A virion contains 3 major polypeptides (VP1, VP2 and VP3), and possibly the fourth, smaller, VP4 polypeptide. The existence of VP4 within the HAV particle is an important question, since VP4 is N-terminally myristylated in each of the other major picornaviral genera (113). The myristylation of VP4 is considered to play a role in the earliest steps of particle assembly involving the formation of 14S pentamer morphogenesis intermediates containing 5 copies of each of the capsid polypeptides (114). Unlike other picornaviruses, however, the N-terminus of the putative HAV VP4 sequence does not contain a consensus myristylation signal. Nonetheless, a myristylation signal does exist internally within the VP4 coding sequence, suggesting the possibility that the VP4 polypeptide might be cleaved into a leader (L) polypeptide such as is present in the aphthoviruses and cardioviruses, albeit one that would be much shorter in length (only 6 amino acids). Whether or not this is the case, the very small size of the VP4 protein (17-23 amino acids) would make its detection in conventional gels extremely difficult. Attempts to demonstrate myristylation of the mature HAV particle have failed (115). These observations suggest that early steps in the morphogenesis of the HAV particle may differ from what is thought to occur in the other major picornaviral genera. The role of the VP4 sequence in the assembly process has recently been studied using recombinant vaccinia viruses expressing HAV precursors that assemble into pentamers and empty capsids. Probst et al. reported that deletion of the VP4 sequence in the capsid protein precursor does not affect pentamer assembly but prevents the formation of empty capsids (89). Despite this, we have found that empty capsids do form in the absence of VP4 (Martin et al., unpublished data). Thus, not only the presence of VP4 in virions, but also its role, if any, in particle assembly remain uncertain.

Additional information supporting the notion that HAV differs from other picornaviruses in its assembly process comes from recent studies of the role of the 2A sequence. As described above, this carboxy-terminal extension of the VP1 sequence is present (as the VP1-2A precursor, the PX protein) in both pentamers as well as some early virion morphogenesis intermediates (116), and has no homology with other picornaviral 2A proteins. We have demonstrated that deletions in the N-terminal third of 2A abolish infectivity of the viurs, whereas deletions in the C-terminal two-thirds result in a reduced replication phenotype. As mentioned above, 2A is not required for RNA synthesis, but is required for the complete life cycle of the virus. We further demonstrated that deletions in the C-terminal part of 2A alter the VP1/2A cleavage, resulting in accumulation of uncleaved VP1-2A precursor in virions and possibly accounting for a delay in the appearance of infectious particles and a slightly decreased particle specific infectivity with these mutants. In addition, our data also indicate that the N-terminal domain of 2A must be present as a C-terminal extension of VP1 for pentamer assembly, after which cleavage at the VP1-2A junction produces the mature VP1 protein (see above) (Cohen, Bénichou, Martin, manuscript in preparation). The relevance of these observations to the virus life cycle *in vivo* is unknown. However, a short deletion in the C-terminal region of 2A had no effect on either the infectivity or virulence of HAV in primates (117). This suggests that HAV particles containing abnormal amounts of 2A as a C-terminal extension of the VP1 capsid protein are fully infectious *in vivo*.

In the case of highly cell culture-adapted, cytopathic HAV strains, cell death may occur due to apoptosis (96, 118). However, in most infected cells there is no cytopathology. Although much of the newly replicated virus remains tightly cell-associated in such cells, there is extensive release of progeny virus into supernatant fluids. In polarized colonic epithelial cell (Caco-2) cultures, this release of virus occurs almost exclusively into apical supernatant fluids (43). Brefeldin A causes a profound inhibition of HAV replication, probably because of disruption of the membranous replication complexes (see above), but also selectively reduced the apical release of virus (43). These results suggest that the release of virus across the apical membrane of polarized cells may involve vectorial cellular vesicular transport mechanisms. Such results are potentially of relevance to the situation in infected hepatocytes, which are also polarized cells of epithelial origin. The apical surface of the hepatocyte forms a well-demarcated groove that encircles the cell and provides access to the biliary canaliculi through which components of bile (including HAV during acute hepatitis A) are secreted from the liver into the feces.

REFERENCES

1. Feinstone S.M., Kapikian A.Z., Purcell R.H. Hepatitis A: detection by immune electron microscopy of a viruslike antigen associated with acute illness. Science 1973; 182:1026-1028.
2. Daemer R.J., Feinstone S.M., Gust I.D., Purcell R.H. Propagation of human hepatitis A virus in African Green Monkey kidney cell culture: primary isolation and serial passage. Infect Immun 1981; 32:388-393.

3. Binn L.N., Lemon S.M., Marchwicki R.H., Redfield R.R., Gates N.L., Bancroft W.H. Primary isolation and serial passage of hepatitis A virus strains in primate cell cultures. J Clin Microbiol 1984; 20:28-33.
4. Provost P.J., Hilleman M.R. Propagation of human hepatitis A virus in cell culture *in vitro*. Proc Soc Exper Biol Med 1979; 160:213-221.
5. Gauss-Muller V., Deinhardt F. Effect of hepatitis A virus infection on cell metabolism *in vitro*. Proc Soc Exper Biol Med 1984; 175:10-15.
6. Lemon S.M., Shapiro C.N. The value of immunization against hepatitis A. Infectious Agents and Disease 1994; 3:38-49.
7. Dienstag J.L., Feinstone S.M., Kapikian A.Z., Purcell R.H., Boggs J.D., Conrad M.E. Faecal shedding of hepatitis-A antigen. Lancet 1975; i:765-767.
8. Asher L.V.S., Binn L.N., Mensing T.L., Marchwicki R.HVassell., R.A., Young G.D. Pathogenesis of hepatitis A in orally inoculated owl monkeys (Aotus trivergatus). J Med Virol 1995; 47:260-268.
9. Siegl G., Weitz M., Kronauer G. Stability of hepatitis A virus. Intervirology 1984; 22:218-226.
10. Lemon S.M., Binn L.N., Marchwicki R., Murphy P.C., Ping L.-H., Jansen R.W., Asher L.V.S., Stapleton J.T., Taylor D.G., LeDuc J.W. In vivo replication and reversion to wild-type of a neutralization-resistant variant of hepatitis A virus. J Infect Dis 1990; 161:7-13.
11. Lemon S.M., Binn L.N. Serum neutralizing antibody response to hepatitis A virus. J Infect Dis 1983; 148:1033-1039.
12. Decker R.H., Overby L.R., Ling C.-M., Frosner G.G., Deinhardt F., Boggs J. Serologic studies of transmission of hepatitis A in humans. J Infect Dis 1979; 139:74-82.
13. Vallbracht A., Gabriel P., Maier K., Hartmann F., Steinhardt H.J., Miller C., Wolf A., Manncke K.H., Flehmig B. Cell-mediated cytotoxicity in hepatitis A virus infection. Hepatology 1986; 6:1308-1314.
14. Vallbracht A., Maier K., Stierhof Y.-D., Wiedmann K.H., Flehmig B., Fleischer B.. Liver-derived cytotoxic T cells in hepatitis A virus infection. J Infect Dis 1989; 160:209-217.
15. Fleischer B., Fleischer S., Maier K., Wiedmann K.H., Sacher M., Thaler H.. Vallbracht A. Clonal analysis of infiltrating T lymphocytes in liver tissue in viral hepatitis A. Immunology 1990; 69:14-19.
16. Vallbracht A., Fleischer B., Busch F.W. Hepatitis A: hepatotropism and influence on myelopoiesis. Intervirology 1993; 35:133-139.
17. Maier K., Gabriel P., Koscielniak E., Stierhof Y.-D., Wiedmann K.H., Flehmig B., Vallbracht A.. Human gamma interferon production by cytotoxic T lymphocytes sensitized during hepatitis A virus infection. J Virol 1988; 62:3756-3763.
18. Coulepis A.G., Locarnini S.A., Lehmann N.I., Gust I.D. Detection of hepatitis A virus in the feces of patients with naturally acquired infections. J Infect Dis 1980; 141:151-156.
19. Skinhoj P., Mikkelsen F., Hollinger F.B. Hepatitis A in Greenland: importance of specific antibody testing in epidemiologic surveillance. Am J Epidemiol 1977; 105:140-147.
20. Rosenblum L.S., Villarino M.E., Nainan O.V., Melish M.E., Hadler S.C., Pinsky P.P., Jarvis W.R., Ott C.E., Margolis H.S. Hepatitis A outbreak in a neonatal intensive care unit: Risk factors for transmission and evidence of prolonged viral excretion among preterm infants. J Infect Dis 1991; 164:476-482.
21. Dienstag, J.L., Feinstone S.M., Purcell R.H., Hoofnagle J.H., Barker L.F., London W.T., Popper H., Peterson J.M., Kapikian A.Z. Experimental infection of chimpanzees with hepatitis A virus. J Infect Dis 1975; 132:532-545.

22. LeDuc J.W., Lemon S.M., Keenan C.M., Graham R.R., Marchwicki R.H., Binn L.N.. Experimental infection of the New World owl monkey (*Aotus trivirgatus*) with hepatitis A virus. Infect Immun 1983; 40:766-772.
23. Lemon S.M., Binn L.N. Antigenic relatedness of two strains of hepatitis A virus determined by cross-neutralization. Infect Immun 1983; 42:418-420.
24. Robertson B.H., Jansen R.W., Khanna B., Totsuka A., Nainan O.V., Siegl G., Widell A., Margolis H.S., Isomura S., Ito K., Ishizu T., Moritsugu Y., Lemon S.M. Genetic relatedness of hepatitis A virus strains recovered from different geographic regions. J Gen Virol 1992; 73:1365-1377.
25. Lemon S.M., Thomas D.L. Vaccines to prevent viral hepatitis. N Engl J Med 1997; 336:196-204.
26. Hughes J.V., Bennett C., Stanton L.W., Linemeyer D.L., Mitra S.W. "Hepatitis-A virus structural proteins: sequencing and ability to induce virus-neutralizing antibody responses." In *Vaccines 85: Molecular and Chemical Basis of Resistance to Parasitic, Bacterial and Viral Diseases,* Lerner R.A., Chanock R.M., Brown F., eds., Cold Spring Harbor Laboratory, New York: Cold Spring Harbor, 1985; pp 255-259.
27. Lemon S.M., Amphlett E., Sangar D. Protease digestion of hepatitis A virus: disparate effects on capsid proteins, antigenicity, and infectivity. J Virol 1991; 65:5636-5640.
28. Hogle J.M., Chow M., Filman D.J. Three-dimensional structure of poliovirus at 2.9 Å resolution. Science 1985; 229:1358-1365.
29. Acharya R., Fry E., Stuart D., Fox G., Rowlands D., Brown F. The three-dimensional structure of foot-and-mouth disease virus at 2.9 Å resolution. Nature 1989; 337:709-716.
30. Ping L.-H., Lemon S.M. Antigenic structure of human hepatitis A virus defined by analysis of escape mutants selected against murine monoclonal antibodies. J Virol 1992; 66:2208-2216.
31. Cheng R.H. Pleomorphic and symmetrical organizations of viral capsids in virus assembly and entry. The Sixth International Symposium on Positive Strand Viruses, sponsored by the Institute Pasteur and the Federation of European Microbiological Societies; Paris, France, 2001; May 28 – June 2.
32. Ticehurst J.R., Racaniello V.R., Baroudy B.M., Baltimore D., Purcell R.H., Feinstone S.M. Molecular cloning and characterization of hepatitis A virus cDNA. Proc Natl Acad Sci USA 1983; 80:5885-5889.
33. Cohen J.I., Ticehurst J.R., Purcell R.H., Buckler-White A., Baroudy B.M. Complete nucleotide sequence of wild-type hepatitis A virus: comparison with different strains of hepatitis A virus and other picornaviruses. J Virol 1987; 61:50-59.
34. Weitz M., Baroudy B.M., Maloy W.L., Ticehurst J.R., Purcell R.H. Detection of a genome-linked protein (VPg) of hepatitis A virus and its comparison with other picornaviral VPgs. J Virol 1986; 60:124-130.
35. Paul, A.V., Van Boom J.H., Filippov D., Wimmer E. Protein-primed RNA synthesis by purified poliovirus RNA polymerase. Nature 1998; 393:280-284.
36. Cohen J.I., Ticehurst J.R., Feinstone S.M., Rosenblum B., Purcell R.H. Hepatitis A virus cDNA and its RNA transcripts are infectious in cell culture. J Virol 1987; 61:3035-3039.
37. Tesar M., Harmon S.A., Summers D.F., Ehrenfeld E. Hepatitis A virus polyprotein synthesis initiates from two alternative AUG codons. Virology 1992; 186:609-618.
38. Emerson S.U., Huang Y.K., McRill C., Lewis M., Purcell R.H. Mutations in both the 2B and 2C genes of hepatitis A virus are involved in adaptation to growth in cell culture. J Virol 1992; 66:650-654.
39. Zhang H.C., Chao S.F., Ping L.H., Grace K., Clarke B., Lemon S.M. An infectious cDNA clone of a cytopathic hepatitis A virus: Genomic regions associated with rapid replication and cytopathic effect. Virology 1995; 212:686-697.

40. Graff J., Kasang C., Normann A., Pfisterer-Hunt M., Feinstone S.M., Flehmig B. Mutational events in consecutive passages of hepatitis A virus strain GBM during cell culture adaptation. Virology 1994; 204:60-68.
41. Kusov Y., Gauss-Muller V. Improving proteolytic cleavage at the 3A/3B site of the hepatitis A virus polyprotein impairs processing and particle formation, and the impairment can be complemented in trans by 3AB and 3ABC. J Virol 1999; 73:9867-9878.
42. Yi M., Lemon S.M. The replication of subgenomic hepatitis A virus RNAs expressing firefly luciferase is enhanced by mutations associated with adaptation of virus to growth in cultured cells. 2001; Submitted for publication.
43. Blank C.A., Anderson D.A., Beard M., Lemon S.M. Infection of polarized cultures of human intestinal epithelial cells with hepatitis A virus: vectorial release of progeny virions through apical cellular membranes. J Virol 2000; 74:6476-6484.
44. Schultheiss T., Sommergruber W., Kusov Y., Gauss-Müller V. Cleavage specificity of purified recombinant hepatitis A virus 3C proteinase on natural substrates. J Virol 1995; 69:1727-1733.
45. Sicinski P., Rowinski J., Warchol J.B., Jarzcabek Z., Gut W., Szczygiel B., Bielecki K., Koch G. Poliovirus type 1 enters the human host through intestinal M cells. Gastroenterology 1990; 98:56-58.
46. Mendelsohn C.L., Wimmer E., Racaniello V.R. Cellular receptor for poliovirus: molecular cloning, nucleotide sequence, and expression of a new member of the immunoglobulin superfamily. Cell 1989; 56:855-865.
47. Dotzauer A., Feinstone S.M., Kaplan G. Susceptibility of nonprimate cell lines to hepatitis A virus infection. J Virol 1994; 68:6064-6068.
48. Frings W., Dotzauer A. Adaptation of primate cell-adapted hepatitis A virus strain HM175 to growth in guinea pig cells is independent of mutations in the 5' nontranslated region. J Gen Virol 2001; 82:597-602.
49. Kaplan G., Totsuka A., Thompson P., Akatsuka T., Moritsugu Y., Feinstone S.M. Identification of a surface glycoprotein on African green monkey kidney cells as a receptor for hepatitis A virus. EMBO J 1996; 15:4282-4296.
50. Silberstein E., Dveksler G., Kaplan G.G. Neutralization of hepatitis A virus (HAV) by an immunoadhesin containing the cysteine-rich region of HAV cellular receptor-1. J Virol 2001; 75:717-725.
51. Feigelstock D., Thompson P., Mattoo P., Zhang Y., Kaplan G.G. The human homolog of HAVcr-1 codes for a hepatitis A virus cellular receptor. J Virol 1998; 72:6621-6628.
52. Dotzauer A., Gebhardt U., Bieback K., Gottke U., Kracke A., Mages J., Lemon S.M., Vallbracht A. Hepatitis A virus-specific immunoglobulin A mediates infection of hepatocytes with hepatitis A virus via the asialoglycoprotein receptor. J. Virol 2000; 74:10950-10957.
53. Bishop N.E., Anderson D.A. Uncoating kinetics of hepatitis A virus virions and provirions. J Virol 2000; 74:3423-3426.
54. Bishop N.E., Anderson D.A. RNA-dependent cleavage of VPO capsid protein in provirions of hepatitis A virus. Virology 1993; 197:616-623.
55. Jackson R.J., Hunt S.L., Gibbs C.L., Kaminski A. Internal initiation of translation of picornavirus RNAs. Mol Biol Rep 1994; 19:147-159.
56. Pelletier J., Sonenberg N. Internal initiation of translation of eukaryotic mRNA directed by a sequence derived from poliovirus RNA. Nature 1988; 334:320-325.
57. Jang S.K., Wimmer E. Cap-independent translation of encephalomyocarditis virus RNA: structural elements of the internal ribosomal entry site and involvement of a cellular 57-kD RNA-binding protein. Genes Dev 1990; 4:1560-1572.
58. Brown E.A., Zajac A.J., Lemon S.M. In vitro characterization of an internal ribosomal entry site (IRES) present within the 5' nontranslated region of hepatitis A

virus RNA: Comparison with the IRES of encephalomyocarditis virus. J Virol 1994; 68:1066-1074.

59. Glass M.J., Jia X.-Y., Summers D.F. Identification of the hepatitis A virus internal ribosome entry site: *in vivo* and *in vitro* analysis of bicistronic RNAs containing the HAV 5' noncoding region. Virology 1993; 193:842-852.

60. Brown E.A., Day S.P., Jansen R.W., Lemon S.M. The 5' nontranslated region of hepatitis A virus: secondary structure and elements required for translation in vitro. J Virol 1991; 65:5828-5838.

61. Le S.-Y., Chen J.-H., Sonenberg N., Maizel J.V., Jr. Conserved tertiary structural elements in the 5' nontranslated region of cardiovirus, aphthovirus and hepatitis A virus RNAs. Nucleic Acids Res 1993; 21:2445-2451.

62. Wimmer E., Hellen C.U.T., Cao X. Genetics of poliovirus. Annu Rev Genet 1993; 27:353-436.

63. Whetter L.E., Day S.P., Elroy-Stein O., Brown E.A., Lemon S.M. Low efficiency of the 5' nontranslated region of hepatitis A virus RNA in directing cap-independent translation in permissive monkey kidney cells. J Virol 1994; 68:5253-5263.

64. Borman A.M., Kean K.M. Intact eukaryotic initiation factor 4G is required for hepatitis A virus internal initiation of translation. Virology 1997; 237:129-136.

65. Borman, A.M., Bailly J.L., Girard M., Kean K.M. Picornavirus internal ribosome entry segments: comparison of translation efficiency and the requirements for optimal internal initiation of translation in vitro. Nucleic Acids Res 1995; 23:3656-3663.

66. Borman A.M., Le Mercier P., Girard M., Kean K.M. Comparison of picornaviral IRES-driven internal initiation of translation in cultured cells of different origins. Nucleic Acids Res 1997; 25:925-932.

67. Kaminski A. Jackson R.J. The polypyrimidine tract binding protein (PTB) requirement for internal initiation of translation of cardiovirus RNAs is conditional rather than absolute. RNA 1998; 4:626-638.

68. Gosert R., Chang K.H., Rijnbrand R., Yi M., Sangar D.V., Lemon S.M. Transient expression of cellular polypyrimidine-tract binding protein stimulates cap-independent translation directed by both picornaviral and flaviviral internal ribosome entry sites in vivo. Mol Cell Biol 2000; 20:1583-1595.

69. Schultz D.E., Hardin C.C., Lemon S.M. Specific interaction of glyceraldehyde 3-phosphate dehydrogenase with the 5' nontranslated RNA of hepatitis A virus. J Biol Chem 1996; 271:14134-14142.

70. Yi M., Schultz D.E., Lemon S.M. Functional significance of the interaction of hepatitis A virus RNA with glyceraldehyde 3-phosphate dehydrogenase (GAPDH): opposing effects of GAPDH and polypyrimidine tract binding protein on internal ribosome entry site function. J Virol 2000; 74:6459-6468.

71. Graff J., Cha J., Blyn L.B., Ehrenfeld E. Interaction of poly(rC) binding protein 2 with the 5' noncoding region of hepatitis A virus RNA and its effects on translation. J Virol 1998; 72:9668-9675.

72. Meerovitch K., Svitkin Y.V., Lee H.S., Lejbkowicz F., Kenan D.J., Chan E.K.L., Agol V.I., Keene J.D., Sonenberg N. La autoantigen enhances and corrects aberrant translation of poliovirus RNA in reticulocyte lysates. J Virol 1993; 67:3798-3807.

73. Svitkin Y.V., Pestova T.V., Maslova S.V., Agol V.I. Point mutations modify the response of poliovirus RNA to a translation initiation factor: a comparison of neurovirulent and attenuated strains. Virology 1988; 166:394-404.

74. Gutierrez A.L., Denova-Ocampo M., Racaniello V.R., del Angel R.M. Attenuating mutations in the poliovirus 5' untranslated region alter its interaction with polypyrimidine tract-binding protein. J Virol 1997; 71:3826-3833.

75. Chang K.H., Brown E.A., Lemon S.M. Cell type-specific proteins which interact with the 5' nontranslated region of hepatitis A virus RNA. J Virol 1993; 67:6716-6725.

76. Day S.P., Murphy P., Brown E.A., Lemon S.M. Mutations within the 5' nontranslated region of hepatitis A virus RNA which enhance replication in BS-C-1 cells. J Virol 1992; 66:6533-6540.
77. Graff J., Normann A., Flehmig B. Influence of the 5' noncoding region of hepatitis A virus strain GBM on its growth in different cell lines. J Gen Virol 1997; 78:1841-1849.
78. Funkhouser A.W., Schultz D.E., Lemon S.M., Purcell R.H., Emerson S.U. Hepatitis A virus translation is rate-limiting for virus replication in MRC-5 cells. Virology 1999; 254:268-278.
79. Funkhouser A.W., Purcell R.H., D'Hondt E., Emerson S.U. Attenuated hepatitis A virus: Genetic determinants of adaptation to growth in MRC-5 cells. J Virol 1994; 68:148-157.
80. Schultz D.E., Honda M., Whetter L.E., McKnight K.L., Lemon S.M. Mutations within the 5' nontranslated RNA of cell culture-adapted hepatitis A virus which enhance cap-independent translation in cultured African green monkey kidney cells. J Virol 1996; 70:1041-1049.
81. Emerson S.U., Huang Y.K., Purcell R.H. 2B and 2C mutations are essential but mutations throughout the genome of HAV contribute to adaptation to cell culture. Virology 1993; 194:475-480.
82. Glass M.J. Summers D.F. Identification of a trans-acting activity from liver that stimulates hepatitis A virus translation *in vitro*. Virology 1993; 193:1047-1050.
83. Palmenberg, A.C. Proteolytic processing of picornaviral polyprotein. Annu Rev Microbiol 1990; 44:603-23.:603-623.
84. Allaire M., Chernala M.M., Malcolm B.A., James M.N.G. Picornaviral 3C cysteine proteinases have a fold similar to chymotrypsin-like serine proteinases. Nature 1994; 369:72-76.
85. Bergmann E.M., Mosimann S.C., Chernaia M.M., Malcolm B.A., James M.N.G.. The refined crystal structure of the 3C gene product from hepatitis A virus: Specific proteinase activity and RNA recognition. J Virol 1997; 71:2436-2448.
86. Martin A., Escriou N., Chao S.-F., Lemon S.M., Girard M., Wychowski C. Identification and site-direct mutagenesis of the primary (2A/2B) cleavage site of the hepatitis A virus polyprotein: Functional impact on the infectivity of HAV transcripts. Virology 1995; 213:213-222.
87. Gosert R., Cassinotti P., Siegl G., Weitz M. Identification of hepatitis A virus non-structural protein 2B and its release by the major virus protease 3C. J Gen Virol 1996; 77:247-255.
88. Martin A., Benichou D., Chao S.F., Cohen L.M., Lemon S.M. Maturation of the hepatitis A virus capsid protein VP1 is not dependent on processing by the 3Cpro proteinase. J Virol 1999; 73:6220-6227.
89. Probst, C., Jecht M., Gauss-Muller V. Intrinsic signals for the assembly of hepatitis A virus particles. Role of structural proteins VP4 and 2A. J Biol Chem 1999; 274:4527-4531.
90. Graff J., Richards O.C., Swiderek K.M., Davis M.T., Rusnak F., Harmon S.A., Jia X.Y., Summers D.F., Ehrenfeld E. Hepatitis A virus capsid protein VP1 has a heterogeneous C terminus. J Virol 1999; 73 :6015-6023.
91. Boege U., Scraba D.G. Mengo virus maturation is accompanied by C-terminal modification of capsid protein VP1. Virology 1989; 168:409-412.
92. Palmenberg A.C., Parks G.D., Hall D.J., Ingraham R.H., Seng T.W., Pallai P.V. Proteolytic processing of the cardioviral P2 region: primary 2A/2B cleavage in clone-derived precursors. Virology 1992; 190:754-762.
93. Probst C., Jecht M., Gauss-Muller V. Processing of proteinase precursors and their effect on hepatitis A virus particle formation. J Virol 1998; 72:8013-8020.

94. Cho M.W., Teterina N., Egger D., Bienz K., Ehrenfeld E. Membrane rearrangement and vesicle induction by recombinant poliovirus 2C and 2BC in human cells. Virology 1994; 202:129-145.

95. Teterina N.L., Bienz K., Egger D., Gorbalenya A.E., Ehrenfeld E. Induction of intracellular membrane rearrangements by HAV proteins 2C and 2BC. Virology 1997; 237:66-77.

96. Gosert R., Egger D., Bienz K. A cytopathic and a cell culture adapted hepatitis A virus strain differ in cell killing but not in intracellular membrane rearrangements. Virology 2000; 266:157-169.

97. Beard M.R., Cohen L., Lemon S.M., Martin A. Characterization of recombinant hepatitis A virus genomes containing exogenous sequences at the 2A/2B junction. J Virol 2001; 75:1414-1426.

98. McKnight K.L., Lemon S.M. Capsid coding sequence is required for efficient replication of human rhinovirus 14 RNA. J Virol 1996; 70:1941-1952.

99. Lobert P.E., Escriou N., Ruelle J., Michiels T. A coding RNA sequence acts as a replication signal in cardioviruses. Proc Natl Acad Sci USA 1999; 96:11560-11565.

100. Goodfellow I., Chaudhry Y., Richardson A., Meredith J., Almond J.W., Barclay W., Evans D.J. Identification of a cis-acting replication element within the poliovirus coding region. J Virol 2000; 74:4590-4600.

101. McKnight K.L., Lemon S.M. The rhinovirus type 14 genome contains an internally located RNA structure that is required for viral replication. RNA 1998; 4:1569-1584.

102. Paul A.V., Rieder E., Kim D.W., Van Boom J.H., Wimmer E. Identification of an RNA hairpin in poliovirus RNA that serves as the primary template in the in vitro uridylylation of VPg. J Virol 2000; 74:10359-10370.

103. Beneduce F., Ciervo A., Kusov Y., Gauss-Muller V., Morace G. Mapping of protein domains of hepatitis A virus 3AB essential for interaction with 3CD and viral RNA. Virology 1999; 264:410-421.

104. Tesar M., Pak I., Jia X.-Y., Richards O.C., Summers D.F., Ehrenfeld E. Expression of hepatitis A virus precursor protein P3 *in vivo* and *in vitro*: Polyprotein processing of the 3CD cleavage site. Virology 1994; 198:524-533.

105. Meredith J.M., Rohll J.B., Almond J.W., Evans D.J. Similar interactions of the poliovirus and rhinovirus 3D polymerases with the 3' untranslated region of rhinovirus 14. J Virol 1999; 73:9952-9958.

106. Todd, S., Towner J.S., Brown D.M., Semler B.L. Replication-competent picornaviruses with complete genomic RNA 3' noncoding region deletions. J Virol 1997; 71:8868-8874.

107. Anderson D.A., Ross B.C., Locarnini S.A. Restricted replication of hepatitis A virus in cell culture: encapsidation of viral RNA depletes the pool of RNA available for replication. J Virol 1988; 62:4201-4206.

108. Andino R., Rieckhof G.E., Achacoso P.L., Baltimore D. Poliovirus RNA synthesis utilizes an RNP complex formed around the 5'-end of viral RNA. EMBO J 1993; 12:3587-3598.

109. Parsley T.B., Towner J.S., Blyn L.B., Ehrenfeld E., Semler B.L. Poly (rC) binding protein 2 forms a ternary complex with the 5'-terminal sequences of poliovirus RNA and the viral 3CD proteinase. RNA 1997; 3:1124-1134.

110. Kusov Y.Y., Gauss-Muller V. In vitro RNA binding of the hepatitis A virus proteinase 3C (HAV 3Cpro) to secondary structure elements within the 5' terminus of the HAV genome. RNA 1997; 3:291-302.

111. Shaffer D.R., Brown E.A., Lemon S.M. Large deletion mutations involving the first pyrimidine-rich tract of the 5' nontranslated RNA of hepatitis A virus define two adjacent domains associated with distinct replication phenotypes. J Virol 1994; 68:5568-5578.

112. Shaffer D.R., Lemon S.M. Temperature-sensitive hepatitis A virus mutants with deletions downstream of the first pyrimidine-rich tract of the 5' nontranslated RNA are impaired in RNA synthesis. J Virol 1995; 69:6498-6506.

113. Chow M., Newman J.F.E., Filman D.J., Hogle J.M., Rowlands D.J., Brown F. Myristylation of picornavirus capsid protein VP4 and its structural significance. Nature 1987; 327:482-486.

114. Marc D., Masson G., Girard M., van der Werf S. Lack of myristoylation of poliovirus capsid polypeptide VP0 prevents the formation of virions or results in the assembly of noninfectious virus particles. J Virol 1990; 64:4099-4107.

115. Tesar M., Jia X.-Y., Summers D.F., Ehrenfeld E. Analysis of a potential myristoylation site in hepatitis A virus capsid protein VP4. Virology 1993; 194:616-626.

116. Anderson D.A., Ross B.C. Morphogenesis of hepatitis A virus: Isolation and characterization of subviral particles. J Virol 1990; 64:5284-5289.

117. Harmon S.A., Emerson S.U., Huang Y.K., Summers D.F., Ehrenfeld E. Hepatitis A viruses with deletions in the 2A gene are infectious in cultured cells and marmosets. J Virol 1995; 69:5576-5581.

118. Brack K., Frings W., Dotzauer A., Vallbracht A. A cytopathogenic, apoptosis-inducing variant of hepatitis A virus. J Virol 1998; 72:3370-3376.

Chapter 3

THE MOLECULAR BIOLOGY OF HEPATITIS B VIRUS

T. S. Benedict Yen

Pathology Service 113B, San Francisco Veterans Affairs Medical Center, 4150 Clement Street, San Francisco, CA 94121, and Department of Pathology, University of California, San Francisco, CA 94143-0506

1. INTRODUCTION

Hepatitis B virus (HBV) is a major infectious cause of morbidity and mortality throughout the world (1). Despite the availability of an effective vaccine for 2 decades, approximately 1.2 million people still die each year from HBV-related diseases. HBV usually causes only mild symptoms upon acute infection, although rarely fatal fulminant hepatitis can occur. The great majority of infected adults successfully clear the virus and acquire life-long immunity to HBV. However, others fail to mount an effective immune response, and the virus replicates for long periods (years to decades) in their livers. Children are even more susceptible to chronic HBV infection, probably because of the immaturity of their immune system. Chronically infected people are at high risk for developing chronic hepatitis, cirrhosis, and hepatocellular carcinoma (HCC). They also serve as the reservoir for further spread of infection. Transmission is by contact with blood and other bodily fluids, although the exact route shows geographic variation and is not always well defined. In Western countries, sexual activity and needle sharing constitute the major routes of transmission, while in Asia mother-to-infant transmission via blood at the time of birth is common. In addition, close contact with infected people is a major risk factor, and indeed is the major route of transmission among children in Africa. Health care workers, especially surgeons, dentists, and nurses, are also highly susceptible to infection.

Recent detailed reviews of both medical and scientific aspects of HBV are available (1, 2). This article will instead give an overview of the molecular biology of HBV and related viruses, and concentrate on new information and aspects of this important virus that are still unknown or under dispute. It is hoped that readers may be spurred to perform research to resolve these issues.

2. PHYLOGENY OF HBV

HBV is the prototype of a family of animal viruses known as the Hepadnaviridae (2). Other members infect birds such as duck and heron, and small mammals such as ground squirrel and woodchuck. All of these viruses are characterized by a predilection for infection of the liver, a limited host range (human beings, chimpanzees and perhaps a few other primate species for HBV), and the ability to cause chronic infections. The virion genetic material is DNA but HBV does not replicate by semi-conservative replication via a DNA-dependent DNA polymerase. Rather, it synthesizes progeny DNA by reverse transcription of a greater-than-genome-length RNA template (3). Thus, its replication strategy is similar to that of retroviruses, and indeed, there is weak but discernable sequence similarity between HBV and retroviruses (4). The HBV genome is only 3.2 kb in size, approximately 3 times smaller than that of retroviruses, giving rise to speculation that hepadnaviruses may represent "degenerate" retroviruses (4). However, sequence comparisons do not rule out the possibility that retroviruses and hepadnaviruses arose independently from cellular retroid elements (5).

The two main differences in the life cycle of retroviruses and hepadnaviruses are the physical state of the viral DNA in the infected host (integrated vs. episomal) and the timing of reverse transcription (upon entry into a newly infected cell vs. prior to release from an infected cell). However, recent data indicate that the foamy retroviruses have an intermediate phenotype in these regards; i.e., the viral DNA is slow to integrate (although integration is necessary for completion of the life cycle), and many viral particles contain DNA rather than RNA (6). Furthermore, like the hepadnaviruses, the foamy retroviruses produce the reverse transcriptase as a primary translation product, rather than as a fusion protein, and bud through the endoplasmic reticulum (ER) (6). Therefore, the simple retroviruses and hepadnaviruses may simply represent 2 extremes of a continuum among viruses that utilize reverse transcription.

3. STRUCTURE OF VIRAL PARTICLES

HBV is an enveloped virus with a nucleocapsid (see Figure 1). The spherical virion measures approximately 42 nm in diameter. The outer envelope comprises host-derived lipids and 3 forms of the virally encoded surface protein. Approximately half of the surface proteins are the small surface protein, 20% are the middle surface protein, and 20% are the large surface protein (7). The large surface protein appears to interact with the cellular receptor(s) and is also necessary for envelopment of the nucleocapsid during morphogenesis (see below). The small surface protein is necessary for the proper formation and secretion of mature virions, although other functions cannot be ruled out. No specific role has been assigned to the middle surface protein, which is dispensable for production of virions but necessary for infectivity (8). Therefore, it may play an accessory role in binding the cellular receptor(s) and/or perform another function during viral entry into the host cell. Indeed, it has been hypothesized that middle surface protein may be endoproteolytically cleaved after virion binding to the cell surface, thereby generating a fusogenic sequence at its N-terminus needed for fusion of the viral envelope with the cell membrane (9, 10).

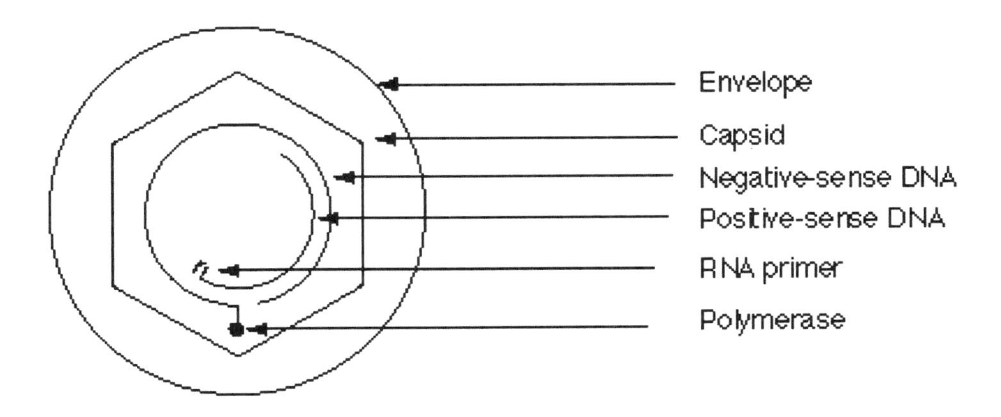

Figure 1. Diagram of HBV virion.

The nucleocapsid is approximately 27 nm in diameter and shows icosahedral symmetry. The protein shell is made up solely of the viral core protein. Inside is the DNA genome and at least 1 copy of the viral polymerase protein, as well as probably a few cellular proteins. The genome has a structure unique among animal viruses, being circular but with a gap in the negative strand and a long discontinuity in the positive strand (Figure 1). Furthermore, the viral polymerase protein is covalently linked to the 5' end of the negative strand via a tyrosine residue (11). This arrangement results from the utilization of reverse transcription for HBV genome replication (see below).

In the sera of infected people, virions actually constitute a small minority of HBV-derived particles. Large numbers of smaller subviral particles are also present, that usually outnumber virions by a ratio of 100:1 or more. These so-called surface antigen (HBsAg) particles are 20 nm spheres or filaments of variable length, and are composed of lipids and the viral surface proteins. Because of the absence of the nucleocapsid, these particles are non-infectious. The spherical forms contain very little large surface protein (7), and hence presumably cannot bind to the cell surface receptor. However, the filamentous forms are estimated to contain approximately half as much large surface protein, on a per weight basis, as virions (7). This presents a puzzle, since these particles should then be able to compete with virions for binding to the cellular receptor and hence block infection. It is possible that the preS1 domain of the large surface protein in the filaments is in the wrong conformation to bind the receptor, but no data are available in this regard. The function, if any, of the subviral particles is unknown. It has been speculated that they may play a role in suppressing the immune response against HBV surface proteins and hence favoring chronic infection. Recently, subviral particles were shown to increase the infectivity of the duck HBV (DHBV) for hepatocytes in culture (12). The relevance of this finding is unclear, since even a single duck HBV particle can productively infect a duckling. Nevertheless, it is possible that this phenomenon may assist in the spread of the virus in the liver. In any case, it is the existence of these particles that allows easy serological diagnosis of active HBV infection by the presence of circulating HBsAg.

Naked nucleocapsid particles can also be secreted by an uncharacterized mechanism from transfected cells in culture, but such particles have not been detected in patients' sera (13). It is difficult to envision a function for such particles, which may reflect an artifact of transfection.

4. VIRAL ENTRY

Because of the lack of information on the identity of the cellular receptor(s) for HBV, the lack of a convenient small animal model, and the lack of an easily reproducible cell culture infection system, little is known about the early steps in HBV infection. There has been somewhat more success with DHBV, which has been found to utilize carboxypeptidase D for entry (14, 15). However, this protein is neither hepatocyte-specific nor sufficient for infection (15). Hence, there must be at least one co-receptor, whose identity is as yet unknown. The N-terminal domain of the DHBV large surface protein (preS1 domain) is the viral determinant of specific binding to carboxypeptidase D (16). After binding the cell surface, DHBV appears to be endocytosed, and then fuses with the membrane of the resulting early (non-acidified) endosome to release the nucleocapsid into the cytosol (17). Presumably, the nucleocapsid then travels to the nucleus to deliver the viral DNA. Indeed, nuclear localization signals have been found in the core proteins of both HBV and DHBV (18, 19). However, it is not clear if the intact nucleocapsid traverses the nuclear pore, or if it disintegrates at the cytosolic face of the nuclear pore and the DNA/polymerase complex then enters the nucleoplasm. In favor of the latter model, indirect data from transgenic mice indicate that empty HBV capsids do not cross the nuclear envelope (20). However, since transgenic mice do not faithfully recapitulate other aspects of the HBV life cycle that require import of viral DNA into the nucleus (i.e., amplification of viral episomal DNA, see below), it is conceivable that murine cells may lack factor(s) needed for import of the nucleocapsid into the nucleus.

Upon gaining entry into the nucleus, the polymerase protein must be hydrolytically cleaved from the viral DNA, which is subsequently repaired into a fully double-stranded, supercoiled covalently closed circular form (CCC DNA). There is evidence that the viral polymerase is not necessary for this process, indirectly implicating host enzymes (21).

5. GENE EXPRESSION

The CCC DNA behaves as a minichromosome, and is transcribed by host RNA polymerase II into 4 major species of mRNA (22). All the HBV genes are encoded on one strand of the DNA, and hence all the mRNAs are of the same polarity. Indeed, they all terminate at the same polyadenylation signal, and differ only in their 5' start sites, which are determined by the respective promoters used (Figure 2).

The core (C) promoter gives rise to 3.5 kb transcripts that contain slightly more than 1 genome length of sequence information. These transcripts show heterogeneity in their 5' ends. The longer ones encompass the entire core open reading frame (ORF) and are translated into the so-called precore protein (Figure 3). This protein is targeted to the lumen of the ER via a signal peptide at its N-terminus, and is subsequently processed and secreted (23). The secreted form of the precore protein is known as the e antigen

(HBeAg), which has long been used as a serological marker for active HBV infection. The function of HBeAg is a mystery, since it is entirely dispensable for viral infectivity and replication and yet is maintained in all hepadnaviruses (24-26). A role in suppression of the immune response against HBV has been posited (27). Indeed, a mutant woodchuck hepatitis virus incapable of synthesizing HBeAg gives rise to chronic infections much less frequently than wildtype virus under laboratory conditions (28), and the HBeAg-negative mutants of HBV are frequently associated with severe forms of hepatitis B (29, 30). These data suggest that HBeAg plays a role in dampening the immune response against the virus. Another possibility, not mutually exclusive, is that HBeAg negatively modulates viral replication and hence indirectly down-regulates the host immune response. This hypothesis is not unreasonable, since there is selective pressure on viruses like HBV that are not easily transmissible to keep the host alive as long as possible. There is evidence from transfected cells and transgenic mice that the precore protein can directly inhibit viral replication (31, 32). The means by which an extracellular protein can affect viral replication is not obvious, but a portion of HBeAg is known to remain cell-associated, apparently because of inefficient translocation of the molecule through the ER membrane (33). As a result, there may be membrane bound and/or intracellular HBeAg that can affect viral or cellular processes.

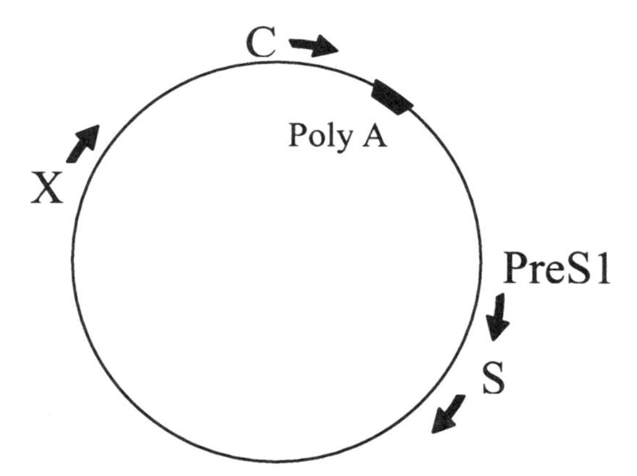

Figure 2. Schematic diagram of the 4 HBV promoters and the single polyadenylation signal. Note that the C promoter is directly upstream of the polyadenylation signal.

The shorter C promoter transcripts are missing the first ATG codon of the core ORF. Hence, translation starts from a downstream in-frame ATG codon, giving rise to the core protein (Figure 3). In addition, a small fraction of scanning ribosomes apparently bypasses this and several down-stream

ATG codons, to initiate translation of the polymerase (Figure 3) (34). The mechanism by which ribosomes can ignore the intervening ATG codons is not known, although it may involve translation of a short intervening ORF (35). Yet a third function of these C transcripts is to serve as the template for reverse transcription (see below), which is the reason they are usually called the pregenomic RNAs. Thus, all components of the progeny nucleocapsids (core protein, polymerase, and pregenomic RNA) arise from the same template.

While the precore and pregenomic transcripts are almost identical, the former cannot be used either for translation of polymerase or for encapsidation. Both situations arise because of the differential usage of ATG codons. The precore ATG codon, unlike the core ATG codon, for unknown reasons, does not allow ribosomes to scan past it to translate the down-stream polymerase ORF. In addition, it appears that when the RNA between the precore and core ATG codons is translated, it changes the secondary structure of this RNA sequence, which is critical for its interaction with polymerase to initiate the encapsidation process (see below). Consequently, the precore mRNA cannot serve as the substrate for reverse transcription and viral replication (36). Thus, even though the precore and pregenomic transcripts are both specified by the C promoter, they serve entirely separate functions. Not surprisingly, therefore, recent data indicate that there can be independent regulation of the 2 types of transcripts (37, 38). Thus, it may be more accurate to view the C promoter as 2 closely spaced but distinct promoters (precore and pregenomic promoters). The details are still being worked out, but interestingly a common C promoter mutant that is associated with severe hepatitis shows a decrease in the ratio of precore to pregenomic RNA synthesis (39), consistent with the hypothesized role of HBeAg in reducing the severity of hepatitis.

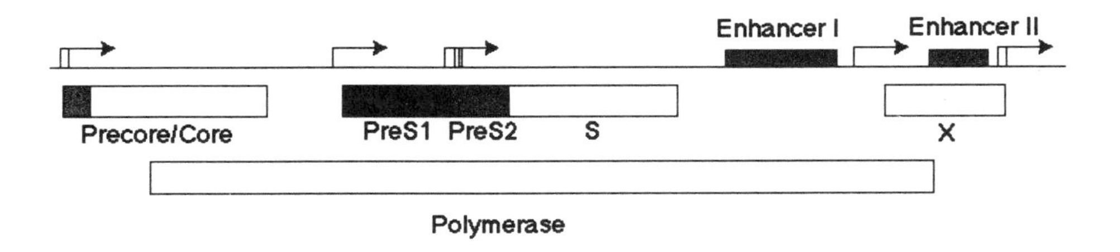

Figure 3. Map of the HBV genome. The DNA is shown as the thin line, the transcriptional start sites of the 4 promoters are shown as bent arrows, the enhancers are shown as thin boxes above the DNA, and the open reading frames are shown as thick boxes under the DNA.

An unusual feature of transcription from the C promoter is that this promoter is almost directly upstream of the single polyadenylation signal in the HBV genome (Figure 2). Therefore, the elongating RNA polymerase II complex must ignore the polyadenylation signal the first time they meet but not the second time. A similar situation occurs in retroviruses, since these viruses must also produce a longer-than-full-length RNA for reverse transcription. For HBV, the reasons for differential polyadenylation are complex. The lack of polyadenylation at the first pass is explained by a non-canonical sequence of the polyadenylation signal. The efficient usage of this same signal at the second pass is partially explained by the presence of sequences up-stream of the C promoter that act at either the DNA or RNA level to increase utilization of the polyadenylation signal (40). In addition, the distance between the 5' end of the nascent pre-mRNA and the polyadenylation signal plays a role, with longer distances leading to greater polyadenylation (41). Both factors are important for full utilization of the polyadenylation signal, but the molecular mechanisms for these effects are unknown. Clearly, the same factors can allow efficient utilization of the polyadenylation signal on the first pass of transcripts originating from the other viral promoters (however, X transcripts may not fully utilize this signal on the first pass; see below).

The C promoter comprises numerous elements that bind different cellular transcription factors, some ubiquitous and some hepatocyte-enriched. Recently, it was shown that a cis-element that binds the hepatocyte-enriched transcription factors HNF4 and PPARα is important for C promoter activity (42). Therefore, this element and the trans-acting factors that bind to it may be one determinant of the hepatotropism of HBV.

Many of the up-stream elements of the C promoter also can function to activate the S promoter (see below) and thus have enhancer function (43). Consequently, this region of the C promoter is also called enhancer II, since it was discovered subsequent to another enhancer up-stream of the X promoter (see below).

The surface ORF is complex. As already mentioned, the major form of the surface protein in subviral particles and virions is the small surface protein, which encompasses the "S" region and starts at an internal ATG codon of this ORF (Figure 3). The middle surface protein starts at the ATG codon immediately up-stream of the S ATG codon, and thus includes an N-terminal extension called the preS2 region. Finally, the large surface protein starts at the most up-stream ATG codon and includes both the preS2 and the preS1 region (Figure 3). While the middle and small surface proteins can theoretically be generated from the large surface protein by proteolysis, in actual fact there is no precursor-product relationship and each surface protein is synthesized as a primary translation product.

The preS1 promoter (also called SpI) is immediately up-stream of the entire surface ORF (Figure 3). Hence, its transcripts are used for the translation of the large surface protein. In transient transfection studies, the function of this promoter is heavily dependent on the hepatocyte-enriched transcription factor HNF1α, that binds to a cis-element in the promoter (44). This property presumably accounts for the liver-enriched expression of large surface protein in transgenic mice. However, transgenic mice with the HNF1α gene knocked out express only slightly lower amounts of preS1 transcripts than wildtype mice (45). This observation raises the possibility that other cis- and trans-acting factors are more important for transcription from this promoter in normal hepatocytes, but it is also possible that in the knockout mice the related transcription factor HNF1β is up-regulated to subsume some of the functions normally performed by HNF1α.

A different promoter (herein called the S promoter, also known as SpII) is used for the synthesis of middle and small surface proteins. This promoter is embedded within the S ORF (Figure 3) and gives rise to transcripts that are heterogeneous at the 5' end. Just as for the C transcripts, the variation in start sites leads to the translation of 2 different proteins. In the case of the S transcripts, the longest mRNA species starts up-stream of an internal ATG codon of the S ORF, and is used for translation of the middle surface protein. The others are shorter and hence are translated into the small surface protein. Because the latter transcripts are more abundant, significantly more small surface protein than middle surface protein is synthesized. The S promoter shows relatively little cell-type specificity, and indeed its function depends mainly on 2 ubiquitous transcription factors, Sp1 and NF-Y (46, 47). However, the S promoter is strongly activated by the 2 viral enhancers, whose function depends largely on hepatocyte-enriched transcription factors (48). Consequently, the S transcripts, like the other HBV transcripts, show the highest level of expression in the liver of transgenic mice.

In infected hepatocytes and cells transfected with the HBV genome, the amount of preS1 transcripts is much lower than that of S transcripts. This difference accounts for the relatively small amount of large surface protein synthesized *in vivo*. However, reporter gene experiments reveal that the preS1 and S promoters are actually of similar strength (48). Thus, other factors must account for the substantial difference in transcript levels. Indeed, the viral enhancers activate the S promoter more than the preS1 promoter (49). Another difference is that the posttranscriptional regulatory element (PRE), an RNA element that overlaps much of the X ORF (see below), increases the export of S transcripts much more than that of preS1 transcripts (50). Finally, there appears to be a block to elongation of the preS1 transcripts through the region of the S promoter. The NF-Y site necessary for S promoter function is also important for this elongation block, although it is not clear if this sequence is functioning in the latter capacity as DNA or RNA (51). Thus, mutation of this element simultaneously leads to a decrease in S mRNA levels and an increase in preS1 mRNA levels. Interestingly, mutants with lesions in the NF-Y site are common in chronically infected people, possibly because of immune pressure. These mutants, when transfected into cultured cells, show the expected derangement in surface protein expression (52).

The fourth HBV promoter is up-stream of the X ORF (Figure 3), and is used for the synthesis of the X protein. This small protein shows no homology to other proteins, and can activate a variety of cellular and viral promoters in transient transfection assays of cultured cells. One pathway by which X protein affects transcription is by interacting directly with host transcription factors, and it is likely that it increases the activity of HBV enhancer I by increasing the ability of CREB and related factors to activate this enhancer via a variant CREB site (53). However, the mechanism by which X protein enhances CREB function is controversial (54, 55). Furthermore, the exact function of X protein during HBV infection remains unclear. In transfection studies of cultured cells, the X protein has at best modest effects on HBV gene expression and replication. Yet woodchuck hepatitis viruses that cannot express a full-length X protein are incapable of establishing infection when injected as naked plasmid DNA into the liver (56, 57). Because the X protein can interact with proteasome subunits in yeast 2-hybrid assays (58), a role in preventing antigen processing and hence immune recognition of infected cells has been proposed. However, woodchucks previously exposed to the X-minus mutant can still be infected by wildtype virus (57), making it unlikely that this is the sole function of X protein. Another activity of X protein in transfected cells is the activation of the Src signal transduction pathway (59), possibly by causing release of calcium from intracellular stores. Src activation can signal to the JAK/STAT and other intracellular pathways, leading to transcriptional activation and other effects. One result of Src activation is to promote cycling of growth-arrested cells into the G1/S transition, which is associated with an increase in cellular deoxyribonucleoside triphosphate (dNTP) levels (59a). This is likely to be an important factor for HBV replication, since in quiescent cells, such as normal hepatocytes, dNTP concentrations are too low to support cytoplasmic reverse transcription, as has been demonstrated for many retroviruses. Indeed, Schneider and colleagues (60) have shown that blockade of Src down-regulates the synthesis of progeny DNA in HBV-transfected cells. This finding may explain why X protein does not have dramatic effects on HBV replication in transfected cells, since cells grown in culture are mitotically active and hence should have ample nucleotide levels. Thus, one attractive hypothesis is that the X protein has 2 separate functions: in the nucleus, it interacts with transcription factors to boost viral gene expression, while in the cytoplasm, it activates signaling pathways to allow viral replication. However, evidence for such activities in the natural setting (infected hepatocytes) is still lacking.

Many other activities have been ascribed to the X protein. These include effects on DNA repair, apoptosis, p53 function, and Smad signaling (61-64). Sometimes, contradictory results have been obtained in different laboratories, and it is difficult at this time to determine the significance of many of these findings. Part of the problem may be ascribed to the use of different cell lines, and it is probably necessary to wait for the availability of a simple small animal model (e.g., murine model) before these questions can be settled. Finally, it is important to point out that X protein has been implicated in carcinogenesis. This topic is discussed in detail in Chapter 9 of this book.

The X ORF is conserved among all mammalian hepadnaviruses, but the avian hepadnaviruses have no apparent orthologue. Since the avian viruses are rather distantly related to the mammalian viruses, it is possible that this reflects a real difference in the biology of these viruses. However, recent data indicate that perhaps there is an avian X protein that is translated from a non-ATG initiation codon (65). If the existence of this protein is confirmed, it may allow greater progress in understanding the role of X protein in the viral life cycle and its molecular mechanism of action, since experimentation with ducks is much easier and cheaper than with woodchucks.

The X promoter has no TATA box, but has a cis-element at the start site of transcription that interacts with an unidentified transcription factor (66). The up-stream elements of this promoter also function as enhancer I to up-regulate the other viral promoters (Figure 3). Some of these elements bind ubiquitous transcription factors, but others bind hepatocyte-enriched factors. This latter property can explain, in part, the liver-enriched expression of HBV genes in transgenic mice, and may also play a role in determining the hepatotropism of HBV.

The X transcripts should be approximately 0.8 kb in size, and indeed mRNA of the expected size have been detected in both infected livers and transfected cells. In addition, 3.9 kb X transcripts have been found in transfected cells (67). They arise from the same promoter but bypass the polyadenylation site on the first encounter, similar to the precore and pregenomic transcripts. This situation presumably results from the relatively short distance between the X promoter and the polyadenylation site. It remains unclear whether in vivo the X protein is synthesized from the 0.8 kb mRNA, 3.9 kb mRNA, or both. In addition, shorter transcripts that initiate from within the X ORF are synthesized in transfected cells, and they are capable of being translated into N-terminally truncated X proteins initiating at internal ATG codons (68). Similar short X proteins have been detected following translation of in vitro transcribed X RNA, although here the mechanism is different, being secondary to leaky scanning of the ribosome through the first X ATG codon (69). Nevertheless, these 2 pieces of data suggest that perhaps multiple protein products are made from the X ORF, just as for the core and surface ORFs. If so, some of the apparent discrepancies in the literature regarding the functions of X protein may also be explained by different amounts of the various X proteins synthesized by the plasmids used by different groups.

Because of the circular nature of the HBV genome and the fact that there are multiple viral promoters but only a single polyadenylation signal, all of the cis-elements important for transcription are also within transcribed regions (Figure 3). Thus, it is not surprising that there may be mutual interactions among the various promoters. One example, the negative posttranscriptional regulation of preS1 transcription by the S promoter has already been cited above. Another example is the apparent "occlusion" of the DHBV S promoter by the C promoter (70), which probably occurs because the elongating RNA polymerase II complexes prevent recognition of the downstream S promoter by trans-acting cellular factors. Undoubtedly other interactions will be discovered in the future, although it is difficult to study them with transfection experiments because it is impossible to duplicate the architecture of the HBV genome with plasmids unless the insert is excised and circularized in vitro before transfection.

Unlike retroviruses, hepadnaviruses are generally believed not to utilize splicing to generate gene products. However, recent data from DHBV indicate that at least 1 splicing event may be important for the viral life cycle. In this virus, the pregenomic pre-mRNA can be spliced to bring the large surface protein ATG codon to the 5' end of the mature mRNA. This mRNA appears to serve a unique function, since ablation of the splice site results in a defective virus, but the nature of the defect is not clear (71). These splice sites are not conserved in the mammalian hepadnaviruses, and thus this splicing event may not be relevant to them. However, a different spliced mRNA produced in HBV-transfected cells has been shown to result in the synthesis of a novel polypeptide fused to the N-terminus of polymerase (72). Antibodies to this protein have been detected in some infected people, and an immunologically related polypeptide of the expected size is present in the liver of infected people, but a role for this protein in the viral life cycle has not been demonstrated. Other spliced RNAs have been detected in infected livers and/or transfected cells, but, again, no data supporting the importance of these RNAs have been forthcoming. However, some of these spliced

RNAs, after reverse transcription, can be secreted in virions (73). These particles have the potential of behaving as defective interfering particles.

In mammalian cells, incompletely spliced nuclear transcripts are usually not exported to the cytoplasm (74). This is presumably a quality-control mechanism to prevent the synthesis of abnormal proteins from mRNAs with retained introns. This failure in export can be partially explained by retention of these pre-mRNAs by the splicing machinery. However, the β-globin cDNA with all introns removed still gives rise to pre-mRNAs that are poorly exported. Thus, it is believed that splicing also plays a positive role in mRNA export, by marking fully spliced pre-mRNAs in some manner that allows efficient export (75). However, a few cellular genes are devoid of introns, and there are other genes showing alternative splicing patterns, resulting in mRNAs that retain splice sites and yet are efficiently exported. Transcripts derived from these genes must utilize an alternative mechanism to access the export pathway. Extensive analysis of both simple and complex retroviruses, most of which give rise to multiple mRNAs by alternative splicing of a single species of primary transcripts, reveals that cis-acting RNA elements are involved (76). The best studied is the human immunodeficiency virus (HIV) Rev response element (RRE), which allows export of incompletely spliced mRNAs by binding to the viral export factor Rev. Some simple retroviruses code for RNA elements known as constitutive transport elements (CTE), which perform a similar function but binds to cellular export factors. Unspliced messages are used for translation of all the HBV structural proteins, and, not surprisingly, an element functionally similar to the RRE and CTE has been discovered in HBV mRNAs (50, 77). This posttranscriptional regulatory element (PRE) overlaps with the X promoter and the up-stream portion of the X ORF, and is necessary for the efficient export of the S mRNA into the cytoplasm. It is also important for the C mRNA but appears to have relatively little effect on preS1 mRNA, for unknown reasons (50, 77). The PRE, like the CTE, depends on cellular rather than viral proteins for its function, but its export pathway is different from those for both the RRE and CTE (78). It contains multiple subelements, each of which shows little or no function unless multimerized (79). These subelements show no obvious sequence or structural homology to each other. This feature has made it difficult to identify the cellular proteins that export PRE-containing RNAs. Glyceraldehyde-3-phosphate dehydrogenase has been shown to bind the PRE (80), but the functional significance of this interaction is not clear. Recent data indicate that the polypyrimidine tract binding protein (PTB) also binds to the PRE (B. Li and T. S. B. Yen, unpublished observations). Mutants of the PRE with decreased PTB binding are defective for export, while increasing the amount of PTB in the cell enhances the function of the PRE. Furthermore, PTB is known to be a nuclear protein that shuttles rapidly between the nucleus and the cytoplasm (81). Therefore, PTB may be directly involved in PRE export function, but additional experiments are needed to confirm this inference. Finally, the PRE may also affect the polyadenylation of viral pre-mRNAs (82), although because all of the posttranscriptional RNA processing steps are tightly linked in mammalian cells, it remains to be determined whether this is an additional function of the PRE or reflects an indirect effect of export on polyadenylation.

6. RNA ENCAPSIDATION AND REPLICATION

The encapsidation of the pregenomic RNA takes place in the cytoplasm, and is initiated by the interaction of the polymerase with a short stem-loop structure near the 5' end of the RNA called ε (83). None of the shorter mRNAs contain a 5' ε, but the slightly longer precore mRNA does. As mentioned above, the precore mRNA is not encapsidated, because the ε region is part of the precore ORF, and translation of e in the precore mRNA by ribosomes appears to preclude recognition by polymerase (36). However, all viral mRNAs contain another copy of ε near the 3' end, because the unique polyadenylation signal is down-stream of ε (Figure 4). This region of the mRNAs is not translated, and hence can potentially be recognized by polymerase for encapsidation. Yet, there is extremely tight exclusion of all viral RNAs other than the pregenomic RNA in nucleocapsids. This discrimination is explained by the finding that polymerase only recognizes ε that is near the 5' of the RNA molecule, apparently because of the need for a nearby cap structure (84). After binding of polymerase to ε, there is a conformational change in both the protein and the RNA (85, 86). This switch allows the polymerase to synthesize a 4-nucleotide cDNA of a non-based-paired region of the ε sequence, using a tyrosine residue in the N-terminal domain of the polymerase as primer (Figure 4, step I) (11). Subsequently, the polymerase moves the attached cDNA to an identical complementary sequence near the 3' end of the pregenomic RNA, within a region called direct repeat 1 (DR1) (Figure 4, step II), and extends the cDNA until it comes to the 5' end of the RNA (Figure 4, steps III and IV). This first DNA strand is called by convention the negative-sense strand, and is terminally redundant by about 8 nucleotides. Simultaneously, the RNase H domain of the polymerase degrades all but the last 15-18 nucleotides of the template RNA (Figure 4, step IV). The location for the last nucleotide to be degraded is determined not by the sequence but by the distance from the 5' end, which may reflect the physical separation of the catalytic center of the reverse transcriptase and RNase domains of the polymerase (87).

Figure 4. The various steps in HBV DNA replication. RNA is shown as thin black lines, while DNA is shown as thicker grey lines. Direct repeat 1 is the open triangle, while direct repeat 2 is the filled-in triangle. R indicates the redundant sequences at the ends of the pregenomic RNA. The black blob is the polymerase. Note that, to maintain clarity, this diagram is not intended to show the correct topology, which in any case is not well defined.

At this point, the polymerase reverses direction, and uses the small fragment of remaining pregenomic RNA as primer to synthesize the second DNA strand (positive-sense strand). However, before this happens, the RNA primer is translocated to the 5' end of the negative-sense DNA strand, which contains a short repeat (DR2) that is complementary to the 3' portion of the RNA (Figure 4, step V). Because of this translocation, only a small amount of the positive-sense strand DNA can be synthesized before the polymerase runs out of template. To avoid this situation, and to make a circular product, the polymerase again switches template to the corresponding site at the 3' end of the negative-sense strand DNA, made possible by the 8-nucleotide terminal redundancy in this DNA. A variable and incomplete length of this DNA strand is synthesized (Figure 1), probably because the nucleocapsid is secreted out of the cell during second strand DNA synthesis, and thus the available nucleotide supply is exhausted before this DNA strand is completed.

All these molecular maneuvers presumably require changes in the secondary and/or tertiary structure of polymerase, and probably explain at least in part why it has been impossible to reproduce even the early steps of HBV replication in vitro. More success has been achieved with the DHBV polymerase, in that the synthesis of the first 4 nucleotides upon polymerase binding to ε has been achieved with in vitro translated polymerase (11). It turns out that cellular cofactors are needed for this process. Among these cofactors are Hsp90 and other chaperone molecules, which bind polymerase and are required for at least the initial stages of replication (88). It is likely that these accessory proteins hold the polymerase in one or more metastable conformations and/or assist in switching from one conformation to another. Detailed understanding of these processes may provide suitable targets for future drug development.

Another possible reason for the difficulty in performing in vitro HBV DNA synthesis is that the whole process takes place inside the nucleocapsid. At some point during or soon following the recognition of pregenomic RNA by polymerase, core protein molecules encapsidate the RNA-protein complex into immature nucleocapsids. The details of this process are unknown, including whether core protein interacts solely with the polymerase or with both the polymerase and RNA. It is clear that core protein cannot bind bare RNA with sequence specificity, since core protein particles formed in the absence of the polymerase protein contains a non-specific mixture of total cytoplasmic RNA rather than only the pregenomic RNA (89).

Because of the complicated multistep process needed to generate the partially double-stranded HBV DNA, it is not surprising that undesired side products can be made. One such product is a linear form of the DNA, that results from a failure of the polymerase to move the RNA primer for positive strand DNA synthesis from the 3' to the 5' end of the negative strand DNA (90). This failure then prevents the circularization of the double-stranded DNA product. These linear DNA molecules, if recycled into the nucleus (see below), can recombine with each other to produce circular DNA that is competent for transcription and hence replication (90). Furthermore, linear HBV DNA molecules in the nucleus appear to be a major source of viral DNA fragments that integrate into the host chromosome (see below) and hence may play a role in carcinogenesis (91).

Just as the polymerase seems to change conformation during viral DNA synthesis, the core protein also undergoes a change. When the polymerase starts to synthesize the positive-sense stand DNA, the nucleocapsid becomes competent (mature) to interact with the surface proteins in the ER. As a result, only DNA-containing nucleocapsids are enveloped and secreted (92). This checkpoint presumably is advantageous for the infectivity of the virus, since DNA is more stable than RNA. The difference between immature and mature nucleocapsids is not known, but there must be a conformational change in the core protein. This inference is strengthened by the observation that a single-residue mutation in the core protein can result in the secretion of immature nucleocapsids in virions (93).

The conformational change may be caused by phosphorylation and dephosophorylation of serine residues in the C-terminal tail of core protein. This region contains multiple residues that can be phosphorylated, and the pattern of phosphorylation is different at different stages of the viral life cycle (94). The responsible kinase(s) have not been positively identified, but there is evidence that at least one cellular kinase is incorporated into the virion (95).

While there must be direct interaction between core protein and surface protein to allow envelopment of the nucleocapsid, the difference between immature and mature nucleocapsids may not reflect a change in the strength of this interaction. Rather, recent data suggest that mature nucleocapsids acquire the ability to interact with intracellular membranes regardless of the presence of surface protein (13). This may reflect an increase in particle hydrophobicity with resultant non-specific binding to lipids, or perhaps specific interaction with a host protein. Presumably, this initial interaction with the membranes brings the nucleocapsid in close enough proximity to surface protein molecules in the ER to allow high-affinity binding and then envelopment.

Envelopment of the nucleocapsid takes place at the ER or ER-Golgi intermediate compartment. Indeed, all forms of the surface protein are synthesized as integral transmembrane ER proteins. The topology and mode of membrane insertion of the small surface protein have been well studied (96). There is no N-terminal leader peptide that is cleaved off. Instead, the first transmembrane domain also acts as a signal to direct the N-terminus into the lumen of the ER. A second hydrophobic domain is also inserted into the membrane, causing the region downstream of it to be intraluminal (Figure 5). The topology of the C-terminal portion of small surface protein is not well defined, but there is evidence that it traverses the membrane twice (Figure 5). Soon after synthesis, disulfide bonding leads to the formation of dimers, which then aggregate and bud into the lumen to form subviral particles, which then are secreted out of the cell by the constitutive secretory pathway (97). Shortly before or during budding, additional disulfide bonds form, resulting in large multimers of surface protein. It should be noted that the term budding may not accurately describe the process of particle formation, since electron microscopy reveals no clear lipid bilayer and the ratio of protein to lipid is much higher in surface antigen particles than in conventional membranes.

A similar budding process presumably takes place during virion morphogenesis, except that all 3 forms of the surface proteins are involved, and the nucleocapsid is included within the outer envelope of surface proteins and lipid. The middle surface protein is not necessary for virion morphogenesis, although it is important for infectivity (see above), but the large surface protein is absolutely required for virion morphogenesis (8).

This fact implies that the unique N-terminal preS1 domain of large surface protein is necessary for interaction with the nucleocapsid. Indeed, binding of a portion of the preS1 domain to core particles has been demonstrated in vitro (98, 99). However, portions of the preS2 and S domains of the large surface protein have also been implicated in binding to core particles.

It had long been assumed that the middle and large surface proteins insert into the ER membrane in a similar fashion, topologically speaking, as the small surface protein, since all these proteins are identical except for N-terminal extensions on the middle and large surface proteins. Such a topology would be necessary for the preS1 domain of large surface protein to bind the cellular receptor, as has been shown to be the case for DHBV, since the luminal surface of the ER is topologically equivalent to the outer surface of the virion particle (form I, Figure 5). Consistent with this view, the preS1 domain can be immunologically detected on intact virions (7). However, the preS1 domain is also thought to be the region of the large surface protein that interacts with the cytosolic nucleocapsid during virion budding. For this purpose, it must be on the opposite (cytosolic) side of the ER. This apparent contradiction became resolved when it was shown that the large surface protein N-terminus, in contrast to that of the small and presumably middle surface protein, is cytosolically disposed upon synthesis (form II, Figure 5) (100, 101). However, in a portion of the molecules, much of the preS1 domain post-translationally apparently flips through the ER membrane to end up in the lumen (form II, Figure 5). This unusual dual localization of the preS1 domain allows the large surface protein to function in both virion formation and the infection process. The large surface protein is also myristilated at its N-terminus. This modification is not necessary for morphogenesis but is required for infectivity (102).

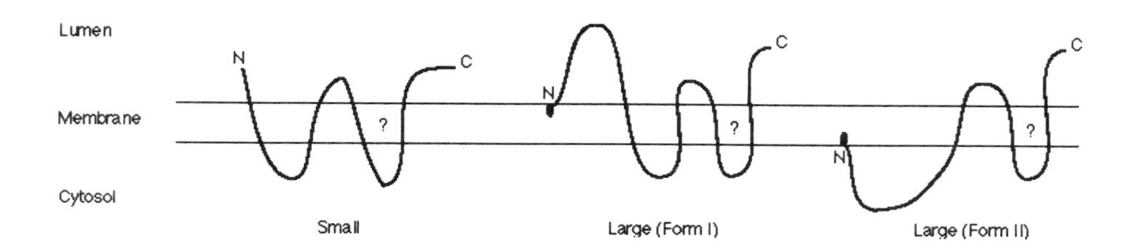

Figure 5. Diagram of the topology of the small and large surface proteins in the ER membrane. The question mark indicates that the exact arrangement of the C-terminal portion is unclear.

Another unusual feature of large surface protein is that if it is overexpressed, relative to the small surface protein, it prevents the secretion of subviral particles (103). At sufficiently high amounts, even virion secretion is blocked (52). Large surface protein by itself is also not secreted. Therefore, its trans-dominant effect on secretion is presumably due to the formation of heteromultimers with small surface protein. Contrary to early speculations, this secretory block is not at the step of budding but during the transport of the intraluminal particles through the ER-Golgi intermediate compartment (104). The molecular basis of the accumulation of large surface protein particles in this compartment is not known, although the fact that the chaperone protein calnexin is stably bound to these particles suggests that somehow they are recognized as improperly folded proteins by the ER quality-control apparatus (104). The significance of the retention of large surface protein is unknown.

Although the accumulation of large surface protein-containing particles within the intermediate compartment presumably serves no useful purpose for the virus, hepatocytes with retained particles are frequently seen during chronic HBV infection. These hepatocytes show enlarged cytoplasm with hyperplastic smooth ER and/or intermediate compartment, and are called "ground glass cells" because of their distinctive appearance by light microscopy . Indeed, prior to the availability of specific immunodiagnostic reagents, such cells were used by pathologists to aid in the diagnosis of chronic hepatitis B. Recent data have shown that large surface protein retention activates the so-called ER stress response to increase the transcription of both ER-resident chaperone proteins and lipids (105, 106), thereby explaining the ER expansion seen in ground glass cells. In addition, overexpression of large surface protein can kill the host cell by apoptosis (N. C. Foo, B. Y. Ahn, X. Ma, W. Hyun, and T. S. B. Yen, unpublished data). It remains to be seen whether this effect plays any role in the pathology of chronic hepatitis B, but it may explain the entity known as fibrosing cholestatic hepatitis B (FCH-B) (107, 108). FCH-B is most frequently seen in organ recipients, but can also occur in other patients who are iatrogenically immunosuppressed (e.g., for treatment of lupus erythematosus). It is marked by rapid course of progression to liver fibrosis and failure. Unlike conventional hepatitis B, the host immune response does not appear to be a causative factor, both based on the histological appearance and the fact that the patients are heavily immunosuppressed. Instead, the hepatocytes show vacuolization and apoptosis. Immunostaining reveals large amounts of surface protein within the hepatocyte cytoplasm, and electron microscopy shows the presence of surface protein particles in smooth vesicles. Thus, the direct toxic effect of large surface protein probably explains the pathogenesis of FCH-B. The cause of large surface protein overexpression in FCH-B remains to be demonstrated.

A final aspect of viral replication is that a portion of the DNA in cytoplasmic nucleocapsids can be recycled into the nucleus. The purpose of this recycling is to boost and maintain the number of CCC DNA in the nucleus, since HBV DNA is not capable of semi-conservative DNA-to-DNA replication. There must be tight control of this recycling process, since each infected hepatocyte usually contains no more than 10-20 copies of CCC DNA. For DHBV, the amount of cellular large surface protein may constitute one mechanism of regulation, since viruses that synthesize mutant large surface proteins give rise to greatly increased number of CCC DNA in the infected cell (109). It seems likely that there is a positive correlation between the amount of large surface protein in the ER membrane and the rate of virion encapsidation. At the early stages of infection, the amount of large surface protein would be low, and hence most of the mature nucleocapsids in the cytosol enter the nucleus by default (note that the mature progeny nucleocapsid is formally equivalent to the nucleocapsid of an infecting virus). When there is a sufficient number of CCC DNA molecules to act as templates for transcription of large surface protein, there would be enough ER-associated large surface protein to efficiently secrete all the available mature nucleocapsids, which then would not be able to travel into the nucleus. Thus, a feedback loop is established to keep the number of CCC DNA in the nucleus constant. Whether such a scenario also applies to mammalian hepadnaviruses remains to be seen. In addition, other types of control, such as at the level of core protein phosphorylation, are also possible, since phosphorylation has been shown to affect the nuclear localization signal of the core protein (110).

During chronic HBV infection, many of the infected cells contain fragments of viral DNA integrated into the chromosome. The source of these fragments is probably linear HBV DNA that results from aberrant viral replication (see above). These integration events play no apparent role in the viral life cycle and are probably an undesired side product of long-term infection. Viral integration can potentially play a role in HBV carcinogenesis, by either inactivating anti-oncogenes or activating protooncogenes. In fact, for the woodchuck hepatitis virus, activation of cellular N-myc genes by insertional activation is known to be the major cause of HCC (see Chapter 9). However, no consistent chromosomal location of HBV DNA integration has been found, although rare insertion events that are potentially oncogenic have been reported (2, 111). The integrated HBV DNA sequences may also cause genomic instability by promoting recombination events (112), and thereby indirectly contribute to carcinogenesis.

7. DISEASE PATHOGENESIS

There is excellent albeit indirect evidence that the liver damage that occurs during acute HBV infection is caused not by the direct action of the virus but by the immune response against viral antigens (1). Thus, the peak of liver enzyme abnormalities occurs after rather than during the peak of viral replication, and there is a positive correlation between the severity of acute symptoms and the likelihood of clearing the virus. Furthermore, cells grown in culture can be stably transfected with the HBV genome and release infectious virions with no measurable effect on cell viability, morphology, or growth (113).

Because of the evidence for immune damage of hepatocytes, it was long believed that viral clearance from the liver resulted from the direct killing of infected hepatocytes by cytotoxic T-lymphocytes. However, elegant work from Chisari and colleagues has clearly demonstrated an alternative pathway in both transgenic mice and infected chimpanzees (114, 115). It appears that cytokines released by lymphocytes and/or macrophages can indirectly control the virus, by blocking both viral gene expression at a posttranscriptional step and by destabilizing cytoplasmic nucleocapsids (116). The molecular mechanisms of these novel effects are still unclear. Presumably, cytokine release of sufficient strength and duration can help clear the virus, assuming that the CCC DNA is unstable and is degraded over time, especially since the antigen-specific cytotoxic T-lymphocytes can mop up any residual infection. This non-cytolytic mechanism explains why the majority of infected people can recover from acute infection with little or no symptoms of hepatitis, despite the fact that typically the majority of hepatocytes are infected. On the other hand, it is possible that this cytokine effect, especially if inadequate in strength or duration, can also be responsible for the establishment of chronic infections, since the decrease or loss of viral antigen expression may allow a small percentage of infected cells to escape killing by cytotoxic T-lymphocytes. In any case, elucidation of the mechanisms involved in this pathway may lead in the future to novel therapies to clear the virus without damaging the host hepatocytes.

It is likely that a significant portion of the liver damage during chronic hepatitis B is also secondary to the immune response. In concert with this view, periods of exacerbation of disease are associated with a broader repertoire of T-lymphocytes directed against viral antigens (117). However, ground glass cells that frequently arise during chronic infection are susceptible to apoptosis (see above) and hence can contribute to liver dysfunction. The cause of ground glass cell formation during chronic infection is uncertain, but there is preliminary data that viral mutants that frequently arise during chronic infection may be involved (118). In addition, it has been demonstrated for DHBV that mutant viruses showing increased CCC DNA accumulation also cause direct cytopathic damage to hepatocytes (119). Therefore, it remains possible that a direct cytopathic effect of mutant viruses that arise during chronic infection can contribute to disease pathogenesis.

The pathogenesis of HCC arising during chronic hepatitis B is an unresolved issue. While it is likely that HBV plays a direct role in carcinogenesis (by cis- and trans-acting effects on the host genome), the chronic inflammation, hepatocyte injury, and hepatocyte turnover also contribute to genetic mutations that lead to HCC. For a detailed discussion, the reader is referred to Chapter 9 of this volume.

8. ACKNOWLEDGEMENTS

Work in my laboratory on hepatitis B virus has been supported by the Department of Veterans Affairs, National Institutes of Health, University of California STAR Program, Immune Response Corporation, Council for Tobacco Research, Formosa Plastics Corporation, and China General Corporation. I apologize that, because of space restrictions, many relevant references cannot be cited.

REFERENCES

1 Hollinger F.B. "Hepatitis B Virus." In *Fields Virology*, B.N. Fields, D.M. Knipe, P.M. Howley, eds. Philadelphia: Lippincott-Raven, 1996.

2 Ganem D. "Hepadnaviridae and Their Replication." In *Fields Virology*, B.N. Fields, D.M. Knipe, P.M. Howley, eds. Philadelphia: Lippincott-Raven, 1996

3 Nassal M., Schaller H. Hepatitis B virus replication--an update. J Viral Hepatitis 1996; 3:217-226.

4 Miller R.H., Robinson W.S. Common evolutionary origin of hepatitis B virus and retroviruses. Proc Natl Acad Sci USA 1986; 83:2531-2535.

5 Doolittle R.F., Feng D.F., Johnson M.S., McClure M.A. Origins and evolutionary relationships of retroviruses. Quarterly Review of Biology 1989; 64:1-30.

6 Lecellier C.H., Saib A. Foamy viruses: between retroviruses and pararetroviruses. Virology 2000; 271:1-8.

7 Heermann K.H., Goldmann U., Schwartz W., Seyffarth T., Baumgarten H., Gerlich W.H. Large surface proteins of hepatitis B virus containing the pre-s sequence. J Virol 1984; 52:396-402.

8 Bruss V., Ganem D. The role of envelope proteins in hepatitis B virus assembly. Proc Natl Acad Sci USA 1991; 88:1059-1063.

9 Lu X., Block T.M., Gerlich W.H. Protease-induced infectivity of hepatitis B virus for a human hepatoblastoma cell line. J Virol 1996; 70:2277-2285.

10 Rodriguez-Crespo I., Nunez E., Yelamos B., Gomez-Gutierrez J., Albar J.P., Peterson D.L., Gavilanes F. Fusogenic activity of hepadnavirus peptides corresponding to sequences downstream of the putative cleavage site. Virology 1999; 261:133-142.

11 Wang G.H., Seeger C. The reverse transcriptase of hepatitis B virus acts as a protein primer for viral DNA synthesis. Cell 1992; 71:663-670.

12 Bruns M., Miska S., Chassot S., Will H. Enhancement of hepatitis B virus infection by noninfectious subviral particles. J Virol 1998; 72:1462-1468.

13 Mabit H., Schaller H. Intracellular hepadnavirus nucleocapsids are selected for secretion by envelope protein-independent membrane binding. J Virol 2000; 74:11472-11478.

14 Kuroki K., Eng F., Ishikawa T., Turck C., Harada F., Ganem D. gp180, a host cell glycoprotein that binds duck hepatitis B virus particles, is encoded by a member of the carboxypeptidase gene family. J Biol Chem 1995; 270:15022-150228.

15 Breiner K.M., Urban S., Schaller H. Carboxypeptidase D (gp180), a Golgi-resident protein, functions in the attachment and entry of avian hepatitis B viruses. J Virol 1998; 72:8098-8104.

16 Urban S., Breiner K.M., Fehler F., Klingmuller U., Schaller H. Avian hepatitis B virus infection is initiated by the interaction of a distinct pre-S subdomain with the cellular receptor gp180. J Virol 1998; 72:8089-8097.

17 Breiner K.M., Schaller H. Cellular receptor traffic is essential for productive duck hepatitis B virus infection. J Virol 2000; 74:2203-2209.

18 Yeh C.T., Liaw Y.F., Ou J.H. The arginine-rich domain of hepatitis B virus precore and core proteins contains a signal for nuclear transport. J Virol 1990; 64:6141-6147.

19 Mabit H., Breiner K.M., Knaust A., Zachmann-Brand B., Schaller H. Signals for bidirectional nucleocytoplasmic transport in the duck hepatitis B virus capsid protein. J Virol 2001; 75:1968-1977.

20 Guidotti L.G., Martinez V., Loh Y.T., Rogler C.E., Chisari F.V. Hepatitis B virus nucleocapsid particles do not cross the hepatocyte nuclear membrane in transgenic mice. J Virol 1994; 68:5469-5475.

21 Kock J., Schlicht H.J. Analysis of the earliest steps of hepadnavirus replication: genome repair after infectious entry into hepatocytes does not depend on viral polymerase activity. J Virol 1993; 67:4867-4874.

22 Yen T.S.B. Regulation of hepatitis B virus gene expression. Seminars Virol 1993; 4:33-42.

23 Ou J.H., Laub O., Rutter W.J. Hepatitis B virus gene function: the precore region targets the core antigen to cellular membranes and causes the secretion of the e antigen. Proc Natl Acad Sci USA 1986; 83:1578-1582.

24 Chen H.S., Kew M.C., Hornbuckle W.E., Tennant B.C., Cote P.J., Gerin J.L., Purcell R.H., Miller R.H. The precore gene of the woodchuck hepatitis virus genome is not essential for viral replication in the natural host. J Virol 1992; 66:5682-5684.

25 Chang C., Enders G., Sprengel R., Peters N., Varmus H.E., Ganem D. Expression of the precore region of an avian hepatitis B virus is not required for viral replication. J Virol 1987; 61:3322-3325.

26 Petrosillo N., Ippolito G., Solforosi L., Varaldo P.E., Clementi M., Manzin A. Molecular epidemiology of an outbreak of fulminant hepatitis B. J Clinical Microbiology 2000; 38:2975-2981.

27 Milich D.R., Jones J.E., Hughes J.L., Price J., Raney A.K., McLachlan A. Is a function of the secreted hepatitis B e antigen to induce immunologic tolerance in utero? Proc Natl Acad Sci USA 1990; 87:6599-6603.

28 Cote P.J., Korba B.E., Miller R.H., Jacob J.R., Baldwin B.H., Hornbuckle W.E., Purcell R.H., Tennant B.C., Gerin J.L. Effects of age and viral determinants on chronicity as an outcome of experimental woodchuck hepatitis virus infection. Hepatology 2000; 31:190-200.

29 Hasegawa K., Huang J.K., Wands J.R., Obata H., Liang T.J. Association of hepatitis B viral precore mutations with fulminant hepatitis B in Japan. Virology 1991; 185:460-463.

30 Yuasa R., Takahashi K., Dien B.V., Binh N.H., Morishita T., Sato K., Yamamoto N., Isomura S., Yoshioka K., Ishikawa T., Mishiro S., Kakumu S. Properties of hepatitis B virus genome recovered from Vietnamese patients with fulminant hepatitis in comparison with those of acute hepatitis. J Med Virol 2000; 61:23-28.

31 Guidotti L.G., Matzke B., Pasquinelli C., Schoenberger J.M., Rogler C.E., Chisari F.V. The hepatitis B virus (HBV) precore protein inhibits HBV replication in transgenic mice. J Virol 1996; 70:7056-7061.

32 Hasegawa K., Huang J., Rogers S.A., Blum H.E., Liang T.J. Enhanced replication of a hepatitis B virus mutant associated with an epidemic of fulminant hepatitis. J Virol 1994; 68:1651-1659.

33 Garcia P.D., Ou J.H., Rutter W.J., Walter P. Targeting of the hepatitis B virus precore protein to the endoplasmic reticulum membrane: after signal peptide cleavage translocation can be aborted and the product released into the cytoplasm. J Cell Biol 1988; 106:1093-1104.

34 Chang L.J., Ganem D., Varmus H.E. Mechanism of translation of the hepadnaviral polymerase (P) gene. Proc Natl Acad Sci USA 1990; 87:5158-5162.

35 Jean-Jean O., Weimer T., de Recondo A.M., Will H., Rossignol J.M. Internal entry of ribosomes and ribosomal scanning involved in hepatitis B virus P gene expression. J Virol 1989; 63:5451-5454.

36 Nassal M., Junker-Niepmann M., Schaller H. Translational inactivation of RNA function: discrimination against a subset of genomic transcripts during HBV nucleocapsid assembly. Cell 1990; 63:1357-1363.

37 Buckwold V.E., Xu Z., Chen M., Yen T.S., Ou J.H. Effects of a naturally occurring mutation in the hepatitis B virus basal core promoter on precore gene expression and viral replication. J Virol 1996; 70:5845-5851.

38 Yu X., Mertz J.E. Differential regulation of the pre-C and pregenomic promoters of human hepatitis B virus by members of the nuclear receptor superfamily. J Virol 1997; 71:9366-9374.

39 Li J., Buckwold V.E., Hon M.W., Ou J.H. Mechanism of suppression of hepatitis B virus precore RNA transcription by a frequent double mutation. J Virol 1999; 73:1239-1244.

40 Russnak R., Ganem D. Sequences 5' to the polyadenylation signal mediate differential poly(A) site use in hepatitis B viruses. Genes and Development 1990; 4:764-776.

41 Cherrington J., Russnak R., Ganem D. Upstream sequences and cap proximity in the regulation of polyadenylation in ground squirrel hepatitis virus. J Virol 1992; 66:7589-7596.

42 Tang H., McLachlan A. Transcriptional regulation of hepatitis B virus by nuclear hormone receptors is a critical determinant of viral tropism. Proc Natl Acad Sci USA 2001;98:1841-1846.

43 Yee J.K. A liver-specific enhancer in the core promoter region of human hepatitis B virus. Science 1989; 246:658-661.

44 Zhou D.X., Yen T.S. The ubiquitous transcription factor Oct-1 and the liver-specific factor HNF-1 are both required to activate transcription of a hepatitis B virus promoter. Mol Cell Biol 1991; 11:1353-1359.

45 Raney A.K., Eggers C.M., Kline E.F., Guidotti L.G., Pontoglio M., Yaniv M., McLachlan A. Nuclear covalently closed circular viral genomic DNA in the liver of hepatocyte nuclear factor 1 alpha-null hepatitis B virus transgenic mice. J Virol 2001; 75:2900-2911.

46 Raney A.K., Le H.B., McLachlan A. Regulation of transcription from the hepatitis B virus major surface antigen promoter by the Sp1 transcription factor. J Virol 1992; 66:6912-6921.

47 Lu C.C., Yen T.S. Activation of the hepatitis B virus S promoter by transcription factor NF-Y via a CCAAT element. Virology 1996; 225:387-394.

48 Antonucci T.K., Rutter W.J. Hepatitis B virus (HBV) promoters are regulated by the HBV enhancer in a tissue-specific manner. J Virol 1989; 63:579-583.

49 Zhou D.X., Yen T.S. Differential regulation of the hepatitis B virus surface gene promoters by a second viral enhancer. J Biol Chem 1990; 265:20731-20734.

50 Huang Z.M., Yen T.S. Hepatitis B virus RNA element that facilitates accumulation of surface gene transcripts in the cytoplasm. J Virol 1994; 68 3193-3199.

51 Lu C.C., Chen M., Ou J.H., Yen T.S. Key role of a CCAAT element in regulating hepatitis B virus surface protein expression. Virology 1995; 206:1155-1158.

52 Xu Z., Yen T.S. Intracellular retention of surface protein by a hepatitis B virus mutant that releases virion particles. J Virol 1996; 70:133-140.

53 Maguire H.F., Hoeffler J.P., Siddiqui A. HBV X protein alters the DNA binding specificity of CREB and ATF-2 by protein-protein interactions. Science 1991; 252:842-844.

54 Barnabas S., Andrisani O.M. Different regions of hepatitis B virus X protein are required for enhancement of bZip-mediated transactivation versus transrepression. Jf Virol 2000; 74:83-90.

55 Pflum M.K., Schneider T.L., Hall D., Schepartz A. Hepatitis B virus X protein activates transcription by bypassing CREB phosphorylation, not by stabilizing bZIP-DNA complexes. Biochemistry 2001; 40:693-703.

56 Chen H.S., Kaneko S, Girones R, Anderson R.W., Hornbuckle W.E., Tennant B.C., Cote P.J., Gerin J.L., Purcell R.H., Miller R.H. The woodchuck hepatitis virus X gene is important for establishment of virus infection in woodchucks. J Virol 1993; 67:1218-1226.

57 Zoulim F., Saputelli J., Seeger C. Woodchuck hepatitis virus X protein is required for viral infection in vivo. J Virol 1994; 68:2026-2030.

58 Zhang Z., Torii N., Furusaka A., Malayaman N., Hu Z., Liang T.J. Structural and functional characterization of interaction between hepatitis B virus X protein and the proteasome complex. J Biol Chem 2000; 275:15157-15165.

59 Klein N.P., Schneider R.J. Activation of Src family kinases by hepatitis B virus HBx protein and coupled signaling to Ras. Mol Cell Biol 1997; 17:6427-6436.

59a. Bouchard M., Giannakopoulos S., Wang E.H., Tanese N., Schneider R.J. Hepatitis B virus HBx protein activation of cyclin A-cyclin-dependent kinase 2 complexes and G1 transit via a Src kinase pathway. J Virol 2001; 75:4247-4257.

60 Klein N.P., Bouchard M.J., Wang L.H., Kobarg C., Schneider R.J. Src kinases involved in hepatitis B virus replication. EMBOP J 1999; 18:5019-5027.

61 Becker S.A., Lee T.H., Butel J.S., Slagle B.L. Hepatitis B virus X protein interferes with cellular DNA repair. J Virol 1998; 72:266-272.

62 Ogden S.K., Lee K.C., Barton M.C. Hepatitis B viral transactivator HBx alleviates p53-mediated repression of alpha-fetoprotein gene expression. J Biol Chem 2000; 275:27806-27814.

63 Shih W.L., Kuo M.L., Chuang S.E., Cheng A.L., Doong S.L. Hepatitis B virus X protein inhibits transforming growth factor-beta -induced apoptosis through the

activation of phosphatidylinositol 3-kinase pathway. J Biol Chem 2000; 275:25858-25864.

64 Su F., Theodosis C.N., Schneider R.J. Role of NF-kappaB and myc proteins in apoptosis induced by hepatitis B virus HBx protein. J Virol 2001; 75:215-225.

65 Chang S.F., Netter H.J., Hildt E., Schuster R., Schaefer S., Hsu Y.C., Rang A., Will H. Duck hepatitis B virus expresses a regulatory HBx-like protein from a hidden open reading frame. J Virol 2001; 5:161-170.

66 Yaginuma K., Nakamura I., Takada S., Koike K. A transcription initiation site for the hepatitis B virus X gene is directed by the promoter-binding protein. J Virol 1993; 67:2559-2565.

67 Guo W.T., Wang J., Tam G., Yen T.S., Ou J.H. Leaky transcription termination produces larger and smaller than genome size hepatitis B virus X gene transcripts. Virology 1991; 181:630-636.

68 Zheng Y.W., Riegler J., Wu J., Yen T.S. Novel short transcripts of hepatitis B virus X gene derived from intragenic promoter. J Biol Chem 1994; 269:22593-22598.

69 Kwee L., Lucito R., Aufiero B., Schneider R.J. Alternate translation initiation on hepatitis B virus X mRNA produces multiple polypeptides that differentially transactivate class II and III promoters. J Virol 1992; 66:4382-4389.

70 Huang M., Summers J. pet, a small sequence distal to the pregenome cap site, is required for expression of the duck hepatitis B virus pregenome. J Virol 1994; 68:1564-1572.

71 Obert S., Zachmann-Brand B., Deindl E., Tucker W., Bartenschlager R., Schaller H. A splice hepadnavirus RNA that is essential for virus replication. EMBO J 1996; 15:2565-2574.

72 Soussan P., Garreau F., Zylberberg H., Ferray C., Brechot C., Kremsdorf D. In vivo expression of a new hepatitis B virus protein encoded by a spliced RNA. J Clinical Investigation 2000; 105:55-60.

73 Rosmorduc O., Petit M.A., Pol S., Capel F., Bortolotti F., Berthelot P., Brechot C., Kremsdorf D. In vivo and in vitro expression of defective hepatitis B virus particles generated by spliced hepatitis B virus RNA. Hepatology 1995; 22:10-19.

74 Nakielny S., Dreyfuss G. Transport of proteins and RNAs in and out of the nucleus. Cell 1999; 99:677-690.

75 Luo M.J., Reed R. Splicing is required for rapid and efficient mRNA export in metazoans. Proc Natl Acad Sci USA 1999; 96:14937-14942.

76 Cullen B.R. Retroviruses as model systems for the study of nuclear RNA export pathways. Virology 1998; 249:203-210.

77 Huang J., Liang T.J. A novel hepatitis B virus (HBV) genetic element with Rev response element-like properties that is essential for expression of HBV gene products. Mol Cell Biol 1993; 13:7476-7486.

78 Zang W.Q., Yen T.S.B. Distinct export pathway utilized by the hepatitis B virus posttranscriptional regulatory element. Virology 1999; 259:299-304.

79 Huang Z.M., Zang W.Q., Yen T.S. Cellular proteins that bind to the hepatitis B virus posttranscriptional regulatory element. Virology 1996; 217:573-581.

80 Zang W.Q., Fieno A.M., Grant R.A., Yen T.S. Identification of glyceraldehyde-3-phosphate dehydrogenase as a cellular protein that binds to the hepatitis B virus posttranscriptional regulatory element. Virology 1998; 248:46-52.

81 Michael W.M., Siomi H., Choi M., Pinol-Roma S., Nakielny S., Liu Q., Dreyfuss G. Signal sequences that target nuclear import and nuclear export of pre-mRNA-binding proteins. Cold Spring Harbor Symposia on Quantitative Biology 1995; 60:663-668.

82 Huang Y., Wimler K.M., Carmichael G.G. Intronless mRNA transport elements may affect multiple steps of pre-mRNA processing. EMBO J 1999; 18:1642-1652.

83 Junker-Niepmann M., Bartenschlager R., Schaller H. A short cis-acting sequence is required for hepatitis B virus pregenome encapsidation and sufficient for packaging of foreign RNA. EMBO J 1990; 9:3389-3396.

84 Jeong J.K., Yoon G.S., Ryu W.S. Evidence that the 5'-end cap structure is essential for encapsidation of hepatitis B virus pregenomic RNA. J Virol 2000; 74:5502-5508.

85 Tavis J.E., Ganem D. Evidence for activation of the hepatitis B virus polymerase by binding of its RNA template. J Virol 1996; 70:5741-5750.

86 Beck J., Nassal M. Formation of a functional hepatitis B virus replication initiation complex involves a major structural alteration in the RNA template. Mol Cell Biol 1998; 18:6265-6272.

87 Loeb D.D., Hirsch R.C., Ganem D. Sequence-independent RNA cleavages generate the primers for plus strand DNA synthesis in hepatitis B viruses: implications for other reverse transcribing elements. EMBO J 1991; 10:3533-3540.

88 Hu J., Toft D.O., Seeger C. Hepadnavirus assembly and reverse transcription require a multi-component chaperone complex which is incorporated into nucleocapsids. EMBO J 1997; 16:59-68.

89 Hatton T., Zhou S., Standring D.N. RNA- and DNA-binding activities in hepatitis B virus capsid protein: a model for their roles in viral replication. J Virol 1992; 66:5232-5241.

90 Yang W., Summers J. Illegitimate replication of linear hepadnavirus DNA through nonhomologous recombination. J Virol 1995; 69:4029-4036.

91 Yang W., Summers J. Integration of hepadnavirus DNA in infected liver: evidence for a linear precursor. J Virol 1999; 73:9710-9717.

92 Gerelsaikhan T., Tavis J.E., Bruss V. Hepatitis B virus nucleocapsid envelopment does not occur without genomic DNA synthesis. J Virol 1996; 70:4269-4274.

93 Yuan T.T., Sahu G.K., Whitehead W.E., Greenberg R, Shih C. The mechanism of an immature secretion phenotype of a highly frequent naturally occurring missense mutation at codon 97 of human hepatitis B virus core antigen. J Virol 1999; 73:5731-5740.

94 Barrasa M.I., Guo J.T., Saputelli J., Mason W.S., Seeger C. Does a cdc2 kinase-like recognition motif on the core protein of hepadnaviruses regulate assembly and disintegration of capsids? J Virol 2001;75: 2024-2028.

95 Kann M., Gerlich W.H. Effect of core protein phosphorylation by protein kinase C on encapsidation of RNA within core particles of hepatitis B virus. Journal of Virology 1994; 68:7993-8000.

96 Eble B.E., MacRae D.R., Lingappa V.R., Ganem D. Multiple topogenic sequences determine the transmembrane orientation of the hepatitis B surface antigen. Mol Cell Biol 1987; 7:3591-3601.

97 Huovila A.P., Eder A.M., Fuller S.D. Hepatitis B surface antigen assembles in a post-ER, pre-Golgi compartment. J Cell Biol 1992; 118:1305-1320.

98 Poisson F., Severac A., Hourioux C., Goudeau A., Roingeard P. Both pre-S1 and S domains of hepatitis B virus envelope proteins interact with the core particle. Virology 1997; 228:115-120.

99 Tan W.S., Dyson M.R., Murray K. Two distinct segments of the hepatitis B virus surface antigen contribute synergistically to its association with the viral core particles. J Mol Biol 1999; 286:797-808.

100 Ostapchuk P., Hearing P., Ganem D. A dramatic shift in the transmembrane topology of a viral envelope glycoprotein accompanies hepatitis B viral morphogenesis. EMBO J 1994; 13:1048-1057.

101 Bruss V., Lu X., Thomssen R., Gerlich W.H. Post-translational alterations in transmembrane topology of the hepatitis B virus large envelope protein. EMBO J 1994; 13:2273-2279.

102 Macrae D.R., Bruss V., Ganem D. Myristylation of a duck hepatitis B virus envelope protein is essential for infectivity but not for virus assembly. Virology 1991; 181:359-363.

103 Persing D.H., Varmus H.E., Ganem D. Inhibition of secretion of hepatitis B surface antigen by a related presurface polypeptide. Science 1986; 234:1388-1390.

104 Xu Z., Bruss V., Yen T.S. Formation of intracellular particles by hepatitis B virus large surface protein. J Virol 1997; 71:5487-5494.

105 Foo N.C., Yen T.S. Activation of promoters for cellular lipogenic genes by hepatitis B virus large surface protein. Virology 2000; 269:420-425.

106 Xu Z., Jensen G., Yen T.S. Activation of hepatitis B virus S promoter by the viral large surface protein via induction of stress in the endoplasmic reticulum. J Virol 1997; 71:7387-7392.

107 Benner K.G., Lee R.G., Keeffe E.B., Lopez R.R., Sasaki A.W., Pinson C.W. Fibrosing cytolytic liver failure secondary to recurrent hepatitis B after liver transplantation [see comments]. Gastroenterology 1992; 103:1307-1312.

108 Lau J.Y., Bain V.G., Davies S.E., O'Grady J.G., Alberti A., Alexander G.J., Williams R. High-level expression of hepatitis B viral antigens in fibrosing cholestatic hepatitis. Gastroenterology 1992; 102:956-962.

109 Lenhoff R.J., Summers J. Coordinate regulation of replication and virus assembly by the large envelope protein of an avian hepadnavirus. J Virol 1994; 68:4565-4571.

110 Liao W., Ou J.H. Phosphorylation and nuclear localization of the hepatitis B virus core protein: significance of serine in the three repeated SPRRR motifs. J Virol 1995; 69:1025-1029.

111 Ganem D., Varmus H.E. The molecular biology of the hepatitis B viruses. Ann Rev Biochem 1987; 56:651-693.

112 Hino O., Tabata S., Hotta Y. Evidence for increased in vitro recombination with insertion of human hepatitis B virus DNA. Proc Natl Acad Sci USA 1991; 88:9248-9252.

113 Sureau C., Romet-Lemonne J.L., Mullins J.I., Essex M. Production of hepatitis B virus by a differentiated human hepatoma cell line after transfection with cloned circular HBV DNA. Cell 1986; 47:37-47.

114 Guidotti L.G., Rochford R., Chung J., Shapiro M., Purcell R., Chisari F.V. Viral clearance without destruction of infected cells during acute HBV infection. Science 1999; 284: 825-829.

115 Guidotti L.G., Ando K., Hobbs M.V., Ishikawa T., Runkel L., Schreiber R.D., Chisari F.V. Cytotoxic T lymphocytes inhibit hepatitis B virus gene expression by a noncytolytic mechanism in transgenic mice. Proc Natl Acad Sci USA 1994; 91:3764-3768.

116 Wieland S.F., Guidotti L.G., Chisari F.V. Intrahepatic induction of alpha/beta interferon eliminates viral RNA-containing capsids in hepatitis B virus transgenic mice. J Virol 2000; 74:4165-4173.

117 Tsai S.L., Chen P.J., Lai M.Y., Yang P.M., Sung J.L., Huang J.H., Hwang L.H., Chang T.H., Chen D.S. Acute exacerbations of chronic type B hepatitis are accompanied by increased T cell responses to hepatitis B core and e antigens. Implications for hepatitis B e antigen seroconversion. J Clinical Investigation 1992; 89:87-96.

118 Fan Y.F., Lu C.C., Chen W.C., Yao W.J., Wang H.C., Chang T.T., Lei H.Y., Shiau A.L., Su I.J. Prevalence and significance of hepatitis B virus (HBV) pre-S mutants

in serum and liver at different replicative stages of chronic HBV infection. Hepatology 2001; 33:277-286.

119 Lenhoff R.J., Luscombe C.A, Summers J. Acute liver injury following infection with a cytopathic strain of duck hepatitis B virus. Hepatology 1999; 29:563-571.

Chapter 4

THE MOLECULAR BIOLOGY OF HEPATITIS C VIRUS

Keril J. Blight, Arash Grakoui, Holly L. Hanson and Charles M. Rice

Center for the Study of Hepatitis C, The Rockefeller University,Box 64, 1230 York Avenue New York, NY 10021

1. INTRODUCTION

Despite the decreased incidence of post-transfusion hepatitis caused by hepatitis C virus (HCV), this chronic viral infection remains a major health concern affecting an estimated 170 million individuals worldwide. The acute phase of HCV infection is generally subclinical, however, approximately 80% of infected individuals fail to clear the virus and develop persistent infections. Although this may result in a healthy carrier state, a high proportion develop chronic liver disease and ~20-30% of chronic carriers progress to cirrhosis. HCV-associated end-stage liver disease is now the leading cause for liver transplantation in the United States, and moreover HCV infection has been epidemiologically linked to the development of hepatocellular carcinoma. There is no vaccine to prevent hepatitis C infection. Prolonged treatment of chronically HCV-infected patients with interferon-α (IFN-α) alone, or in combination with the nucleoside analogue ribavirin is the only currently approved therapy, although poor response rates often accompany these treatment regimens.

The genome for HCV was first cloned in 1989, and has been classified as the sole member in the genus *Hepacivirus* within the *Flaviviridae* family also comprising the classical flaviviruses (e.g. yellow fever virus) and the animal pestiviruses (e.g. bovine viral diarrhea virus). HCV displays considerable genetic heterogeneity and thus far, HCV has been grouped into at least 6 major genotypes comprising numerous subtypes. Genotypes 1a and 1b are the most prevalent worldwide. Moreover, HCV exists within a single infected individual as a pool of closely related variants.

Although genetic studies of HCV replication have been impeded by its limited replication in cell culture and the lack of small animal models, significant progress has been made in understanding, at least in part, the molecular virology of HCV. Studies defining the physical properties of the virus and examining HCV replication have been restricted to clinical samples from human patients or experimentally-inoculated chimpanzees. Surrogate *in vitro* cell culture expression systems have aided in understanding the polyprotein processing scheme as well as characterizing the individual viral proteins. A summary of HCV protein function and potential host interactions is outlined in Table 1.

Table 1. Summary of HCV protein functions and potential host interactions

	Function	**Host interactions**
5'NTR	cap-independent translation; contains IRES	eIF3 (27); PTB (29, 30); La (31); hnRNP L (32); PCBP-2 (33, 34)
Core	forms internal viral nucleocapsid	members of the TNF receptor superfamily (65, 66); DEAD box protein DDX3 (213); p21 binding (214); p53 binding (215); 14-3-3- protein binding (216)
E1	glycoprotein	BiP, calnexin, calreticulin; (78, 80)
E2	glycoprotein	BiP, calnexin, calreticulin (78, 80); CD81 (86)
NS2	autoprotease with N-terminus of NS3	
NS3	serine protease (N terminus); NTPase/helicase (C terminus)	p53 (125, 126); PKA (128); PKC (129); histones (131)
NS4A	cofactor for NS3 serine protease	
NS4B	Unknown	
NS5A	interferon responsiveness (?)	PKR (155); adaptor protein 2 (167); hVAP-33 (168); human karyopherin beta 3 (169); SRCAP (170); p53 (166)
NS5B	RNA-dependent RNA polymerase	
3'NTR	contains poly (U/UC), X-tail	PTB (6-9); La autoantigen (10); GAPDH (13); HuR (12); hnRNP C (11, 12)

2. HCV GENOME STRUCTURE

The HCV genome RNA is ~9.6 kb in length, and consists a 5' non-translated region (NTR), a long open reading frame (ORF) encoding the viral polyprotein and a 3' NTR (Figure 1). The 5' NTR is ~341 nucleotides in length, highly conserved and contains an internal ribosome entry site (IRES), mediating cap-independent translation of the ORF of ~3011 amino acids. The 3' NTR consists of a short (~28-42 nt) variable sequence which is poorly conserved among different genotypes, and a polyuridine/polypyrimidine [poly(U/UC)] tract, followed by a highly conserved sequence of 98 bases, also designated the X-tail (1, 2). Biochemical probing has demonstrated that the last 46 nts fold into a stable stem loop structure (3). The sequence and structural conservation within these 98 bases suggest it may have functional importance in the viral life cycle, a hypothesis that was supported by the inability of HCV RNAs lacking this region to replicate in chimpanzees (4, 5). Moreover, the poly(U/UC) tract is also essential for RNA infectivity in the chimpanzee model, whereas the variable region is dispensable (5). Besides an array of unknown cellular proteins, polypyrimidine tract binding protein (PTB; 6-9), the La autoantigen (10), hnRNP C (11, 12), glyceraldehyde-3-phosphate dehydrogenase (GAPDH; 13) and HuR (12) have been demonstrated to specifically interact with the 3' NTR, all of which exhibit an affinity for the pyrimidine-rich region. The function(s) of the 3' NTR and its interaction with cellular proteins still remains speculative. However, PTB binding to the 3' NTR has been implicated in enhancing HCV IRES-dependent translation (14).

3. RANSLATION AND POLYPROTEIN PROCESSING

The 5' NTR folds into a highly ordered RNA structure, comprising four major structural domains (I-IV) and a psuedoknot formed by base pairing between domains III and IV (for review see 15). The IRES activity appears to require most of the 5' NTR, with the possible exception of a small hairpin in the first 20 nucleotides of the 5' NTR (stem loop I). Deletion of this stem loop appears to enhance translation (16-18) suggesting a possible role for this structure in RNA replication. In contrast, the downstream domains, including the pseudoknot, are important for IRES-mediated translation (19). The initiating AUG is located within the single stranded segment of stem loop IV which is formed in part by the capsid protein-coding sequence (20-22). Reports addressing the importance of the capsid-coding sequence for IRES-mediated translation have been contradictory. However, recently it was shown that for both HCV and the pestiviruses there is no strict requirement for a specific nucleotide sequence immediately downstream of the initiating AUG, but rather a need for the absence of stable RNA structure in this region (23, 24).

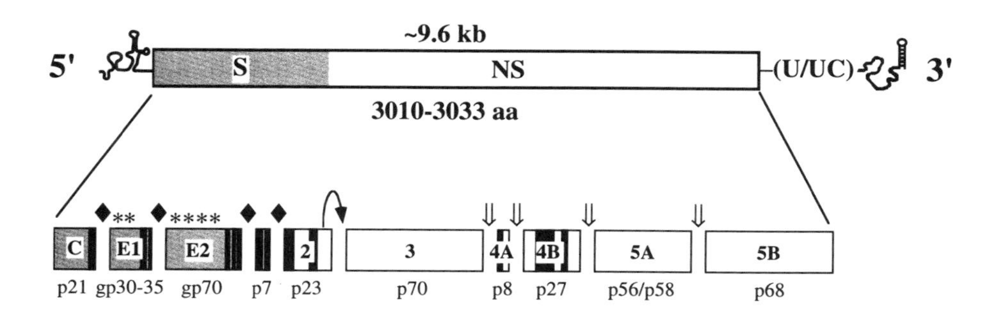

Figure 1. HCV genome structure and polyprotein processing. At the top the viral genome is depicted with the structural (S) and non-structural (NS) protein coding regions, the 5' and 3' NTRs, and the putative 3' secondary structure. Boxes below the genome represent proteins generated by the proteolytic processing cascade. Putative S proteins are indicated by shaded boxes and the NS proteins by open boxes. Contiguous stretches of uncharged amino acids are shown by black bars. Asterisks denote proteins with N-linked glycans but do not necessarily indicate the position or number of sites utilized. Cleavage sites for host signalase (♦), the NS2-3 proteinase (curved arrow), and the NS3-4A serine protease (⇓) are also marked.

Both eukaryotic initiation factor 3 (eIF3) and the 40S ribosomal subunit, the two largest components of the 43S particle, interact directly with the HCV IRES (25-27), and the high affinity interaction of IRES RNA with the 40S subunit drives formation of the IRES RNA-40S-eIF3 ternary complex (25). In contrast to IRES elements of other RNA viruses, this ternary complex forms in the absence of any other canonical eukaryotic translation factors (26). Cryo-electron microscopy studies have demonstrated that binding of the HCV IRES RNA to the 40S ribosomal subunit induces a conformational change in the 40S subunit suggesting IRES RNA actively manipulates the host cell's translation machinery to orchestrate the assembly of the pre-initiation complex (28). Moreover, stable structure immediately downstream of the initiating codon does not inhibit the ability of the 40S subunit to bind to the HCV IRES (24), suggesting the formation of subsequent interactions essential for translation may be impaired. Additional cellular proteins, such as PTB (29, 30), La (31), heterogeneous nuclear protein L (32), poly(rC)-binding protein-2 (PCBP-2; 33, 34), and proteins of 25 kDa (35) and 120 kDa (36), interact specifically with the 5' NTR and typically enhance translation, except PCBP-2. These interacting proteins are not required for 43S binding to the complex, and thus far the importance of these RNA-protein interactions for HCV translation and/or replication remains speculative. Translational activity of the HCV IRES is affected by the cell cycle, where IRES-dependent translation was greatest during the mitotic phases (37). Moreover, a cell type dependence for HCV translation was recently demonstrated (38, 39).

IRES-driven translation yields a polyprotein precursor of more than 3000 amino acids that is processed by a combination of host and viral-encoded proteases to yield at least ten major polypeptides. As illustrated in Figure 1, the order of the cleavage products is: NH_2-C-E1-E2-p7-NS2-NS3-NS4A-NS4B-NS5A-NS5B-COOH. The structural (S) proteins are located in the N-terminal portion followed by the non-structural (NS) proteins in the remainder. Cleavages at the C/E1, E1/E2 and E2/p7 and p7/NS2 junctions are thought to be mediated by a host signal peptidase residing in the lumen of the endoplasmic reticulum (ER). Production of the mature C protein may involve an additional cell-mediated, membrane-dependent cleavage event at its COOH-terminus (40, 41). Processing is delayed at the E2/NS2 site and inefficient at the E2/p7 site, leading to the accumulation of uncleaved E2-p7 (42-44).

Processing in the NS region is mediated by two overlapping HCV-encoded proteases; the zinc-stimulated autoprotease responsible for *cis* (intramolecular) cleavage at the NS2/3 site, or the NS3 serine protease which utilizes NS4A as a cofactor for efficient processing at the NS3/4A, NS4A/4B, NS4B/5A and NS5A/5B junctions. Catalytic activity of the serine protease is not required for cleavage at the NS2/3 site, and inactivation of the NS2-3 protease has minimal effect on processing at downstream sites, indicating the proteolytic activities of these two overlapping proteases are distinct.

4. STRUCTURAL PROTEINS

C protein. The capsid protein is a highly conserved basic protein of 19-21 kDa (45-49). It is membrane-associated and primarily localizes to the cytoplasmic surface of the ER (41, 46-52). With respect to particle assembly, C has been reported to bind to the HCV 5' NTR (53), and more recently found to interact with 31 nts in the IIId loop domain (54). The homotypic interaction of the C protein has been shown to require the N-terminal 115 aa for multimerization (55). Using the yeast two-hybrid system, Nolandt and coworkers have shown that the C-terminal hydrophobic portion (aa 122-172) is incapable of interacting with itself (56). Moreover, it was suggested that this domain may block multimerization by interacting in *trans* with the tryptophan-rich segment (aa 82-102) of C. Coprecipitation experiments have further suggested an interaction of C and E1, but not E2 (57). In most studies C is only translocated to the nucleus when the C-terminal hydrophobic domain responsible for ER membrane association is deleted (48-50, 58-60). Transport to the nucleus of a minor 16 kDa truncated form of C has also been reported (60). C has been shown to have weak transforming potential (61, 62) and may inhibit (63) or enhance (64) apoptosis induced by certain stimuli. Recently, full-length C has been shown to interact with the C-terminal region of some members of the TNF receptor superfamily and increase sensitivity to TNF-mediated apoptosis (65, 66). For a detailed review of the C protein refer to reference 67.

Recently, a potential overlapping reading frame in the C protein coding region was identified using a computer-based sequence analysis (68). Four peptides derived from the alternative reading frame were used in western blots to screen serum from both healthy and HCV-infected patients. Antibodies to three of the four peptides were detected in the HCV-infected patients suggesting the presence of alternative HCV core antigens in chronic infection.

E1 and E2 glycoproteins. HCV E1 and E2 are heavily modified by N-linked glycosylation and believed to be type I transmembrane glycoproteins (45, 51, 58, 69-75). For both E1 and E2, removal of the C-terminal hydrophobic region can lead to secretion rather than ER retention (76). Both a non-covalently linked heterodimer of E1E2 and a heterogeneous disulfide-linked aggregate were initially described (77, 78). Further studies suggest that the noncovalent heterodimer may represent the prebudding subunit of the HCV glycoproteins whereas the disulfide-linked aggregates represent misfolded proteins. Utilization of conformation-dependent monoclonal antibodies, as well as protease sensitivity assays, indicated that formation of putative properly assembled E1E2 oligomers is inefficient, accounting for only ~5% of the E1E2 complexes (79). The limiting steps in assembly have been studied by monitoring the formation of intramolecular disulfide bonds

(76, 78), the formation of conformation-sensitive epitopes (79), and interactions with ER chaperones (78, 80). Both HCV glycoproteins have highly conserved cysteine residues likely to be involved in disulfide bond formation in the oxidizing environment of the ER lumen. Intramolecular disulfide bond formation is slow for E1 (~60 min) (78) but appears to be complete for E2 and E2-p7 by the time cleavage of the E2-NS2 precursor has occurred (~15 min) (78). In the absence of E2, E1 never assumes its oxidized form suggesting that E2 is required for proper folding, perhaps acting as a chaperone. Further, the efficient glycosylation of E1 does not occur in the absence of E2 but instead depends on downstream sequences of the viral polyprotein (81). Recent studies have shown that BiP, calnexin, and calreticulin interact with the HCV glycoproteins in the ER. Calreticulin and BiP are preferentially associated with aggregates of misfolded proteins, whereas calnexin is associated with newly synthesized glycoproteins, oxidized monomeric forms and noncovalent heterodimers (78, 80). However, vaccinia virus-mediated overexpression of these three chaperones individually or in combination, does not increase the efficiency of productive E1E2 folding (80). Other, as yet unidentified, chaperones or foldases may be limiting for HCV glycoprotein folding.

Hypervariable regions (HVR) have been identified in the E1 and E2 genes with the most sequence variation occurring in HVR 1, located within the N-terminal portion of E2 (82-84). Although this dynamic variation in the glycoproteins may be important for the establishment and maintenance of persistent infection, definitive evidence is not yet available. Recently it was shown that E2 HVR 1 was not essential for replication or chronic infection in chimpanzees, although deletion of HVR 1 attenuated the virus (85). A cell-surface molecule, CD81 (or TAPA-1), interacts with the ectodomain of the HCV E2 protein (86). Interestingly, the interaction between the HCV E2 protein and human CD81 can be blocked by serum from protected chimpanzee vaccinees consistent with the idea that HCV may utilize CD81 during infection. Using fold recognition methods and the envelope protein E of tick borne encephalitis virus as a template, Yagnik and coworkers described a structural model for HCV E2 (87). Mapping experimental data onto this model allowed the prediction of a composite interaction site between E2 and CD81 and suggested the interaction may be mediated by the second hypervariable region of E2. However, CD81 alone does not seem to be sufficient to mediate viral attachment, entry and replication. Cell lines such as the human T cell line Molt-4 (88) allow HCV E2 binding to surface-expressed CD81, but do not support robust HCV replication. Furthermore, the HCV-H E2 protein interacts more efficiently with the CD81 molecule from tamarins, which are not susceptible to HCV infection, than with human CD81 (89, 90) confirming that the interaction between the HCV E2 glycoprotein and the CD81 molecule, by itself, is not predictive of susceptibility to HCV infection.

5. NON-STRUCTURAL PROTEINS

NS2. NS2 is a hydrophobic protein with apparent molecular mass of 23 kDa. One of the major functions of NS2 appears to be the NS2-3 protease responsible for cleavage at its own COOH-terminus, in conjunction with the serine protease domain of NS3 (91, 92). NS2-3 protease activity is necessary for the *in vivo* infectivity of full-length HCV genomes (4). However, in cell culture NS2 is dispensable for replication of subgenomic RNAs (93, 94), and no function is known for NS2 other than the NS2/3 cleavage. NS2-3 activity, which is distinct from the serine protease activity, is stimulated by zinc and inhibited by metal chelators initially leading to the suggestion that the NS2-3 protease is a metalloprotease (91, 92, 95). However the absence of motifs typical of the active center of other known metalloproteases, the identification of His-952 and Cys-993 in NS2 as possible catalytic residues (91, 92, 95), and comparative modeling with other viral cysteine proteases (96) seem to be more consistent with the hypothesis that it is a cysteine protease. However it still remains unclear whether zinc plays a catalytic or structural role in NS2/3 processing. Crystallographic (97-99) and biochemical (100, 101) analyses identified Cys-1123, Cys-1125, Cys-1171 and His-1175 in NS3 as a tetrahedrally coordinated zinc-binding site, suggesting the zinc ion stabilizes the structure of the NS3 component of the NS2-3 protease. Additional mechanistic studies of the cleavage reactions and structural studies of NS2-3 are required to determine the mechanism of NS2-3 autoproteolysis. The hydrophobicity of NS2 coupled with the autocatalytic nature of the NS2/3 cleavage pose major barriers for purification of soluble, intact protease. However, the discovery that cleavage at the NS2/3 site can be post-translationally activated in rabbit reticulocyte lysates by the addition of detergents has permitted the analysis of some biochemical properties of the NS2-3 protease, such as its sensitivity to various protease inhibitors (102).

Mutational analysis of the region surrounding the NS2/3 cleavage site revealed that the NS2-3 protease is remarkably tolerant to amino acid substitutions in the P5-P3' positions; except substitutions such as proline with a high potential for disruption of the conformation of the region encompassing the NS2/3 cleavage site (103, 104). Hence, the global conformation, rather than primary amino acid sequence, appears to be an important determinant in recognition of the NS2/3 site.

NS3. HCV NS3 is a 70 kDa protein that encodes a serine protease domain in its N-terminus and an NTPase/helicase domain in the C-terminus. It is a member of DECH-subfamily of DEAD-box helicases. The serine protease is responsible for the cleavages at the NS3/4A, NS4A/5B, NS4B/5A, and NS5A/5B sites (45, 92, 105-109) and its enzymatic activity is required for viral replication *in vivo* (4). While other downstream cleavages can occur in *trans*, the NS3/4A cleavage occurs in *cis* (109-112). The cleavage site for the serine protease is conserved and has the following amino acid sequence: (D/G)XXXX(C/T)↓(S/A). NS4A acts as a cofactor for the NS3 protease activity and is critical for all cleavages downstream except for the NS5A/5B site (111-114). The cofactor activity requires stable NS3-4A complex formation (115-118) that further stabilizes NS3 and anchors it to the cellular membranes (95, 119).

The structures of the NS3 serine protease domain either alone (98), in complex with an NS4A-derived cofactor peptide (97, 99), or full-length NS3 with NS4A (120) have been determined by X-ray crystallography. The structures indicate a critical role for the zinc ion for proper folding of the NS3 protease and reveal that the active site residues and substrate binding pocket are located in a cleft separating two β barrel domains, similar to other members of the trypsin superfamily. The structure of HCV helicase reveals distinct NTPase and RNA binding domains (121). After the recognition of RNA substrate by a conserved arginine-rich sequence on the RNA binding domain, a conformational change induces the rotation of this domain coupled to NTP hydrolysis. NS3 can unwind RNA:RNA, RNA:DNA and DNA:DNA duplexes in a 3' to 5' direction. The helicase activity requires a divalent cation, Mg^{2+} or Mn^{2+}, and an NTP, preferably ATP, although other NTPs can be utilized (122-124). Although a definitive role for the RNA helicase in replication is not yet known, mutations in NS3 that disrupt helicase activity abrogate HCV infectivity *in vivo* (4).

NS3 may have functions beyond those pertaining to HCV polyprotein processing and RNA replication. NS3 has been observed to co-localize with the cellular tumor suppressor gene product p53 (125, 126) and its serine protease domain has weak transforming activity in NIH-3T3 cells (127). A truncated NS3 containing a region with similarity to an inhibitor of PKA can interact with the catalytic subunit of PKA and inhibit its forskolin-stimulated nuclear translocation and PKA-catalyzed phosphorylation (128). Additionally, a peptide derived from the NS3 has been shown to be a selective substrate for protein kinase C (PKC) suggesting that NS3 may disrupt PKC-mediated signal transduction (129, 130). Finally, an interaction between NS3 and core histones H2B and H4 has been demonstrated (131).

NS4A and NS4B. NS4A is a small, hydrophobic polypeptide of approximately 8 kDa. The central domain of NS4A contains the NS3 serine proteinase cofactor activity (114, 115, 117, 132, 133). Synthetic peptides mimicking this region of NS4A bind NS3 with 1:1 stoichiometry and activate its proteolytic activity *in vitro* (134, 135). NS4A anchors NS3 and other NS proteins (136, 137) to cellular membranes, presumably through the N-terminal hydrophobic domain (114, 138). On the other hand, NS4B (27 kDa), also a hydrophobic protein, has not been assigned a function(s). NS4B is a cytoplasmically oriented integral ER membrane protein co-localizing with NS3, NS4A, NS5A and NS5B, suggesting NS4B is a component of the viral replication complex (139).

NS5A. The NS5A protein is differentially phosphorylated generating at least two forms of NS5A with apparent molecular masses of 56 and 58 kDa (140, 141). Phosphorylation occurs predominantly on serine residues and to a much lesser extent on threonine residues (140, 142). Ser-2321, located within a proline-rich amino acid sequence, was identified as the major site of phosphorylation for the genotype 1a HCV-H strain (143), however this site is not conserved among different HCV genotypes. On the other hand, for a genotype 1b isolate, a major phosphate acceptor site was mapped to Ser-2194 (144). Deletion analyses suggest that hyperphosphorylation sites reside in a conserved, central region of NS5A (141). Site-directed mutagenesis of the nine conserved serine residues in this region tentatively identified Ser-2197, Ser-2201 and Ser-2204 as sites of p58 phoshorylation (141). Although NS5A has been shown to be phosphorylated in the absence of other viral proteins (145), NS4A appears to enhance p58 production for the HCV-J isolate (140, 141) through association with NS5A (146). Recent studies suggest that the upstream NS proteins influence differential NS5A phosphorylation; expression of NS5A from an HCV NS polyprotein was required for p58 production (147-149). However, in the context of subgenomic RNA replicons, NS5A hyperphosphorylation was not essential for replication (93)

The cellular kinase responsible for phosphorylation of NS5A has not been identified, but is tightly associated with NS5A. The effects of various protein kinase inhibitors on NS5A phosphorylation were consistent with a kinase activity belonging to the casein kinase II/mitogen-activated protein kinase/glycogen synthase kinase 3 (CMGC) group of proline-directed serine-threonine kinases (142). Although the significance of these observations is unclear given the inhibitor profile of the NS5A-associated kinase(s), NS5A could be phosphorylated *in vitro* by a purified cAMP-dependent protein kinase A (PKA)-α catalytic subunit (150) and casein kinase II (151).

Although the function of NS5A in viral replication is ill-defined, NS5A has been implicated in determining the susceptibility of the virus to treatment with IFN. A cluster of amino acid mutations within the central region of NS5A (amino acids 2209-2248 of the polyprotein) have been identified in Japanese patients infected with HCV genotype 1b which appear to correlate with effectiveness of IFN treatment (152, 153). This sequence was coined the interferon sensitivity determining region (ISDR). However, this correlation is substantially weaker or lacking in patients infected with genotype 1a strains or European patients infected with strains from genotype 1b, 2b, or 3a (reviewed in 154). The mechanism by which NS5A may mediate IFN resistance is unclear, although it has been demonstrated that NS5A is an inhibitor of the IFN-induced double-stranded RNA-activated protein kinase (PKR). Evidence suggests that NS5A interacts with PKR in an ISDR-dependent manner subsequently blocking PKR-dependent phosphorylation of eIF2-α, and this is believed to be one mechanism of HCV-mediated IFN resistance (155). Moreover, cells expressing NS5A derived from IFN-resistant isolates prevented the antiviral action of IFN, thus allowing the rescue of IFN-sensitive viruses, encephalomyocarditis virus (EMCV) and vesicular stomatitis virus (VSV) (156-159). However, these effects have also been observed in the absence of the ISDR (158). Furthermore, replication of a subgenomic HCV replicon carrying a 47 amino acid deletion encompassing the ISDR is sensitive to the antiviral action of IFN (93). For an overview of NS5A and its potential involvement in IFN response see reference 160.

N-terminal truncated forms of NS5A fused to the DNA-binding domain of the yeast Gal4 protein activate transcription of reporter genes under the control of promoters containing Gal4 binding sites (161-163). However, since full-length NS5A lacks this *trans*-activating ability and localizes primarily to the cytoplasm, the biological significance of these findings is unclear. However, NS5A contains a putative nuclear localization signal in its C-terminal portion, and recent work provided evidence for caspase-mediated cleavages in NS5A, possibly in an apoptotic-dependent manner (164). The cleaved forms of NS5A tended to relocate to the nucleus and function as transcriptional activators regulated by PKA (164). In addition, NS5A has been shown to modulate cell growth, through regulation of p21/waf1 gene expression in a p53-dependent manner (165, 166). However, these reports are contradictory with respect to whether NS5A acts as a transcriptional activator or repressor of p21/waf1 gene expression.

Additional functions for NS5A have been proposed based on its interactions with several cellular proteins. It has been suggested that NS5A may interfere with signal transduction by interacting with growth factor receptor-bound adaptor protein 2 (167). NS5A has also been reported to interact with a SNARE-like protein, human vesicle-associated membrane protein (hVAP-33), possibly serving as another membrane anchor for the RNA replicase (168). NS5A also interacts with a nuclear import machinery component, human karyopherin beta 3 (169), and with cellular transcriptional factors including a novel cellular transcription factor SRCAP

(170) and p53 (166). However, more data is needed to establish the importance of these interactions in the life cycle and pathogenesis of HCV.

NS5B. NS5B, the C-terminal protein in the HCV polyprotein, is approximately 68 kDa in size, and represents the RNA dependent RNA polymerase (RdRp) subunit of the viral replicase. It contains motifs shared by other RdRp's, such as the Gly-Asp-Asp motif which is involved in binding the Mg^{2+} ions essential for polymerase activity. Indeed, mutation of this active site or deletion of this motif ablates infectivity of infectious clones *in vivo* (4) and replication of subgenomic RNA replicons in cell culture (93, 94). Mg^{2+}-dependent RdRp activity has been demonstrated *in vitro* using recombinant NS5B expressed in *Escherichia coli* (171-173) and in insect cells from recombinant baculoviruses (174, 175). Although recombinant NS5B utilizes HCV RNA as a template, it also readily polymerizes other RNA templates, including homopolymeric and heteropolymeric RNAs, in a primer-dependent manner. NS5B is capable of copying genome-length HCV RNA with an estimated elongation rate of 150-200 nucleotides per minute at 22°C (176, 177), and this rate was independent of NS5B concentration (176), indicating a high degree of enzyme processivity. Nevertheless, HCV NS5B alone appears to lack specificity for HCV RNA, possibly reflecting the requirement for additional viral or host factors for specific recognition. Although NS5B is capable of initiating RNA synthesis in a primer-dependent manner or by a "copy-back" mechanism, in the presence of high concentrations of ATP or GTP, *de novo* initiation of RNA synthesis has recently been demonstrated ([178-181).

The C-terminal 21 amino acids have been identified as the putative NS5B membrane-anchoring domain (172). Deletion of this hydrophobic region of NS5B allows the production of a highly soluble and enzymatically active RdRp (172, 175, 179, 181, 182). This finding enabled high resolution determinations of NS5B by X-ray crystallography (183-185). NS5B is structurally similar to other polymerases, adopting a "right hand" conformation with the palm subdomain containing active site residues, and discernable fingers and thumb subdomains. However, extensive interactions between the finger and thumb polymerase subdomains completely surround the NS5B active site. The thumb subdomain also contains structural similarity to "armadillo" repeats, which could be involved in mediating protein-protein interactions required for replicase assembly and RNA replication.

6. HCV REPLICATION AND EXPERIMENTAL SYSTEMS

Detection of HCV-encoded proteins and positive and negative sense HCV RNA in liver tissue from HCV-infected individuals (reviewed in 186), suggest hepatocytes are the major site of HCV replication. However, the precise intracellular steps of HCV RNA amplification, virion assembly and release are poorly understood due to the lack of a suitable cell culture system. In stable cell lines supporting replication of subgenomic replicons the viral proteins have been found in association with membranes derived from the ER, suggesting this is the site of RNA amplification (187). By analogy with other members of the *Flaviviridae*, HCV is thought to replicate via a negative sense RNA intermediate that serves as a template for the synthesis of additional positive sense RNAs for translation, replication, and packaging into progeny virus. During flavivirus replication the negative strand is present at a level of approximately one-tenth that of the positive strand. In the case of HCV, negative strand RNA has been detected at 5-10 fold lower levels than positive sense RNA in cell cultures supporting replication of subgenomic RNA replicons (94), and in liver tissue from HCV-infected patients (186).

Mathematical models based on measuring viral production during interferon therapy (188) or plasmapheresis (189) suggest a virion half-life of approximately 3 hours and a production rate of 10^{12} virions per day. Contrary to initial reports of low level replication in infected individuals, HCV replication is a very dynamic process, which is capable of continuously generating variants.

Animal Models. Thus far, the only animal that can be infected with HCV reproducibly is the chimpanzee. The chimpanzee has served as a critical model for viral hepatitis research with the advantage of being more than 98.5% genetically identical to humans. All five hepatitis viruses, A, B, C, D, and E are able to infect chimpanzees, and this model has been essential for the development of the currently licensed vaccines for hepatitis A and B (190-195). HCV infection in the chimpanzee follows a clinical course similar to that seen in human patients. Circulating HCV RNA is detectable within days after exposure. The acute viremic phase is often followed by the development of elevated liver transaminanses, and HCV-specific immune responses. About 50% of infected animals resolve the infection. The chimpanzee has proved particularly valuable for the identification of infectious cDNA clones for HCV genotypes 1a, 1b and 2a (196-201). The availability of functional cDNA clones has permitted verification of HCV-encoded enzymes and RNA conserved elements in the 3' NTR important for HCV replication *in vivo* (4, 5). Moreover, clonal infections of chimpanzees are being used for studies on HCV evolution, pathogenesis and the host immune response, that are relevant to understanding the factors that determine viral clearance versus chronic infection.

Attempts to establish a small animal model have thus far been unsuccessful. A Chinese subspecies of the tupaia, *Tupaia belangeri chinensis,* was susceptible to HCV infection, but these animals developed either transient or intermittent viremia with low titers (202). Limited HCV replication has been detected in irradiated mice that were rescued with SCID bone marrow cells and engrafted with human liver tissue (203). For an overview of experimental models refer to reference 204.

Cell culture systems. Cell culture studies have focused on the ability of HCV to replicate in numerous cell lines following transfection with HCV RNA transcribed from cloned cDNA or infection with virus-containing inoculum (for review see 205). However, only low levels of HCV replication have been observed, and consequently such systems are not readily applicable for genetic analysis of HCV and for antiviral screening and evaluation. Two groups (206, 207) have reported HCV replication after transfection of hepatoma cell lines with transcribed RNAs lacking the highly conserved 3' terminal 98 base sequence. This is surprising given that this terminal sequence is essential for replication *in vivo* (see above). A major breakthrough was the development of bicistronic subgenomic RNA replicons, where the HCV structural region was replaced by the neomycin phosphotransferase gene and translation of HCV proteins NS2 or NS3 to NS5B was directed by the EMCV IRES (94). Following transfection into a human hepatoma cell line, Huh-7, G418-resistant cell colonies, harboring high levels of autonomous HCV replication, were selected at low frequency. Recently, a spectrum of adaptive mutations in the HCV NS proteins was identified increasing the replication efficiency. For instance, a replicon harboring a single amino acid substitution in NS5A allowed efficient initiation of RNA replication in as many as 10% of transfected Huh-7 cells (100,000 fold improvement; 93). In another study, a point mutation in NS5B (208), and more recently, adaptive mutations in NS3 and NS5A act synergistically to increase the efficiency of colony formation (209).

Stable cell lines harboring autonomously replicating subgenomic RNAs have begun to provide information on polyprotein processing kinetics, protein half lives and viral-host interactions. For example, cleavages at the NS3/4A and NS5A/5B site are rapid, whereas the NS4A-4B-5A precusor was processed at a slower rate (187). The mature proteins had half lives ranging from 10 to 16 hours, except the hyperphosphorylated form of NS5A which appeared less stable. No obvious ultrastructural changes were evident in these cells, suggesting subgenomic RNA replication and NS3-5B are not cytopathic. However, the level of RNA replication is dictated by the growth stage of the cells (187).

Thus far, HCV replication is restricted to Huh-7 cells, suggesting a favorable cellular environment. Data suggest Huh-7 cells may be inherently defective in their IFN response mechanism (210, 211), thus allowing the establishment and persistence of HCV replication. Furthermore, replication of HCV subgenomic RNAs is dramatically inhibited by interferon-α in a dose-dependent manner (93, 212). The mechanism by which IFN inhibits HCV replication is ill-defined, although inhibition is independent of the IFN-induced GTPase, MxA (212).

7. CONCLUDING REMARKS

Knowledge of the HCV genome organization, polyprotein processing, protein function and structure has been accumulating. The establishment of a subgenomic replicon-based cell culture system not only provides a workable system for the identification of specific inhibitors of HCV replication, but also enables functional genetic and biochemical analyses of HCV replication *in vitro*. However, extensive work remains to be done with respect to HCV virion structure, mode of entry into hepatocytes, processes of uncoating, assembly, and pathogenesis. Hence, systems capable of virus assembly and export, cell-free systems and alternative animal models are still required.

8. ACKNOWLEDGEMENTS

Supported in part by grants from the Public Health Service to C.M.R. (CA57973 and AI40034) and the Greenberg Medical Foundation. A.G. is supported by a Cancer Research Institute fellowship.

REFERENCES

1 Kolykhalov A.A., Feinstone S.M., Rice C.M. Identification of a highly conserved sequence element at the 3' terminus of hepatitis C virus genome RNA. J Virol 1996; 70:3363-3371.

2 Tanaka T., Kato N., Cho M.-J., Shimotohno K. A novel sequence found at the 3' terminus of hepatitis C virus genome. Biochem Biophys Res Comm 1995; 215:744-749.

3 Blight K.J., Rice C.M. Secondary structure determination of the conserved 98-base sequence at the 3' terminus of hepatitis C virus genome RNA. J Virol 1997; 71:7345-7352.

4 Kolykhalov A.A., Mihalik K., Feinstone S.M., Rice C.M. Hepatitis C virus-encoded enzymatic activities and conserved RNA elements in the 3' nontranslated region are essential for virus replication in vivo. J Virol 2000; 74:2046-2051.

5 Yanagi M., St Claire M., Emerson S.U., Purcell R.H., Bukh J. In vivo analysis of the 3' untranslated region of the hepatitis C virus after in vitro mutagenesis of an infectious cDNA clone. Proc Natl Acad Sci USA 1999; 96:2291-5.

6 Chung R.T., Kaplan L.M. Heterogeneous nuclear ribonucleoprotein I (hnRNP-I/PTB) selectively binds the conserved 3' terminus of hepatitis C viral RNA. Biochem Biophys Res Commun 1999; 254:351-362.

7 Gontarek R.R., Gutshall L.L., Tsai J., Sathe G.M., Mao J.Y., Prescott C.D., Vecchio A.M. Interaction of polypyrimidine tract-binding protein with the 3' non-translated region of the hepatitis C virus genome. Nucleic Acids Symp Ser 1997; 36:146-149.

8 Ito T., Lai M.M.C. Determination of the secondary structure of and cellular protein binding to the 3'-untranslated region of the hepatitis C virus RNA genome. J Virol 1997; 71:8698-8706.

9 Tsuchihara K., Tanaka T., Hijikata M., Kuge S., Toyoda H., Nomoto A., Yamamoto N., Shimotohno K. Specific interaction of polypyrimidine tract-binding protein with the extreme 3'-terminal structure of the hepatitis C virus genome, the 3'X. J Virol 1997; 71:6720-6726.

10 Spangberg K., Goobar-Larsson L., Wahren-Herlenius M., Schwartz S. The La protein from human liver cells interacts specifically with the U-rich region in the hepatitis C virus 3' untranslated region. J Hum Virol 1999; 2:296-307.

11 Gontarek R.R., Gutshall L.L., Herold K.M., Tsai J., Sathe G.M., Mao J., Prescott C., Del Vecchio A.M. hnRNP C and polypyrimidine tract-binding protein specifically interact with the pyrimidine-rich region within the 3'NTR of the HCV RNA genome. Nucleic Acids Research 1999; 27:1457-1463.

12 Spangberg K., Wiklund L., Schwartz S. HuR, a protein implicated in oncogene and growth factor mRNA decay, binds to the 3' ends of hepatitis C virus RNA of both polarities. Virology 2000; 274:378-390.

13 Petrik J., Parker H., Alexander G.J. Human hepatic glyceraldehyde-3-phosphate dehydrogenase binds to the poly(U) tract of the 3' non-coding region of hepatitis C virus genomic RNA. J Gen Virol 1999; 80:3109-13.

14 Ito T., Tahara S.M., Lai M.M.C. The 3'-untranslated region of hepatitis C virus RNA enhances translation from an internal ribosomal entry site. J Virol 1998; 72:8789-8796.

15 Rijnbrand R.C.A., Lemon S.M. "Internal ribosome entry site-mediated translation in hepatitis C virus replication". In *Hepatitis C virus*, C. Hagedorn, C. M. Rice eds. Berlin: Springer-Verlag, 2000.

16 Honda M., Ping L.H., Rijnbrand R.C., Amphlett E., Clarke B., Rowlands D., Lemon S.M. Structural requirements for initiation of translation by internal ribosome entry within genome-length hepatitis C virus RNA. Virology 1996; 222:31-42.

17 Rijnbrand R., Bredenbeek P.J., Van Der Straaten T., Whetter L., Inchauspe G., Lemon S., Spaan W. Almost the entire 5' non-translated region of hepatitis C virus is required for cap-independent translation. FEBS Lett. 1995; 365:115-119.

18 Yoo B.J., Spaete R.R., Geballe A.P., Selby M., Houghton M., Han J.H. 5' end-dependent translation initiation of hepatitis C viral RNA and the presence of putative positive and negative translational control elements within the 5' untranslated region. Virology 1992; 191:889-899.

19 Wang C., Siddiqui A. Structure and function of the hepatitis C virus internal ribosome entry site. Curr Top Microbiol Immunol 1995; 203:99-115.

20 Honda M., Brown E.A., Lemon S.M. Stability of a stem-loop involving the initiator AUG controls the efficiency of internal initiation of translation on hepatitis C virus RNA. RNA 1996; 2:955-968.

21 Reynolds J.E., Kaminski A., Carroll A.R., Clarke B.E., Rowlands D.J., Jackson R.J. Internal initiation of translation of hepatitis C virus RNA: the ribosome entry site is at the authentic initiation codon. RNA 1996; 2:867-878.

22 Rijnbrand R.C., Abbink T.E., Haasnoot P.C., Spaan W.J., Bredenbeek P.J. The influence of AUG codons in the hepatitis C virus 5' nontranslated region on translation and mapping of the translation initiation window. Virology 1996; 226:47-56.

23 Myers T.M., Kolupaeva V.G., Mendez E., G. B.S., Frolov I., Hellen C.U.T., Rice C.M. Efficient translation initiation is required for replication of bovine viral diarrhea virus subgenomic replicons. J Virol 2001; 75:4226-4238.

24 Rijnbrand R., Bredenbeek P.J., Haasnoot P.C., Kieft J.S., Spaan W.J.M., Lemon S.M. The influence of downstream protein-coding sequence on internal ribosome entry on hepatitis C virus amnd other flavivirus RNAs. RNA 2001; 7:585-597.

25 Kieft J.S., Zhou K., Jubin R., Doudna J.A. Mechansim of ribosome recruitment by hepatitis C IRES RNA. RNA 2001; 7:194-206.

26 Pestova T.V., Shatsky I.N., Fletcher S.P., Jackson R.J., Hellen C.U. A prokaryotic-like mode of cytoplasmic eukaryotic ribosome binding to the initiation codon during internal translation initiation of hepatitis C and classical swine fever virus RNAs. Genes Dev 1998; 12:67-83.

27 Sizova D.V., Kolupaeva V.G., Pestova T.V., Shatsky I.N., Hellen C.U.T. Specific interaction of eukaryotic translation initiation factor 3 with the 5' nontranslated regions of hepatitis C virus and classical swine fever virus RNAs. J Virol 1998; 72:4775-4782.

28 Spahn C.M.T., Kieft J.S., Grassucci R.A., Penczek P.A., Zhou K., Doudna J.A., Frank J. Hepatitis C virus IRES RNA-induced changes in the conformation of the 40S ribosomal subunit. Science 2001; 291:1959-1962.

29 Ali N., Siddiqui A. Interaction of polypyrimidine tract-binding protein with the 5' noncoding region of the hepatitis C virus RNA genome and its functional requirement in internal initiation of translation. J. Virol. 1995; 69:6367-6375.

30 Kaminski A., Hunt S.L., Patton J.G., Jackson R.J. Direct evidence that polypyrimidine tract binding protein (PTB) is essential for internal initiation of translation of encephalomyocarditis virus RNA. RNA 1995; 1:924-938.

31 Ali N., Siddiqui A. The La antigen binds 5' noncoding region of the hepatitis C virus RNA in the context of the initiator AUG codon and stimulates internal ribosome entry site-mediated translation. Proc Natl Acad Sci USA 1997; 94:2249-2254.

32 Hahm B., Kim Y.K., Kim J.H., Kim T.Y., Jang S.K. Heterogeneous nuclear ribonucleoprotein L interacts with the 3' border of the internal ribosomal entry site of hepatitis C virus. J Virol 1998; 72:8782-8788.

33 Fukushi S., Okada M., Kageyama T., Hoshino F.B., Nagai K., Katayama K. Interaction of poly(rC)-binding protein 2 with the 5'-terminal stem loop of the hepatitis C-virus genome. Virus Res. 2001; 73:67-79.

34 Spangberg K., Schwartz S. Poly(C)-binding protein interacts with the hepatitis C virus 5' untranslated region. J Gen Virol 1999; 80:1371-1376.

35 Fukushi S., Kurihara C., Ishiyama N., Hoshino F.B., Oya A., Katayama K. The sequence element of the internal ribosome entry site and a 25-kilodalton cellular protein contribute to efficient internal initiation of translation of hepatitis C virus RNA. J Virol 1997; 71:1662-1666.

36 Yen J.-H., Chang S.C., Hu C.-R., Chu S.-C., Lin S.-S., Hsieh Y.-S., Chang M.-F. Cellular proteins specifically bind to the 5'-noncoding region of hepatitis C virus RNA. Virology 1995; 208:723-732.

37 Honda M., Kaneko S., Matsushita E., Kobayashi K., Abell G.A., Lemon S.M. Cell cycle regulation of hepatitis C virus internal ribosome entry site-directed translation. Gastroenterology 2000; 118:152-162.

38 Laporte J., Malet I., Andrieu T., Thibault V., Toulme J.J., Wychowski C., Pawlotsky J.M., Huraux J.M., Agut H., Cahour A. Comparative analysis of translation efficiencies of hepatitis C virus 5' untranslated regions anong intraindividual quasispecies present in chronic infection: opposite behaviors depending on cell type. J Virol 2000; 74:10827-10833.

39 Lerat H., Shimizu Y.K., Lemon S.M. Cell type-specific enhancement of hepatitis C virus internal ribosome entry site-directed translation due to 5' nontranslated region substitutions selected during passage of virus in lymphoblastoid cells. J Virol 2000; 74:7024-7031.

40 Hüssy P., Langen H., Mous J., Jacobsen H. Hepatitis C virus core protein: carboxy-terminal boundaries of two processed species suggest cleavage by a signal peptide peptidase. Virology 1996; 224:93-104.

41 Santolini E., Migliaccio G., La Monica N. Biosynthesis and biochemical properties of the hepatitis C virus core protein. J Virol 1994; 68:3631-3641.

42 Lin C., Lindenbach B.D., Prágai B., Mccourt D.W., Rice C.M. Processing of the hepatitis C virus E2-NS2 region: Identification of p7 and two distinct E2-specific products with different C termini. J Virol 1994; 68:5063-5073.

43 Mizushima H., Hijikata H., Asabe S.-I., Hirota M., Kimura K., Shimotohno K. Two hepatitis C virus glycoprotein E2 products with different C termini. J Virol 1994; 68:6215-6222.

44 Selby M.J., Glazer E., Masiarz F., Houghton M. Complex processing and protein:protein interactions in the E2:NS2 region of HCV. Virology 1994; 204:114-122.

45 Grakoui A., Wychowski C., Lin C., Feinstone S.M., Rice C.M. Expression and identification of hepatitis C virus polyprotein cleavage products. J Virol 1993; 67:1385-1395.

46 Harada S., Watanabe Y., Takeuchi K., Suzuki T., Katayama T., Takebe Y., Saito I., Miyamura T. Expression of processed core protein of hepatitis C virus in mammalian cells. J Virol 1991; 65:3015-3021.

47 Moradpour D., Englert C., Wakita T., Wands J.R. Characterization of cell lines allowing tightly regulated expression of hepatitis C virus core protein. Virology 1996; 222:51-63.

48 Ravaggi A., Natoli G., Primi D., Albertini A., Levrero M., Cariani E. Intracellular localization of full-length and truncated hepatitis C virus core protein expressed in mammalian cells. J Hepatol 1994; 20:833-836.

49 Suzuki R., Matsuura Y., Susuki T., Ando A., Chiba J., Harada S., Saito I., Miyamura T. Nuclear localization of the truncated hepatitis C virus core protein with its hydrophobic C terminus deleted. J Gen Virol 1995; 76:53-61.

50 Chang S.C., Yen J.-H., Kang H.-Y., Jang M.-H., Chang M.-F. Nuclear localization signals in the core protein of hepatitis C virus. Biochem Biophys Res Comm 1994; 205:1284-1290.

51 Hijikata M., Kato N., Ootsuyama Y., Nakagawa M., Shimotohno K. Gene mapping of the putative structural region of the hepatitis C virus genome by *in vitro* processing analysis. Proc Natl Acad Sci USA 1991; 88:5547-5551.

52 Kim D.W., Suzuki R., Harada T., Saito I., Miyamura T. *Trans*-suppression of gene expression by hepatitis C viral core protein. Jpn J Med Sci Biol 1994; 47:211-220.

53 Hwang S.B., Lo S.-Y., Ou J.-H., Lai M.M.C. Detection of cellular proteins and viral core protein interacting with the 5' untranslated region of hepatitis C virus RNA. J Biomed Sci 1995; 2:227-236.

54 Tanaka Y., Shimoike T., Ishii K., Suzuki R., Suzuki T., Ushijima H., Matsuura Y., Miyamura T. Selective binding of hepatitis C virus core protein to synthetic oligonucleotides corresponding to the 5' untranslated region of the viral genome. Virology 2000; 270:229-36.

55 Matsumoto M., Hwang S.B., Jeng K.-S., Zhu N., Lai M.M.C. Homotypic interaction and multimerization of hepatitis C virus core protein. Virology 1996; 218:43-51.

56 Nolandt O., Kern V., Muller H., Pfaff E., Theilmann L., Welker R., Krausslich H.G. Analysis of hepatitis C virus core protein interaction domains. J Gen Virol 1997; 78:1331-1340.

57 Lo S.-Y., Selby M.J., Ou J.-H. Interaction between hepatitis C virus core protein and E1 envelope protein. J Virol 1996; 70:5177-5182.

58 Lanford R.E., Notvall L., Chavez D., White R., Frenzel G., Simonsen C., Kim J. Analysis of hepatitis C virus capsid, E1, and E2/NS1 proteins expressed in insect cells. Virology 1993; 197:225-235.

59 Lo S.-Y., Selby M., Tong M., Ou J.-H. Comparative studies of the core gene products of two different hepatitis C virus isolates: Two alternative forms determined by a single amino acid substitution. Virology 1994; 199:124-131.

60 Lo S.-Y., Masiarz F., Hwang S.B., Lai M.M.C., Ou J.-H. Differential subcellular localization of hepatitis C virus core gene products. Virology 1995; 213:455-461.

61 Chang J., Yang S.-H., Cho Y.-G., Hwang S.B., Hahn Y.S., Sung Y.C. Hepatitis C virus core from two different genotypes has an oncogenic potential but is not sufficient for transforming primary rat embryo fibroblasts in cooperation with the H-*ras* oncogene. J Virol 1998; 72:3060-3065.

62 Ray R.B., Lagging L.M., Meyer K., Ray R. Hepatitis C virus core protein cooperates with *ras* and transforms primary rat embryo fibroblasts to tumorigenic phenotype. J Virol 1996; 70:4438-4443.

63 Ray R.B., Meyer K., Ray R. Suppression of apoptotic cell death by hepatitis C virus core protein. Virology 1996; 226:176-182.

64 Ruggieri A., Harada T., Matsuura Y., Miyamura T. Sensitization to Fas-mediated apoptosis by hepatitis C virus core protein. Virology 1997; 229:68-76.

65 Matsumoto M., Hsieh T.-Y., Zhu N., Vanarsdale T., Hwang S.B., Jeng K.-S., Gorbalenya A.E., Lo S.-Y., Ou J.-H., Ware C.F., Lai M.M.C. Hepatitis C virus core protein interacts with the cytoplasmic tail of lymphotoxin-b receptor. J Virol 1997; 71:1301-1309.

66 Zhu N., Khoshnan A., Schneider R., Matsumoto M., Dennert G., Ware C., Lai M.M.C. Hepatitis C virus core protein binds to the cytoplasmic domain of tumor necrosis factor (TNF) receptor 1 and enhances TNF-induced apoptosis. J Virol 1998; 72:3691-3697.

67 Choi J., Lu W., Ou J.H. Structure and functions of hepatitis C virus core protein. Recent Res Devel Virol 2001; 3:105-120.

68 Walewski J.L., Keller T.R., Stump D.D., Branch A.D. Evidence for a new hepatitis C virus antigen encoded in an overlapping reading frame. RNA 2001; 7:710-21.

69 Hsu H.H., Donets M., Greenberg H.B., Feinstone S.M. Characterization of hepatitis C virus structural proteins with a recombinant baculovirus expression system. Hepatology 1993; 17:763-771.

70 Kohara M., Tsukiyama-Kohara K., Maki N., Asano K., Yamaguchi K., Miki K., Tanaka S., Hattori N., Matsuura Y., Saito I., Miyamura T., Nomoto A. Expression and characterization of glycoprotein gp35 of hepatitis C virus using recombinant vaccinia virus. J Gen Virol 1992; 73:2313-2318.

71 Koike K., Moriya K., Ishibashi K., Matsuura Y., Suzuki T., Saito I., Iino S., Kurokawa K., Miyamura T. Expression of hepatitis C virus envelope proteins in transgenic mice. J Gen Virol 1995; 76:3031-3038.

72 Matsuura Y., Harada S., Suzuki R., Watanabe Y., Inoue Y., Saito I., Miyamura T. Expression of processed envelope protein of hepatitis C virus in mammalian and insect cells. J Virol 1992; 66:1425-1431.

73 Matsuura Y., Suzuki T., Suzuki R., Sato M., Aizaki H., Saito I., Miyamura T. Processing of E1 and E2 glycoproteins of hepatitis C virus expressed in mammalian and insect cells. Virology 1994; 205:141-150.

74 Ryu W.-S., Choi D.-Y., Yang J.-Y., Kim C.-H., Kwon Y.-S., So H.-S., Cho J.M. Characterization of the putative E2 envelope glycoprotein of hepatitis C virus expressed in stably transformed Chinese hamster ovary cells. Molecules & Cells 1995; 5:563-568.

75 Spaete R.R., Alexander D., Rugroden M.E., Choo Q.-L., Berger K., Crawford K., Kuo C., Leng S., Lee C., Ralston R., Thudium K., Tung J.W., Kuo G., Houghton M. Characterization of the hepatitis E2/NS1 gene product expressed in mammalian cells. Virology 1992; 188:819-830.

76 Michalak J.P., Wychowski C., Choukhi A., Meunier J.C., Ung S., Rice C.M., Dubuisson J. Characterization of truncated forms of the hepatitis C virus glycoproteins. J Gen Virol 1997; 78:2299-2306.

77 Dubuisson J., Hsu H.H., Cheung R.C., Greenberg H., Russell D.R., Rice C.M. Formation and intracellular localization of hepatitis C virus envelope glycoprotein complexes expressed by recombinant vaccinia and Sindbis viruses. J Virol 1994; 68:6147-6160.

78 Dubuisson J., Rice C.M. Hepatitis C virus glycoprotein folding: Disulfide bond formation and association with calnexin. J Virol 1996; 70:778-786.

79 Deleersnyder V., Pillez A., Wychowski C., Blight K., Xu J., Hahn Y.S., Rice C.M., Dubuisson J. Formation of native hepatitis C virus glycoprotein complexes. J Virol 1997; 71:697-704.

80 Chookhi A., Ung S., Wychowski C., Dubuisson J. Involvement of endoplasmic reticulum chaperones in the folding of hepatitis C virus glycoproteins. J Virol 1998; 72:3851-3858.

81 Dubuisson J., Duvet S., Meunier J.C., Op De Beeck A., Cacan R., Wychowski C., Cocquerel L. Glycosylation of the hepatitis C virus envelope protein E1 is dependent on the presence of a downstream sequence on the the viral polyprotein. J Biol Chem 2000; 275:30605-30609.

82 Martell M., Esteban J.I., Quer J., Genesca J., Weiner A., Esteban R., Guardia J., Gomez J. Hepatitis C virus (HCV) circulates as a population of different but closely related genomes: Quasispecies nature of the HCV genome distribution. J Virol 1992; 66:3225-3229.

83 Klenerman P., Lechner F., Kantzanou M., Ciurea A., Hengartner H., Zinkernagel R. Viral escape and the failure of cellular immune responses. Science 2000; 289:2003.

84 Farci P., Shimoda A., Coiana A., Diaz G., Peddis G., Melpolder J.C., Strazzera A., Chien D.Y., Munoz S.J., Balestrieri A., Purcell R.H., Alter H.J. The outcome of acute hepatitis C predicted by the evolution of the viral quasispecies. Science 2000; 288:339-44.

85 Forns X., Thimme R., Govindarajan S., Emerson S.U., Purcell R.H., Chisari F.V., Bukh J. Hepatitis C virus lacking the hypervariable region 1 of the second envelope glycoprotein is infectious and causes acute resolving or persistent infection in chimpanzees. Proc Natl Acad Sci USA 2000; 97:13318-23.

86 Pileri P., Uematsu Y., Compagnoli S., Galli G., Falugi F., Petracca R., Weiner A.J., Houghton M., Rosa D., Grandi G., Abrignani S. Binding of hepatitis C virus to CD81. Science 1998; 282:938-941.

87 Yagnik A.T., Lahm A., Meola A., Roccasecca R.M., Ercole B.B., Nicosia A., Tramontano A. A model for the hepatitis C virus envelope glycoprotein E2. Proteins 2000; 40:355-66.

88 Shimizu Y.K., Iwamoto A., Hijikata M., Purcell R.H., Yoshikura H. Evidence for *in vitro* replication of hepatitis C virus genome in a human T-cell line. Proc Natl Acad Sci USA 1992; 89:5477-5481.

89 Allander T., Forns X., Emerson S.U., Purcell R.H., Bukh J. Hepatitis C virus envelope protein E2 binds CD81 of tamarins. Virology 2000; 277:358-67.

90 Meola A., Sbardellati A., Bruni Ercole B., Cerretani M., Pezzanera M., Ceccacci A., Vitelli A., Levy S., Nicosia A., Traboni C., Mckeating J., Scarselli E. Binding of hepatitis C virus E2 glycoprotein to CD81 does not correlate with species permissiveness to infection. J Virol 2000; 74:5933-8.

91 Grakoui A., Mccourt D.W., Wychowski C., Feinstone S.M., Rice C.M. A second hepatitis C virus-encoded proteinase. Proc Natl Acad Sci USA 1993; 90:10583-10587.

92 Hijikata M., Mizushima H., Akagi T., Mori S., Kakiuchi N., Kato N., Tanaka T., Kimura K., Shimotohno K. Two distinct proteinase activities required for the processing of a putative nonstructural precursor protein of hepatitis C virus. J Virol 1993; 67:4665-4675.

93 Blight K.J., Kolykhalov A.A., Rice C.M. Efficient initiation of HCV RNA replication in cell culture. Science 2000; 290:1972-1974.

94 Lohmann V., Korner F., Koch J.O., Herian U., Theilmann L., Bartenschlager R. Replication of subgenomic hepatitis C virus RNAs in a hepatoma cell line. Science 1999; 285:110-113.

95 Hijikata M., Mizushima H., Tanji Y., Komoda Y., Hirowatari Y., Akagi T., Kato N., Kimura K., Shimotohno K. Proteolytic processing and membrane association of putative nonstructural proteins of hepatitis C virus. Proc Natl Acad Sci USA 1993; 90:10773-10777.

96 Gorbalenya A.E., Snijder E.J. Viral cysteine proteinases. Perspectives in Drug Discovery and Design 1996; 6:64-86.

97 Kim J.L., Morgenstern K.A., Lin C., Fox T., Dwyer M.D., Landro J.A., Chambers S.P., Markland W., Lepre C.A., O'malley E.T., Harbeson S.L., Rice C.M., Murcko M.A., Caron P.R., Thomson J.A. Crystal structure of the hepatitis C virus NS3 protease domain complexed with a synthetic NS4A cofactor peptide. Cell 1996; 87:343-355.

98 Love R.A., Parge H., Wickersham J.A., Hostomsky Z., Habuka N., Moomaw E.W., Adachi T., Hostomska Z. The crystal structure of hepatitis C virus NS3 proteinase reveals a trypsin-like fold and a structural zinc binding site. Cell 1996; 87:331-342.

99 Yan Y., Li Y., Munshi S., Sardana V., Cole J.L., Sardana M., Steinkuehler C., Tomei L., De Francesco R., Kuo L.C., Chen Z. Complex of NS3 protease and NS4A peptide of BK strain hepatitis C virus: a 2.2 Å resolution structure in a hexagonal crystal form. Protein Sci 1998; 7:837-847.

100 De Francesco R., Urbani A., Nardi M.C., Tomei L., Steinkuhler C., Tramontano A. A zinc binding site in viral serine proteinases. Biochemistry 1996; 35:13282-13287.

101 Stempniak M., Hostomska Z., Nodes B.R., Hostomsky Z. The NS3 proteinase domain of hepatitis C virus is a zinc-containing enzyme. J Virol 1997; 71:2881-2886.

102 Pieroni L., Santolini E., Fipaldini C., Pacini L., Migliaccio G., La Monica N. In vitro study of the NS2-3 protease of hepatitis C virus. J Virol 1997; 71:6373-6380.

103 Hirowatari Y., Hijikata M., Tanji Y., Nyunoya H., Mizushima H., Kimura K., Tanaka T., Kato N., Shimotohno K. Two proteinase activities in HCV polypeptide expressed in insect cells using baculovirus vector. Arch. Virol. 1993; 133:349-356.

104 Reed K.E., Grakoui A., Rice C.M. The hepatitis C virus NS2-3 autoproteinase: cleavage site mutagenesis and requirements for bimolecular cleavage. J Virol 1995; 69:4127-4136.

105 Bartenschlager R., Ahlborn-Laake L., Mous J., Jacobsen H. Nonstructural protein 3 of the hepatitis C virus encodes a serine-type proteinase required for cleavage at the NS3/4 and NS4/5 junctions. J Virol 1993; 67:3835-3844.

106 Grakoui A., Mccourt D.W., Wychowski C., Feinstone S.M., Rice C.M. Characterization of the hepatitis C virus-encoded serine proteinase: determination of proteinase-dependent polyprotein cleavage sites. J Virol 1993; 67:2832-2843.

107 Eckart M.R., Selby M., Masiarz F., Lee C., Berger K., Crawford K., Kuo C., Kuo G., Houghton M., Choo Q.-L. The hepatitis C virus encodes a serine protease involved in processing of the putative nonstructural proteins from the viral polyprotein precursor. Biochem Biophys Res Comm 1993; 192:399-406.

108 Manabe S., Fuke I., Tanishita O., Kaji C., Gomi Y., Yoshida S., Mori C., Takamizawa A., Yoshida I., Okayama H. Production of nonstructural proteins of hepatitis C virus requires a putative viral protease encoded by NS3. Virology 1994; 198:636-644.

109 Tomei L., Failla C., Santolini E., Defrancesco R., La Monica N. NS3 is a serine protease required for processing of hepatitis C virus polyprotein. J Virol 1993; 67:4017-4026.

110 Tanji Y., Hijikata M., Hirowatari Y., Shimotohno K. Hepatitis C virus polyprotein processing: kinetics and mutagenic analysis of serine proteinase-dependent cleavage. J Virol 1994; 68:8418-8422.

111 Lin C., Prágai B., Grakoui A., Xu J., Rice C.M. Hepatitis C virus NS3 serine proteinase: *trans*-cleavage requirements and processing kinetics. J Virol 1994; 68:8147-8157.

112 Bartenschlager R., Ahlborn-Laake L., Mous J., Jacobsen H. Kinetic and structural analyses of hepatitis C virus polyprotein processing. J Virol 1994; 68:5045-5055.

113 Failla C., Tomei L., Defrancesco R. Both NS3 and NS4A are required for proteolytic processing of hepatitis C virus nonstructural proteins. J Virol 1994; 68:3753-3760.

114 Tanji Y., Hijikata M., Satoh S., Kaneko T., Shimotohno K. Hepatitis C virus-encoded nonstructural protein NS4A has versatile functions in viral protein processing. J Virol 1995; 69:1575-1581.

115 Bartenschlager R., Lohmann V., Wilkinson T., Koch J.O. Complex formation between the NS3 serine-type proteinase of the hepatitis C virus and NS4A and its importance for polyprotein maturation. J Virol 1995; 69:7519-7528.

116 Failla C., Tomei L., Defrancesco R. An amino-terminal domain of the hepatitis C virus NS3 protease is essential for interaction with NS4A. J Virol 1995; 69:1769-1777.

117 Lin C., Thomson J.A., Rice C.M. A central region in the hepatitis C virus NS4A protein allows formation of an active NS3-NS4A serine proteinase complex in vivo and in vitro. J Virol 1995; 69:4373-4380.

118 Satoh S., Tanji Y., Hijikata M., Kimura K., Shimotohno K. The N-terminal region of hepatitis C virus nonstructural protein 3 (NS3) is essential for stable complex formation with NS4A. J Virol 1995; 69:4255-4260.

119 Tanji Y., Hijikata M., Hirowatari Y., Shimotohno K. Identification of the domain required for trans-cleavage activity of hepatitis C viral serine proteinase. Gene 1994; 145:215-219.

120 Yao N., Reichert P., Taremi S.S., Prosise W.W., Weber P.C. Molecular views of viral polyprotein processing revealed by the crystal structure of the hepatitis C virus bifunctional protease-helicase. Structure Fold Des 1999; 7:1353-63.

121 Yao N., Hesson T., Cable M., Hong Z., Kwong A.D., Le H.V., Weber P.C. Structure of the hepatitis C virus RNA helicase domain. Nat Struct Biol 1997; 4:463-467.

122 Gwack T., Kim D.W., Hang J.H., Choe J. Characterization of RNA binding activity and RNA helicase activity of the hepatitis C virus NS3 protein. Biochem Biophys Res Comm 1996; 225:654-659.

123 Gwack Y., Kim D.W., Han J.H., Choe J. DNA helicase activity of the hepatitis C virus nonstructural protein 3. Eur J Biochem 1997; 250:47-54.

124 Tai C.-L., Chi W.-K., Chen D.-S., Hwang L.-H. The helicase activity associated with hepatitis C virus nonstructural protein 3 (NS3). J Virol 1996; 70:8477-8484.

125 Ishido S., Muramatsu S., Fujita T., Iwanaga Y., Tong W.Y., Katayama Y., Itoh M., Hotta H. Wild-type, but not mutant-type, p53 enhances nuclear accumulation of the NS3 protein of hepatitis C virus. Biochem Biophys Res Commun 1997; 230:431-436.

126 Muramatsu S., Ishido S., Fujita T., Itoh M., Hotta H. Nuclear localization of the NS3 protein of hepatitis C virus and factors affecting the localization. J Virol 1997; 71:4954-4961.

127 Sakamuro D., Furukawa T., Takegami T. Hepatitis C virus nonstructural protein NS3 transforms NIH 3T3 cells. J Virol 1995; 69:3893-3896.

128 Borowski P., Oehlmann K., Heiland M., Laufs R. Nonstructural protein 3 of hepatitis C virus blocks the distribution of the free catalytic subunit of cyclic AMP-dependent protein kinase. J Virol 1997; 71:2838-2843.

129 Borowski P., Zur Wiesch J.S., Resch K., Feucht H., Laufs R., Schmitz H. Protein kinase C recognizes the protein kinase A-binding motif of nonstructural protein 3 of hepatitis C virus. J Biol Chem 1999; 274:30722-30728.

130 Borowski P., Resch K., Schmitz H., Heiland M. A synthetic peptide derived from the non-structural protein 3 of hepatitis C virus serves as a specific substrate for PKC. Biol Chem 2000; 381:19-27.

131 Borowski P., Kuhl R., Laufs R., Schulze Zur Wiesch J., Heiland M. Identification and characterization of a histone binding site of the non-structural protein 3 of hepatitis C virus. J Clinical Virol 1999; 13:61-69.

132 Shimizu Y., Yamaji K., Masuho Y., Yokota T., Inoue H., Sudo K., Satoh S., Shimotohno K. Identification of the sequence on NS4A required for enhanced cleavage of the NS5A/5B site by hepatitis C virus NS3 protease. J Virol 1996; 70:127-132.

133 Tomei L., Failla C., Vitale R.L., Bianchi E., Defrancesco R. A central hydrophobic domain of the hepatitis C virus NS4A protein is necessary and sufficient for the activation of the NS3 protease. J Gen Virol 1996; 77:1065-70.

134 Bianchi E., Steinkuhler C., Taliani M., Urbani A., Francesco R.D., Pessi A. Synthetic depsipeptide substrates for the assay of human hepatitis C virus protease. Anal Biochem 1996; 237:239-44.

135 Steinkuhler C., Urbani A., Tomei L., Biasiol G., Sardana M., Bianchi E., Pessi A., Defrancesco R. Activity of purified hepatitis C virus protease NS3 on peptide substrates. J Virol 1996; 70:6694-6700.

136 Ishido S., Fujita T., Hotta H. Complex formation of NS5B with NS3 and NS4A proteins of hepatitis C virus. Biochem Biophys Res Commun 1998; 244:35-40.

137 Lin C., Wu J.W., Hsiao K., Su M.S. The hepatitis C virus NS4A protein: interactions with the NS4B and NS5A proteins. J Virol 1997; 71:6465-71.

138 Wölk B., Sansonno D., Kräusslich H.-G., Dammacco F., Rice C.M., Blum H.E., Moradpour D. Subcellular localization, stability and trans-cleavage of hepatitis C virus NS3-4A complex expressed in tetracyclin-regulated cell lines. J Virol 2000; 74:2293-2304.

139 Hugle T., Fehrmann F., Bieck E., Kohara M., Krausslich H.-G., Rice C.M., Blum H.E., Moradpour D. The hepatitis C virus nonstructural protein 4B is an intergral endoplasmic reticulum membrane protein. Virology 2001; 284:70-81.

140 Kaneko T., Tanji Y., Satoh S., Hijikata M., Asabe S., Kimura K., Shimotohno K. Production of two phosphoproteins from the NS5A region of the hepatitis C viral genome. Biochem Biophys Res Commun 1994; 205:320-326.

141 Tanji Y., Kaneko T., Satoh S., Shimotohno K. Phosphorylation of hepatitis C virus-encoded nonstructural protein NS5A. J Virol 1995; 69:3980-3986.

142 Reed K.E., Xu J., Rice C.M. Phosphorylation of the hepatitis C virus NS5A protein *in vitro* and *in vivo*: properties of the NS5A-associated kinase. J Virol 1997; 71:7187-7197.

143 Reed K.E., Rice C.M. Identification of the major phosphorylation site of the hepatitis C virus H strain NS5A protein as serine 2321. J Biol Chem 1999; 274:28011-28018.

144 Katze M.G., Kwieciszewski B., Goodlett D.R., Blakely C.M., Neddermann P., Tan S.L., Aebersold R. Ser^{2194} is a highly conserved major phosphorylation site of the hepatitis C virus nonstructural protein NS5A. Virology 2000; 278:501-513.

145 Hirota M., Satoh S., Asabe S., Kohara M., Tsukiyama-Kohara K., Kato N., Hijikata M., Shimotohno K. Phosphorylation of nonstructural 5A protein of hepatitis C virus: HCV group-specific hyperphosphorylation. Virology 1999; 257:130-7.

146 Asabe S.-I., Tanji Y., Satoh S., Kaneko T., Kimura K., Shimotohno K. The N-terminal region of hepatitis C virus-encoded NS5A is important for NS4A-dependent phosphorylation. J Virol 1997; 71:790-796.

147 Koch J.O., Bartenschlager R. Modulation of hepatitis C virus NS5A hyperphosphorylation by nonstructural proteins NS3, NS4A, and NS4B. J Virol 1999; 73:7138-7146.

148 Liu Q., Bhat R.A., Prince A.M., Zhang P. The hepatitis C virus NS2 protein generated by NS2-3 autocleavage is required for NS5A phosphorylation. Biochem Biophys Res Commun 1999; 254:572-7.

149 Neddermann P., Clementi A., De Francesco R. Hyperphosphorylation of the hepatitis C virus NS5A protein requires an active NS3 protease, NS4A, NS4B, and NS5A encoded on the same polyprotein. J Virol 1999; 73:9984-9991.

150 Ide Y., Tanimoto A., Sasaguri Y., Padmanabhan R. Hepatitis C virus NS5A protein is phosphorylated in vitro by a stably bound protein kinase from HeLa cells and by cAMP-dependent protein kinase A-a catalytic subunit. Gene 1997; 201:151-158.

151 Kim J., Lee D., Choe J. Hepatitis C virus NS5A protein is phosphorylated by casein kinase II. Biochem Biophys Res Commun 1999; 257:777-781.

152 Enomoto N., Sakuma I., Asahina Y., Kurosaki M., Murakami T., Yamamoto C., Izumi N., Marumo F., Sato C. Comparison of full-length sequences of interferon-sensitive and resistant hepatitis C virus 1b. J Clinical Invest 1995; 96:224-230.

153 Enomoto N., Sakuma I., Asahina Y., Kurosaki M., Murakami T., Yamamoto C., Ogura Y., Izumi N., Marumo F., Sato C. Mutations in the nonstructural protein 5A gene and response to interferon in patients with chronic hepatitis C virus 1b infection. N Engl J Med 1996; 334:77-81.

154 Pawlotsky J.M. Hepatitis C virus (HCV) NS5A protein: role in HCV replication and resistance to interferon-alpha. J Viral Hepatitis 1999; 6 (Suppl) 1:47-8.

155 Gale Jr. M.J., Korth M.J., Tang N.M., Tan S.-L., Hopkins D.A., Dever T.E., Polyak S.J., Gretch D.R., Katze M.G. Evidence that hepatitis C virus resistance to interferon is mediated through repression of the PKR protein kinase by the nonstructural 5A protein. Virology 1997; 230:217-227.

156 Gale Jr. M., Kwieciszewski B., Dossett M., Nakao H., Katze M.G. Antiapoptotic and oncogenic potentials of hepatitis C virus are linked to interferon resistance by viral repression of the PKR protein kinase. J Virol 1999; 73:6506-16.

157 Paterson M., Laxton C.D., Thomas H.C., Ackrill A.M., Foster G.R. Hepatitis C virus NS5A protein inhibits interferon antiviral activity, but the effects do not correlate with clinical response. Gastroenterology 1999; 117:1187-97.

158 Polyak S.J., Paschal D.M., Mcardle S., Gale M.J., Jr., Moradpour D., Gretch D.R. Characterization of the effects of hepatitis C virus nonstructural 5A protein expression in human cell lines and on interferon-sensitive virus replication. Hepatology 1999; 29:1262-71.

159 Song J., Fujii M., Wang F., Itoh M., Hotta H. The NS5A protein of hepatitis C virus partially inhibits the antiviral activity of interferon. J Gen Virol 1999; 80:879-86.

160 Tan S.L., Katze M.G. How hepatitis C virus counteracts the interferon response: the jury is still out on NS5A. Virology 2001; 284:1-12.

161 Chung K.M., Song O.K., Jang S.K. Hepatitis C virus nonstructural protein 5A contains potential transcriptional activator domains. Mol Cells 1997; 7:661-667.

162 Kato N., Lan K.-H., Ono-Nita S.K., Shiratori Y., Omata M. Hepatitis C virus nonstructural region 5A protein is a potent transcriptional activator. J Virol 1997; 71:8856-8859.

163 Tanimoto A., Ide Y., Arima N., Sasaguri Y., Padmanabhan R. The amino terminal deletion mutants of hepatitis C virus nonstructural protein NS5A function as transcriptional activators in yeast. Biochem Biophys Res Commun 1997; 236:360-364.

164 Satoh S., Hirota M., Noguchi T., Hijikata M., Handa H., Shimotohno K. Cleavage of hepatitis C virus nonstructural protein 5A by a caspase-like protease(s) in mammalian cells. Virology 2000; 270:476-87.

165 Arima N., Kao C.Y., Licht T., Padmanabhan R., Sasaguri Y., Padmanabhan R. Modulation of cell growth by the hepatitis C virus nonstructural protein NS5A. J Biol Chem 2001; 276:12675-12684.

166 Majumder M., Ghosh A.K., Steele R., Ray R., Ray R.B. Hepatitis C virus NS5A physically associates with p53 and regulates p21/waf1 gene expression in a p53-dependent manner. J Virol 2001; 75:1401-1407.

167 Tan S.-L., Nakao H., He Y., Vijaysri S., Neddermann P., Jacobs B.L., Mayer B.J., Katze M.G. NS5A, a non-structural protein of hepatitis C virus, binds growth factor receptor-bound protein 2 adaptor protein in a src homology 3 domain/ligand-dependent manner and perturbs mitogenic signaling. Proc Natl Acad Sci USA 1999; 96:5533-5538.

168 Tu H., Gao L., Shi S.T., Taylor D.R., Yang T., Mircheff A.K., Wen Y., Gorbalenya A.E., Hwang S.B., Lai M.M.C. Hepatitis C virus RNA polymerase and NS5A complex with a SNARE-like protein. Virology 1999; 263:30-41.

169 Chung K.M., Lee J., Kim J.E., Song O.K., Cho S., Lim J., Seedorf M., Hahm B., Jang S.K. Nonstructural protein 5A of hepatitis C virus inhibits the function of karyopherin beta3. J Virol 2000; 74:5233-41.

170 Ghosh A.K., Majumder M., Steele R., Yaciuk P., Chrivia J., Ray R., Ray R.B. Hepatitis C virus NS5A protein modulates transcription through a novel cellular transcription factor SRCAP. J Biol Chem 2000; 275:7184-7188.

171 Al R.H., Xie Y., Wang Y., Hagedorn C.H. Expression of recombinant hepatitis C virus non-structural protein 5B in Escherichia coli. Virus Res 1998; 53:141-149.

172 Yamashita T., Kaneko S., Shirota Y., Qin W., Nomura T., Kobayashi K., Murakami S. RNA-dependent RNA polymerase activity of the soluble recombinant hepatitis C virus NS5B protein truncated at the C-terminal region. J Biol Chem 1998; 273:15479-15486.

173 Yuan Z.-H., Kumar U., Thomas H.C., Wen Y.-M., Monjardino J. Expression, purification, and partial characterization of HCV RNA polymerase. Biochem Biophys Res Commun 1997; 232:231-235.

174 Behrens S.E., Tomei L., Defrancesco R. Identification and properties of the RNA-dependent RNA polymerase of hepatitis C virus. EMBO J. 1996; 15:12-22.

175 Lohmann V., Körner F., Herian U., Bartenschlager R. Biochemical properties of hepatitis C virus NS5B RNA-dependent RNA polymerase and identification of

amino acid sequence motifs essential for enzymatic activity. J Virol 1997; 71:8416-8428.

176 Lohmann V., Roos A., Korner F., Koch J.O., Bartenschlager R. Biochemical and kinetic analyses of NS5B RNA-dependent RNA polymerase of the hepatitis C virus. Virology 1998; 249:108-118.

177 Oh J.W., Ito T., Lai M.M. A recombinant hepatitis C virus RNA-dependent RNA polymerase capable of copying the full-length viral RNA. J Virol 1999; 73:7694-702.

178 Kao C.C., Yang X., Kline A., Wang Q.M., Barket D., Heinz B.A. Template requirements for RNA synthesis by recombinant hepatitis C virus RNA-dependent RNA polymerase. J Virol 2000; 74:11121-11128.

179 Luo G., Hamatake R.K., Mathis D.M., Racela J., Rigat K.L., Lemm J., Colonno R.J. De novo initiation of RNA synthesis by the RNA-dependent RNA polymerase (NS5B) of hepatitis C virus. J Virol 2000; 74:851-63.

180 Oh J.W., Sheu G.T., Lai M.M. Template requirement and initiation site selection by hepatitis C virus polymerase on a minimal viral RNA template. J Biol Chem 2000; 275:17710-7.

181 Zhong W., Uss A.S., Ferrari E., Lau J.Y., Hong Z. De novo initiation of RNA synthesis by hepatitis C virus nonstructural protein 5B polymerase. J Virol 2000; 74:2017-22.

182 Ferrari E., Wright-Minogue J., Fang J.W., Baroudy B.M., Lau J.Y., Hong Z. Characterization of soluble hepatitis C virus RNA-dependent RNA polymerase expressed in Escherichia coli. J Virol 1999; 73:1649-54.

183 Ago H., Adachi T., Yoshida A., Yamamoto M., Habuka N., Yatsunami K., Miyano M. Crystal structure of the RNA-dependent RNA polymerase of hepatitis C virus. Structure Fold Des 1999; 7:1417-26.

184 Bressanelli S., Tomei L., Roussel A., Incitti I., Vitale R.L., Mathieu M., De Francesco R., Rey F.A. Crystal structure of the RNA-dependent RNA polymerase of hepatitis C virus. Proc Natl Acad Sci U S A 1999; 96:13034-9.

185 Lesburg C.A., Cable M.B., Ferrari E., Hong Z., Mannarino A.F., Weber P.C. Crystal structure of the RNA-dependent RNA polymerase from hepatitis C virus reveals a fully encircled active site. Nature Structural Biology 1999; 6:937-943.

186 Blight K.J., Gowans E.J. *In situ* hybridization and immunohistochemical staining of hepatitis C virus products. Viral Hepatitis Rev 1995; 1:143-155.

187 Pietschmann T., Lohmann V., Rutter G., Kurpanek K., Bartenschlager R. Characterization of cell lines carrying self-replicating hepatitis C virus RNAs. J Virol 2001; 75:1252-1264.

188 Neumann A.U., Lam N.P., Dahari H., Gretch D.R., Wiley T.E., Layden T.J., Perelson A.S. Hepatitis C viral dynamics in vivo and the antiviral efficacy of interferon-alpha therapy. Science 1998; 282:103-7.

189 Ramratnam B., Bonhoeffer S., Binley J., Hurley A., Zhang L., Mittler J.E., Markowitz M., Moore J.P., Perelson A.S., Ho D.D. Rapid production and clearance of HIV-1 and hepatitis C virus assessed by large volume plasma apheresis. Lancet 1999; 354:1782-5.

190 Alter H.J., Purcell R.H., Holland P.V., Popper H. Transmissible agent in non-A, non-B hepatitis. Lancet 1978; 1:459-463.

191 Barker L.F., Maynard J.E., Purcell R.H., Hoofnagle J.H., Berquist K.R., London W.T. Viral hepatitis, type B, in experimental animals. Am J Med Sci 1975; 270:189-195.

192 Karron R.A., Daemer R., Ticehurst J., D'hondt E., Popper H., Mihalik K., Phillips J., Feinstone S., Purcell R.H. Studies of prototype live hepatitis A virus vaccines in primate models. J Infect Dis 1988; 157:338-345.

193 Maynard J.E., Lorenz D., Bradley D.W., Feinstone S.M., Krushak D.H., Barker L.F., Purcell R.H. Review of infectivity studies in nonhuman primates with virus-like particles associated with MS-1 hepatitis. Am J Med Sci 1975; 270:81-85.

194 Purcell R.H., Gerin J.L. Hepatitis B subunit vaccine: a preliminary report of safety and efficacy tests in chimpanzees. Am J Med Sci 1975; 270:395-399.

195 Rizzetto M., Canese M.G., Gerin J.L., London W.T., Sly D.L., Purcell R.H. Transmission of the hepatitis B virus-associated delta antigen to chimpanzees. J Infect Dis 1980; 141:590-602.

196 Beard M.R., Abell G., Honda M., Carroll A., Gartland M., Clarke B., Suzuki K., Lanford R., Sangar D.V., Lemon S.M. An infectious molecular clone of a Japanese genotype 1b hepatitis C virus. Hepatology 1999; 30:316-324.

197 Hong Z., Beaudet-Miller M., Lanford R.E., Guerra B., Wright-Minogue J., Skelton A., Baroudy B.M., Reyes G.R., Lau J.Y.N. Generation of transmissible hepatitis C virions from a molecular clone in chimpanzees. Virology 1999; 256:36-44.

198 Kolykhalov A.A., Agapov E.V., Blight K.J., Mihalik K., Feinstone S.M., Rice C.M. Transmission of hepatitis C by intrahepatic inoculation with transcribed RNA. Science 1997; 277:570-574.

199 Yanagi M., Purcell R.H., Emerson S.U., Bukh J. Transcripts from a single full-length cDNA clone of hepatitis C virus are infectious when directly transfected into the liver of a chimpanzee. Proc Natl Acad Sci USA 1997; 94:8738-8743.

200 Yanagi M., Purcell R.H., Emerson S.U., Bukh J. Hepatitis C virus: an infectious molecular clone of a second major genotype (2a) and lack of viability of intertypic 1a and 2a chimeras. Virology 1999; 262:250-263.

201 Yanagi M., St. Claire M., Shapiro M., Emerson S.U., Purcell R.H., Bukh J. Transcripts of a chimeric cDNA clone of hepatitis C virus genotype 1b are infectious in vivo. Virology 1998; 244:161-172.

202 Xie Z.C., Riezu-Boj J.I., Lasarte J.J., Guillen J., Su J.H., Civeira M.P., Prieto J. Transmission of hepatitis C virus infection to tree shrews. Virology 1998; 244:513-20.

203 Galun E., Burakova T., Ketzinel M., Lubin I., Shezen E., Kahana Y., Eid A., Ilan Y., Rivkind A., Pizov G., Et A. Hepatitis C virus viremia in SCID-->BNX mouse chimera. J Infect Dis 1995; 172:25-30.

204 Grakoui A., Hanson H.L., Rice C.M. Bad time for Bonzo? Experimental models of HCV infection, replication and pathogenesis. Hepatology 2001:in press.

205 Bartenschlager R., Lohmann V. Replication of hepatitis C virus. J Gen Virol 2000; 81:1631-1648.

206 Dash S., Halim A.-B., Tsuji H., Hiramatsu N., Gerber M.A. Transfection of HepG2 cells with infectious hepatitis C virus genome. Am J Pathol 1997; 151:363-373.

207 Yoo B.J., Selby M., Choe J., Suh B.S., Choi S.H., Joh J.S., Nuovo G.J., Lee H.-S., Houghton M., Han J.H. Transfection of a differentiated human hepatoma cell line (Huh7) with in vitro-transcribed hepatitis C virus (HCV) RNA and establishment of a long-term culture persistently infected with HCV. J Virol 1995; 69:32-38.

208 Lohmann V., Korner F., Dobierzewska A., Bartenschlager R. Mutations in hepatitis C virus RNAs conferring cell culture adaptation. J Virol 2001; 75:1437-1449.

209 Krieger N., Lohmann V., Bartenschlager R. Enhancement of hepatitis C virus RNA replication by cell culture-adaptive mutations. J Virol 2001; 75:4614-4624.

210 Keskinen P., Nyqvist M., Sareneva T., Pirhonen J., Melen K., Julkunen I. Impaired antiviral response in human hepatoma cells. Virology 1999; 263:364-375.

211 Melen K., Keskinen P., Lehtonen A., Julkunen I. Interferon-induced gene expression and signaling in human hepatoma cell lines. J Hepatol 2000; 33:764-772.

212 Frese M., Pietschmann T., Moradpour D., Haller O., Bartenschlager R. Interferon-a inhibits hepatitis C virus subgenomic RNA replication by an MxA-independent pathway. J Gen Virol 2001; 82:723-733.

213 Owsianka A.M., Patel A.H. Hepatitis C virus core protein interacts with a human DEAD box protein DDX3. Virology 1999; 257:330-40.

214 Wang F., Yoshida I., Takamatsu M., Ishido S., Fujita T., Oka K., Hotta H. Complex formation between hepatitis C virus core protein and p21Wafl/Cip1/Sdi1. Biochem Biophys Res Commun 2000; 273:479-84.

215 Lu W., Lo S.Y., Chen M., Wu K.J., Fung Y.K., Ou J.-H. Activation of p53 tumor suppressor by hepatitis C virus core protein. Virology 1999; 264:134-141.

216 Aoki H., Hayashi J., Moriyama M., Arakawa Y., Hino O. Hepatitis C virus core protein interacts with 14-3-3 protein and activates the kinase Raf-1. J Virol 2000; 74:1736-41.

Chapter 5

THE MOLECULAR BIOLOGY OF HEPATITIS DELTA VIRUS

Thomas B. Macnaughton and Michael M. C. Lai

Department of Molecular Microbiology and Immunology, and Howard Hughes Medical Institute, University of Southern California Keck School of Medicine, 2011 Zonal Avenue, Los Angeles, CA 90033

1. INTRODUCTION

Hepatitis delta virus (HDV) infection was first detected as a novel nuclear antigen in the hepatocytes of HBV-infected patients (1). This new antigen, termed delta antigen by Mario Rizzetto, was initially thought to represent a previously unrecognized HBV antigen. Subsequently, it was demonstrated to be a novel and distinct transmissible agent, composed of the HBV surface antigens (HBsAg), delta antigen (HDAg), and a unique small circular RNA (2). The production and transmission of HDV requires a concurrent HBV infection to supply HBsAg, thus explaining the initial confusion and obligatory association between these two viral infections. HDV infection has been reported worldwide, and is particularly prevalent in the Mediterranean basin, Middle East, South America, West Africa and South Pacific Islands (3, 4). Its infection frequently leads to severe acute and chronic hepatitis and accounts for a large proportion of fulminant hepatitis cases in many parts of the world (5). However, a significant decline in the incidence of HDV infections has been noted worldwide in recent years. Despite the welcome drop in the clinical significance of delta hepatitis, HDV remains a treasure trove of exciting molecular biological phenomena, many of which are yet to be explored. Recent years have also witnessed significant modifications in the HDV replication model, as new data generated by novel experimental approaches provided a new understanding of the HDV replication cycle. Thus, the molecular biology of HDV is entering an exciting phase.

HDV is considered a subviral particle (6) because it relies on HBV as a helper virus to make part (envelope) of the viral particle. However, it can replicate autonomously even in the absence of the helper virus and does not share sequence with HBV. Its circular RNA resembles that of the subviral plant pathogens, viroids and virusoids. No other animal viruses are related to HDV. It is classified as a lone member of a floating genus, Deltavirus, without any family designation (7).

2. VIRION STRUCTURE

The HDV particle is an enveloped virus of 36 nm, distinct from the 42-nm particles of HBV or 22-nm HBsAg particles (2, 8). The viral envelope consists exclusively of HBsAg and a lipid bilayer. The three known forms of HBsAg (L, M and S forms) are all present in the virion although the relative ratio of the three forms in HDV is slightly different from that in HBV (9). Within the virion is an internal nucleocapsid consisting of HDAg and viral RNA. The HDAg is usually composed of two related protein species, a large form (L-HDAg) and small form (S-HDAg), in varying ratio (9-11). These HDAg molecules may form dimers through antiparallel coiled coils, which further interact to form octamers (12). The structure of this basic building block for the HDV nucleocapsid is distinct from that of the other known viral nucleocapsids; however, the exact nucleocapsid structure of HDV has not been determined, although it appears to be icosahedral. It has been reported that the nucleocapsid derived from the nuclei of the infected cells has a slightly different structure from that of the virion in terms of the number of the HDAg molecules (13). However, this distinction is yet to be confirmed by direct structural determination.

3. THE RNA GENOME

The HDV genome is a single-stranded, circular RNA of 1700 nucleotides (14, 15), which is the smallest and the only circular RNA among the genomes of animal viruses. Almost 70% of the nucleotides in the RNA are complementary to one another such that the HDV RNA assumes a double-stranded rod-like structure, which is visible by electron microscopy under native conditions (16). Once denatured, the HDV RNA clearly appears as a covalently closed circle.

Sequence analysis of various HDV isolates has shown that HDV can be divided into three genotypes, each of which shares approximately 60-80% sequence similarity (17). Genotype I is the most common and is present worldwide; it is associated with hepatitis of various severity. Genotype II has been detected mainly in Asia and is associated with relatively mild hepatitis (18, 19). Genotype III has so far been found only in South America and is associated frequently with fulminant hepatitis (17). The genetic basis for the different disease pathogenicity is not clear. For purposes of genomic sequence organization in view of the circular nature of the RNA, a Hind III restriction site present in the cDNA of the prototype HDV RNA was designated nucleotide 1 (14). Based on such demarcation, the HDV RNA can be divided into two functional domains (20) (Figure 1). One is termed the ribozyme domain (nt 615-950), which is distantly related to plant viroid RNAs in primary sequence (21) and correspondingly contains a ribozyme activity. The remaining three-quarters of the genome is called the protein-coding domain, which contains an open reading frame (ORF) encoding HDAg on the complementary (antigenomic) strand. It has been proposed that these two domains may have been derived by recombination between a viroid-like RNA and a cellular mRNA (22). Indeed, a putative cellular homologue of HDAg, termed delta-interacting protein A (DIP-A), has been reported (23) but its evolutionary relationship with the HDAg has been questioned (24).

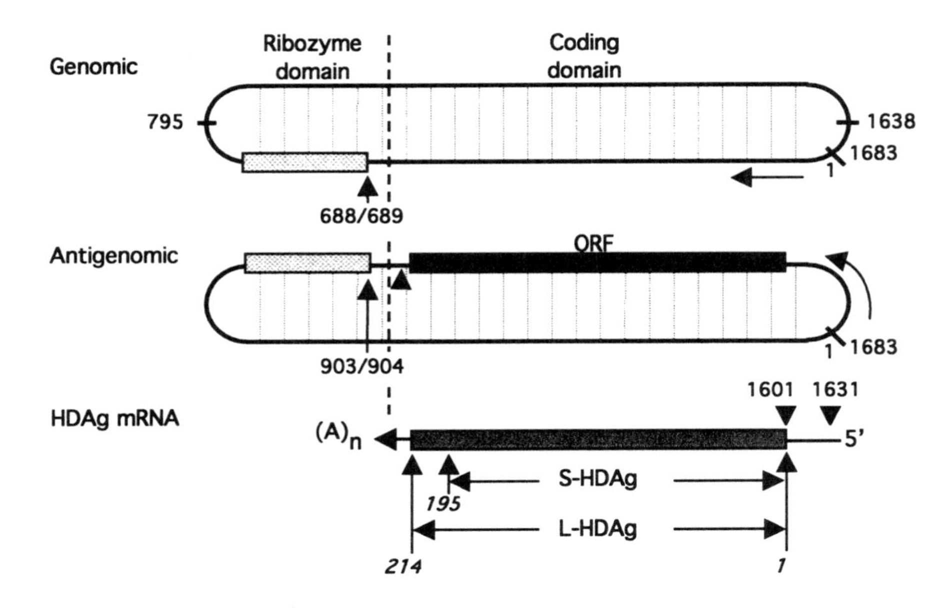

Figure 1. Schematic structures of HDV genomic and antigenomic RNA and HDAg-encoding mRNA. The minimum ribozymes are indicated by light boxes. The italicized numbers for HDAg are amino acid residues. All other numbers are nucleotide positions on the genomic-sense RNA.

The ribozyme structure of HDV RNA has been extensively studied. The minimum HDV ribozyme consists of no more than 87 nucleotides in the ribozyme domain (nt 683-770) of HDV RNA (25). Similar ribozyme activities exist in both the genomic strand and the corresponding region of the complementary antigenomic RNA strand (26-29). These stretches of RNA can cleave themselves at a specific site (nt 688/689 on the genomic strand and the corresponding complementary site on the antigenomic strand) in the presence of magnesium ions at physiological pH (29-31). It has recently been shown that HDV ribozyme also can undergo base catalysis in the absence of divalent cations, a novel property among ribozymes in general (32). The ribozyme can fold into five helical segments connected as a double pseudoknot (33); this structure is distinct from that of the other known ribozymes, such as the hammerhead or hairpin/paperclip ribozymes (34, 35). The ribozyme activity and the RNA cleavage mediated by it are required for HDV RNA replication, for the loss of the ribozyme activity or mutations of the cleavage site sequences almost invariably abolished the replication ability of the RNA (36, 37).

The ribozyme domain has also been shown to have a self-ligation activity (38, 39). However, this activity could be detected only under very artificial conditions. Its biological significance has become increasingly questionable, as recent studies indicated that HDV RNA ligation may be carried out by cellular RNA ligase, instead of RNA catalysis (40).

4. HEPATITIS DELTA ANTIGEN

HDAg is the only protein encoded by HDV RNA. It normally consists of two species: the S-HDAg (195 amino acids, 24 kDa) and L-HDAg (214 amino acids, 27 kDa), which are identical in primary amino acid sequence except that the L-HDAg has an additional 19 amino acids at the C-terminus (Figure 2). These two protein species are present in the virion in variable ratios. Both of them are phosphorylated, but the extent of phosphorylation and the phosphorylation patterns of S-HDAg and L-HDAg appear to be different (41-44). Both of them contain the following domains: a coiled-coil structure (45, 46), which mediates the multimerization of S- and L-HDAg; a bipartite nuclear localization signal (47, 48), which directs HDV RNA to the nucleus; and several RNA-binding domains, which enable HDAg to bind to HDV RNA and possibly other RNAs as well (49, 50). HDAg shows some binding preference for HDV RNA although the specificity is not absolute. In addition, a helix-loop-helix domain may also contribute to the oligomerization of HDAg or its interaction with other proteins (51). The L-

HDAg contains an isoprenylation signal within the unique C-terminal 19 amino acids (52, 53). The resultant modification enables L-HDAg to interact with HBsAg to form virus particles (54-57).

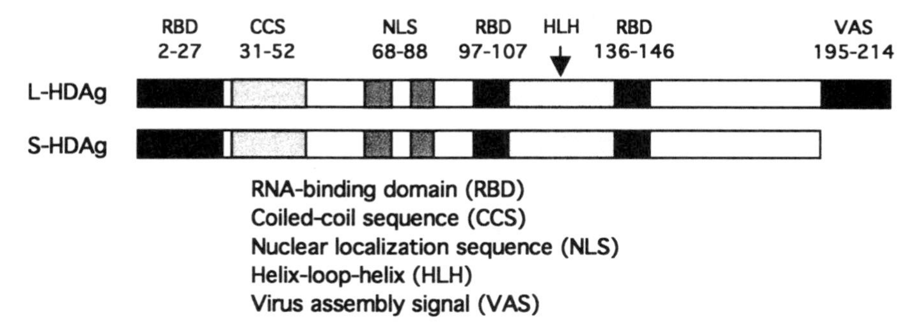

Figure 2. Schematic diagram of the functional domains of HDAg.

Both S- and L-HDAg are localized in the nucleus of HDV-replicating cells. Despite their overall structural similarity, they have distinct biological functions. S-HDAg is required for initiating RNA synthesis from the HDV genomic or antigenomic RNA (58), whereas L-HDAg has been reported to inhibit HDV RNA replication, even when S-HDAg is present in large excess (46, 59, 60). In addition, L-HDAg is required for virus assembly (61, 62). The ability of L-HDAg to inhibit the functions of S-HDAg in RNA replication is dependent on the interaction between these two proteins (46). The ability of S-HDAg to initiate RNA replication appears to be genotype-specific, as S-HDAg of genotype I cannot support genotype III RNA replication and vice versa (63). The molecular basis for such specificity is not clear. Both forms of HDAg also have been reported to stabilize HDV RNA (64) and enhance its ribozyme activities (65). In addition, S-HDAg has RNA chaperon activities (66). Some other reported activities of HDAg, such as inhibition of poly(A) addition signals (67, 68), are likely the artifacts of experimental systems (see VIRAL REPLICATION below).

5. VIRAL REPLICATION

In natural infections, HDV infects only hepatocytes; no extrahepatic infection has been reported. Although HDV uses HBsAg as its envelope protein, HDV apparently has a slightly different host range; for example,

HBV has been reported to infect pancreas and peripheral blood cells, whereas HDV is strictly hepatotropic. On the other hand, HDV can infect woodchucks (69), whereas HBV can not. The slight difference in host range between HDV and HBV could be attributed to the difference in their relative ratio of the three forms of HBsAg (9). HDV RNA is able to replicate in many different cells once the RNA is introduced into cells by transfection. Therefore, there is no intrinsic cell-specific limitation on HDV RNA replication, unlike the replication of HBV DNA, which requires liver-specific factors. However, HDV RNA can not replicate in cells other than mammalian cells, such as avian cells (70). It has been claimed that avian cells can support HDV RNA replication, but HDV replication induces apoptosis in the cells that support HDV replication (70). However, the validity of this claim is not yet certain. Conceivably, HDV RNA replication may require mammalian cell-specific factors.

5.1 In Vitro and In Vivo Systems for Studying HDV Replication

HDV cannot infect cultured cell lines, and can infect only very inefficiently primary hepatocytes from woodchucks or chimpanzees (71-73). Consequently, these cells are of very limited value for studying HDV replication. The most useful approach for studying HDV replication is the transfection of HDV cDNA or RNA into cultured cells, which can lead to robust HDV RNA replication. Several versions of this approach have been developed. The first is the use of HDV cDNA, usually in the form of dimer or trimer cDNA under the control of a foreign promoter. HDV RNA transcripts produced from these constructs can lead to RNA replication in the transfected cells. Even the circularized monomer HDV cDNA without a foreign promoter can lead to RNA replication, suggesting the possible presence of an endogenous promoter in the HDV cDNA (74, 75). These approaches have proven so powerful that most of our current knowledge of HDV replication has come from them. Unfortunately, these approaches introduce an artificial requirement for DNA-dependent transcription, which may have distorted the true characteristics of HDV replication.

The second approach is HDV RNA transfection. However, transfection of the multimer HDV RNA by itself does not lead to RNA replication in the transfected cells (76, 77). It has to be cotransfected with either purified S-HDAg protein (78-80) or a mRNA encoding S-HDAg (77), suggesting that pre-existing S-HDAg is required for the initiation of HDV RNA synthesis. These RNA transfection protocols should be the systems of choice for studying HDV replication because they obviate the need for DNA-dependent transcription. Alternatively, transfection of the HDV multimer

RNA alone into cells that constitutively express S-HDAg (76, 81) can lead to RNA replication, but this approach also requires DNA-dependent transcription to produce the mRNA for HDAg. As will be clear in the following sections, the HDV RNA-HDAg co-transfection approach has led to recent revision of several concepts in the original HDV RNA replication model, which had been derived from cDNA transfection studies.

Animal models for HDV infection include chimpanzees and woodchucks (69, 82, 83), which are valuable for studying HDV pathogenesis but not practical for studying viral replication. Several mouse strains have also been shown to be susceptible to HDV infection by intrahepatic inoculation, but only a small percentage of hepatocytes can be infected and the infection is only transient (84). HDV RNA expressed in transgenic mice containing the HDV dimer cDNA also can replicate (85); surprisingly, HDV RNA replicates most robustly in the muscle, not liver, of these mice. Furthermore, intramuscular injection of HDV cDNA can lead to HDV RNA replication (86). These studies suffer from the same drawbacks as the HDV cDNA transfection in cultured cells, because all of these approaches require DNA-dependent RNA transcription to initiate HDV RNA replication.

5.2 Machinery of HDV RNA Replication

Because HDV uses HBsAg as its envelope proteins, HDV probably uses the same cellular receptor as does HBV. The receptor has not been identified, and the mechanism of virus entry is not yet clear. Once inside the cells, HDV can replicate in the absence of HBV. Moreover, HDV replication can occur in nonhepatic cells. Once the incoming virion is uncoated, the viral nucleocapsid is transported to the nucleus by virtue of the nuclear localization signal of the HDAg (48) and probably remains there throughout the replication cycle (87). Therefore, HDV RNA replication is thought to occur exclusively in the nucleus; however, this concept will likely be amended, as recent data showed that HDV RNA is actually transported to the cytoplasm immediately after synthesis (Macnaughton and Lai, unpublished). Thus, some parts of the RNA replication process may take place in the cytoplasm. Replication proceeds by RNA-dependent RNA synthesis; no DNA intermediate is involved. Since HDV RNA does not encode a polymerase, HDV RNA replication is most likely mediated by cellular enzymes, either a known DNA-dependent polymerase (e.g., pol I, II or III), which is converted to utilize RNA as templates, or a heretofore unknown RNA-dependent RNA polymerase in the mammalian cells. Several *in vitro* replication studies using permeabilized isolated nuclei or nuclear extracts have shown that HDV RNA synthesis is sensitive to very low concentrations of α-amanitin, suggesting that RNA polymerase II, which is responsible for the transcription of cellular mRNAs and some small nuclear RNAs, is responsible for the replication of HDV RNA (88-90). However, the relevance of these *in vitro* studies to HDV RNA replication in cells is questionable, since the metabolic requirements or the nature of the RNA products in these studies did not reflect those of the natural HDV replication. For example, HDAg was not required for, and did not have any effect on, these *in vitro* reactions. Recent studies based on RNA transfection methods indicate that pol II is probably involved only in the transcription of the mRNA encoding HDAg, since the production of this mRNA is sensitive to a low concentration of α-amanitin (91) (see below); in contrast, replication of the HDV genomic-length RNA is resistant to α-amanitin and is thus most likely carried out by other cellular enzymes (91). The nature of these enzymes remains to be identified.

5.3 Double Rolling Circle Model of HDV RNA Replication

HDV RNA replication is thought to occur by a rolling-circle mechanism similar to that proposed for plant viroid or virusoid RNAs (92, 93) (Figure 3, right half). Antigenomic RNA is first synthesized from the circular genomic RNA template. The initial RNA product is likely the dimer or multimer antigenomic RNA, as the polymerase rolls around the template without termination. These intermediate RNA products are then processed into monomer-length RNA by the ribozyme associated with HDV RNA. The monomer RNA is ligated into a circular antigenomic RNA, which is then used as the template for the synthesis of the genomic-sense RNA by another round of rolling circle replication. This double rolling circle replication model is consistent with the findings that the dimer and multimer HDV RNAs of both genomic and antigenomic sense are detectable in HDV RNA-replicating cells. Furthermore, the ribozyme activities associated with both the genomic and antigenomic RNAs are required for HDV RNA replication (36). Thus, the synthesis of both genomic and antigenomic RNA strands appears to involve rolling circle replication. In this replication model, the ribozyme is proposed to cleave the multimer HDV RNA intermediate *in cis*. This model, however, encounters a conceptual quandary since the multimer HDV RNA is expected to form a double-stranded, rod-like structure, which will prevent the formation of the ribozyme-active structure, which consists of multiple helices and pseudoknots, and is necessary for its activity *in vitro* and *in vivo* (33, 37). Thus, other factors, such as HDAg, may be involved in the formation of the ribozyme-active structure in the multimer HDV RNA. Indeed, HDAg can serve as an RNA chaperone (66) and can enhance the ribozyme activity *in vivo* (65).

Ligation of the linear RNA into circular RNA is probably mediated by cellular RNA ligases (40), instead of autocatalytic ligation as previously proposed (38). Another unanswered question is where RNA replication initiates on either genomic or antigenomic RNA. The antigenomic RNA synthesis appears to initiate from a point (nt 1631) near one end of the circular genomic RNA (77); the sequences around this initiation site indeed have a transcription promoter-like activity (74, 94). However, it is not clear whether this represents the initiation of mRNA transcription or genomic-length RNA replication. So far, there have been no data on the location of the initiation site of genomic RNA synthesis.

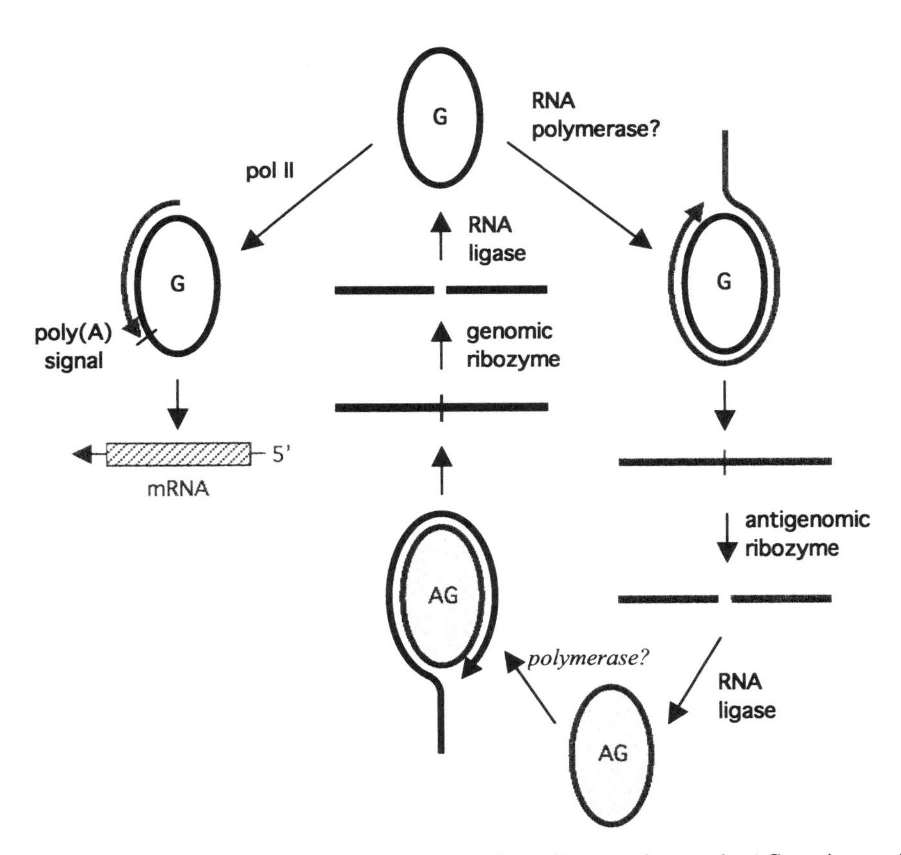

Figure 3. Proposed model of HDV RNA replication. G: genomic strand. AG: antigenomic strand. The HDAg-encoding mRNA is synthesized by pol II from the genomic RNA template independently of RNA replication. The enzymes for genomic and antigenomic RNA replication are not yet known.

Several recent studies have revealed new insights into the mechanism of HDV RNA replication. Genomic and antigenomic RNA synthesis are likely carried out by different mechanisms, since genomic and antigenomic RNA synthesis have different sensitivities to α-amanitin: the antigenomic RNA synthesis is resistant to α-amanitin to as high as 100 μg/ml, whereas the genomic RNA synthesis is sensitive to inhibition at a much lower concentration (Modahl and Lai, unpublished). Moreover, antigenomic RNA synthesis is much more resistant to the inhibitory effects of L-HDAg than genomic RNA synthesis (95), and the recombinant S-HDAg from *E. coli* can initiate RNA synthesis when co-transfected with the HDV genomic RNA but not the antigenomic RNA, suggesting that the genomic and antigenomic RNA synthesis require different forms of HDAg (80). These observations indicate that the metabolic requirements of genomic and antigenomic RNA synthesis are different. It is even possible that they are carried out by different polymerases or take place in different subcellular compartments.

5.4 Transcription of the mRNA for HDAg

In addition to the genomic- and antigenomic-sense RNAs of the genomic length (1.7 kb), a smaller HDV RNA of antigenomic sense can be detected in the HDV-replicating cells. This is a polyadenylated RNA of 0.8 kb, which contains the ORF for HDAg (96, 97). The 5'-end of this RNA starts at nt 1631 and its 3'-end is mapped at approximately 76 nucleotides downstream from the termination codon of the ORF (77, 97). This is the mRNA for translation of HDAg. Despite the fact that this mRNA is present at very low levels in the HDV-replicating cells, it appears that all the HDAg in the cells is translated from this mRNA, and not from the full-length antigenomic RNA (98), which also includes the ORF for HDAg. It has been proposed that the synthesis of the 0.8-kb mRNA is part of the replication process of antigenomic RNA; when antigenomic RNA replication is terminated at a polyadenylation signal, the 0.8-kb mRNA is produced. When the termination is suppressed, the replicating RNA goes on to produce multimer HDV RNA (99). The polyadenylation signal was thought to be suppressed by HDAg, which is translated from the 0.8-kb mRNA, a mechanism akin to feedback inhibition (67). Thus, this mRNA was proposed to be synthesized only early in the HDV replication cycle; when a sufficient level of HDAg is reached, subsequent mRNA synthesis would be inhibited. This model has now been shown to be incorrect. In fact, the 0.8-kb mRNA is synthesized continuously throughout the replication cycle, even when an abundant amount of HDAg is present (77). Furthermore, mRNA transcription and genomic RNA replication can be uncoupled under a variety of experimental conditions. The current thinking is that this mRNA is synthesized by a different mechanism from genomic RNA replication, and is likely mediated by pol II (77) (Fig. 3, left half). It is conceivable that mRNA transcription and genomic RNA replication are carried out in two different compartments in the nucleus.

5.5 RNA Editing

Sometime during HDV RNA replication, a specific nucleotide conversion occurs at the termination codon of the S-HDAg ORF, so that the ORF is extended for additional 19 amino acids (100). This specific mutation (termed RNA editing) results in the generation of the mRNA for L-HDAg. The editing is carried out by a cellular double-stranded RNA-adenosine deaminase (101) and takes place on the antigenomic strand, which, after RNA replication, will result in the conversion of U to C at nucleotide 1015 of the genomic strand (102). [Earlier reports concerning the genomic or antigenomic strand as the substrate of RNA editing were contradictory (100, 103, 104).] Since L-HDAg is required for virus assembly (61, 62), this RNA editing, which generates the mRNA for L-HDAg, is important for the production of virus particles. The editing product usually is not detected until a few days after the initiation of RNA replication in cell culture; this timing is consistent with the fact that virus assembly does not take place until late in the viral life cycle. The mechanism of temporal regulation of this editing is not clear. Nor is the mechanism for selection of the editing site known. [This selectivity is important since the adenosine deaminase can edit at many sites on the HDV RNA *in vitro* (105)]. It appears that the specific sequences around the editing site (on both the editing strand and the complementary site) are required for the regulation of the editing (102), and that the editing can be suppressed by HDAg (105). Both the edited and unedited genomic RNAs are incorporated into the virus particle. Thus, in every clinical sample, HDV virions contain both RNA species capable of synthesizing S- and L-HDAg (106).

5.6 The Role of HDAg in HDV RNA Replication

S-HDAg is required for HDV RNA replication. This is best demonstrated by the ability of S-HDAg to complement *in trans* the replication defects of mutant HDV RNAs that cannot synthesize a functional S-HDAg. Various mutations affecting the phosphorylation site, nuclear localization signal, coiled-coil domain, RNA-binding motifs or C-terminal domain of the S-HDAg rendered the HDAg unable to complement the replication defects of mutant HDV RNAs that did not encode a functional S-

HDAg (44, 47, 49, 51, 56, 107). Therefore, S-HDAg is required for HDV RNA replication in a *trans*-acting manner. This function appears to be at an early step, probably at the initiation, of RNA synthesis. This conclusion was based on the experiments using RNA transfection technique, in which HDV genomic RNA was unable to replicate unless an mRNA for S-HDAg or a recombinant S-HDAg protein was provided, even though the genomic RNA itself contains a functional ORF for S-HDAg (77, 80). Even when a plasmid DNA encoding an S-HDAg was cotransfected with the HDV RNA, no RNA replication occurred (81), indicating that S-HDAg has to be present together with the HDV RNA to initiate RNA replication. Several possible roles for S-HDAg in HDV RNA replication are conceivable: (1) S-HDAg may transport HDV RNA to the appropriate site in the nucleus for replication. Although both the S-HDAg and L-HDAg are transported to the nucleus, they may be localized in different subnuclear compartments during RNA replication (108-110). (2) S-HDAg may serve as a transcription factor by forming a transcription complex with cellular polymerases and other transcription factors. HDAg (both S- and L-HDAg) has been shown to interact with several cellular proteins, including nucleolin (111). HDAg has also been shown to modulate (either enhance or suppress) pol II transcription, suggesting its ability to interact with cellular transcription machineries (112, 113). The ability of HDAg to mediate protein-protein interactions is important for its role in HDV RNA replication (47). (3) S-HDAg may convert the conformation of the double-stranded rod-like structure of HDV RNA into a double-stranded DNA conformation, so that the cellular DNA-dependent RNA polymerases can recognize HDV RNA as a template. (4) HDAg may be required to stabilize HDV RNA (107). It should be noted that it is still not known whether S-HDAg is continuously needed once RNA replication is initiated.

On the other hand, L-HDAg has been reported to inhibit RNA replication (59). This activity has been shown to be due to its ability to complex with S-HDAg, presumably resulting in the alteration of the conformation of the S-HDAg multimers (46). The ability of L-HDAg to inhibit RNA replication appears to make sense as it complements its role in triggering virus assembly, which occurs only after a sufficient amount of HDV RNA has already accumulated. Thus, the appearance of L-HDAg will shut off additional RNA synthesis and simultaneously allow virus assembly to take place. Several potential mechanisms can account for the inhibition of RNA synthesis by L-HDAg: (1) L-HDAg transports HDV RNA to a site not conducive for RNA replication. (2) L-HDAg disrupts the conformation of the functional transcription complex. This is supported by the finding that L-HDAg can affect pol II-mediated transcription of cellular genes (112, 113). A puzzling question in this scenario of L-HDAg functions arises in view of the facts that L-HDAg is part of the virion nucleocapsid (which contains both

S-HDAg and L-HDAg in variable ratios) (8, 10, 11) and that HDV virion contains RNAs capable of encoding both S-HDAg and L-HDAg (106). Thus, L-HDAg would be expected to inhibit initiation of viral replication if L-HDAg indeed is capable of trans-dominant inhibition of viral RNA replication. Studies using cDNA transfection showed that a small fraction of L-HDAg (as low as S-HDAg:L-HDAg = 10:1) is sufficient to shut off HDV RNA replication (59). This puzzle has recently been solved: L-HDAg inhibits primarily only genomic RNA synthesis but not antigenomic RNA synthesis (95). This finding again suggests that genomic and antigenomic RNAs are synthesized by different mechanisms. This differential sensitivity will allow the incoming HDV genomic RNA to synthesize antigenomic RNA after infection, despite the presence of L-HDAg in the virion. However, the reported inhibitory activity of L-HDAg needs to be re-examined, as HDV RNA does not decrease appreciably late in the viral replication cycle, when a large amount of L-HDAg has accumulated and virus particle assembly is actively taking place (Macnaughton and Lai, unpublished). It is possible that L-HDAg interferes with RNA replication only at the initial step of RNA replication.

5.7 Virus Assembly

The complete HDV particle can only be formed when HBV is also present, which supplies HBsAg to form the envelope proteins of HDV particle. However, experimentally, HDV particles can be assembled if HBsAg alone is expressed in the HDV-replicating cells. Further, the co-presence of L-HDAg and HBsAg only are sufficient to trigger the formation of empty virus-like particles (61, 62, 114). Thus, the minimum requirements for HDV particle assembly are L-HDAg and HBsAg, which have been shown to interact with each other *in vitro* (54). Such an interaction requires the presence of isoprenylates in the C-terminus of L-HDAg (54, 57). The C-terminal 19 amino acids of L-HDAg constitute the minimum virus assembly signal (115). S-HDAg alone, which does not have the unique C-terminal amino acid residues of L-HDAg, can not trigger virus assembly but can be incorporated into the virus-like particles probably as a result of its interaction with L-HDAg through the coiled-coil domain (45). The relative ratio of the L- and S-HDAg in the virions is variable. Viral RNA is packaged as a result of its interaction with either L- or S-HDAg; the presence of S-HDAg enhances the efficiency of RNA packaging into the virion (116). Only the genomic sense, but not the antigenomic-sense, RNA is packaged (116), even though both RNAs interact with HDAg *in vitro* equally well (117). The basis for the selectivity of genomic RNA packaging is not clear. One possibility is that the genomic RNA is preferentially exported to the cytoplasm where

HBsAg is located. The infectivity of the complete virus particle depends on the presence of the large form of HBsAg (118), since the pre-S1 domain in the large HBsAg contains the receptor-binding sequence.

6. PERSPECTIVES

HDV is a unique pathogen with features not seen in conventional viruses. Most intriguingly, it relies on cellular machineries to carry out its RNA replication, and yet cells do not have the natural ability to perform RNA-dependent RNA synthesis. How HDV usurps the cellular machineries to replicate its own RNA remains the paramount question in HDV molecular biology. This property suggests that normal cells have a hidden capability to replicate RNA under certain conditions. The understanding of this mechanism will bring an exciting frontier to the molecular biology of cells. The cross-talks between HDV and cells likely contribute to HDV pathogenesis. Because natural infection of HDV in cultured cells or small animals is inefficient, the current understanding of the framework of HDV life cycle has been largely deduced from the studies based on the transfection of HDV cDNA into tissue culture cells. Recent findings using RNA transfection techniques, however, suggest that many of the currently accepted features of the HDV replication model (93, 99) are not correct. We are at an exciting time of HDV molecular studies. In the next few years, a brand new understanding of HDV molecular biology is likely to emerge.

REFERENCES

1. Rizzetto M., Canese M.G., Arico S., Crivelli O., Trepo C., Bonino F., Verme G. Immunofluorescence detection of a new antigen-antibody system (delta/anti-delta) associated with hepatitis B virus in liver and serum of HBsAg carrier. Gut 1977; 18:997-1003.
2. Rizzetto M., Hoyer B., Canese M.G., Shih J.W.-K., Purcell R.H., Gerin J.L. Delta agent: Association of δ antigen with hepatitis B surface antigen and RNA in serum of d-infected chimpanzees. Proc Natl Acad Sci USA 1980; 77:6124-6128.
3. Rizzetto M., Purcell R.H., Gerin J.L. Epidemiology of HBV-associated delta agent: Geographical distribution of anti-delta and prevalence in polytransfused HBsAg carrier. *Lancet*, 1980; 1:1215-1218.
4. Ponzetto A., Forzani B., Parravicini P.P., Hele C., Zanetti A., Rizzetto M. Epidemiology of hepatitis delta virus infection. Eur J Epidemiol. 1985; 1:257-263.
5. Hadler S.C., de Monzon M.A., Rivero D., Perez M., Bracho A., Fields H. Epidemiology and long-term consequences of hepatitis delta virus. Am J Epidemiol 1992; 136:1507-1516.

6. Diener T.O. Pruisner S.B. "The recognition of subviral pathogens". In *Subviral Pathogens of Plants and Animals. Viroids and Prions*, K. Maramorosch, J.J.J. McKelvey, eds. Orlando: Academic Press, Inc, 1985.
7. Murphy F.A. "Virus Taxonomy." In *Sixth Report of the International Committee on Taxonomy of Viruses*." Wien; New York: Springer-Verlag. 1995.
8. Bonino F., Hoyer B., Shih J.W.-K., Rizzetto M., Purcell R.H., Gerin J.L. Delta hepatitis agent: structural and antigenic properties of the delta-associated particles. Infect Immunity 1984; 43:1000-1005.
9. Bonino F., Heermann K.H., Rizzetto M., Gerlich W.H. (1986) Hepatitis delta virus: protein composition of delta antigen and its hepatitis B virus-derived envelope. J Virol 1986; 58:945-950.
10. Bergmann K.F. Gerin J.L. Antigens of hepatitis delta virus in the liver and serum of humans and animals. J Infect Dis 1986; 154:702-706.
11. Pohl C., Baroudy B.M., Bergmann K.F., Cote P.J., Purcell R.H., Hoofnagle J., Gerin J.L. A human monoclonal antibody that recognizes viral polypeptides and in vitro translation products of the genome of the hepatitis D virus. J Infect Dis 1987; 156:622-629.
12. Zuccola H.J., Rozzelle J.E., Lemon S.M., Erickson B.W., Hogle J.M. Structural basis of the oligomerization of hepatitis delta antigen. Structure 1998; 6:821-830.
13. Ryu W.-S., Netter H.J., Bayer M., Taylor J. Ribonucleoprotein complexes of hepatitis delta virus. J Virol 1993; 67:3281-3287.
14. Wang K.S., Choo Q.L., Weiner A.J., Ou J.H., Najarian R.C., Thayer R.M., Mullenbach G.T., Denniston K.J., Gerin J.L., Houghton M. Structure, sequence and expression of the hepatitis delta viral genome. Nature 1986; 323:508-514.
15. Makino S., Chang M.-F., Shieh C.-K., Kamahora T., Vannier D.M., Govindarajan S., Lai M.M.C. "Molecular biology of a human hepatitis delta virus RNA." In *Hepadnaviruses*, W. Robinson, K. Koike, H. Will, eds., New York: Alan Liss, Inc. 1987.
16. Kos A., Dijkema R., Arnberg A.C., van der Merde P.H., Schellekens H. The hepatitis delta (d) virus possesses a circular RNA. Nature 1986; 323:558-560.
17. Casey J.L., Brown T.L., Colan E.J., Wignall F.S., Gerin J.L. A genotype of hepatitis D virus that occurs in northern South America. Proc Natl Acad Sci USA 1993; 90:9016-9020.
18. Imazeki F., Omata M., Ohto M. Complete nucleotide sequence of hepatitis delta virus RNA in Japan. Nucleic Acids Res 1991; 19:5439.
19. Wu J.-C., Choo K.-B., Chen C.-M., Chen T.-Z., Hue T.-C., Lee S.-D. Genotyping of hepatitis D virus by restriction-fragment length polymorphism and relation to outcome of hepatitis D. Lancet 1995; 346:939-941.
20. Branch A.D., Benenfeld B.J., Baroudy B.M., Wells F.V., Gerin J.L., Robertson H.D. An ultraviolet-sensitive RNA structural element in a viroid-like domain of the hepatitis delta virus. Science 1989; 243:649-652.
21. Elena S.F., Dopazo J., Flores R., Diener T.O., Moya A.. Phylogeny of viroids, viroidlike satellite RNAs, and the viroidlike domain of hepatitis δ virus RNA. Proc Natl Acad Sci USA 1991; 88:5631-5634.
22. Robertson H.D. How did replicating and coding RNAs first get together? Science 1996; 274:66-67.
23. Brazas R., Ganem D. A cellular homolog of hepatitis delta antigen: implications for viral replication and evolution. Science 1996; 274:90-94.
24. Long M., de Souza S.J., Gilbert W. Delta-interacting protein A and the origin of hepatitis delta antigen. Science 1996; 274:90-94.

25. Wu H.-N., Wang Y.-J., Hung C.-F., Lee H.-J., Lai M.M.C. Sequence and structure of the catalytic RNA of hepatitis delta virus genomic RNA. J Mol Biol.1992; 223:233-245.
26. Kuo M.Y.-P., Sharmeen L., Dinter-Gottlieb G., Taylor J. Characterization of self-cleaving RNA sequences on the genome and antigenome of human hepatitis delta virus. J Virol 1988 62:4439-4444.
27. Wu H.-N. Lai M.M.C. Reversible cleavage and ligation of hepatitis delta virus RNA. Science 1989; 243:652-654.
28. Sharmeen L., Kuo M.Y.-P., Dinter-Gottlieb G., Taylor J. The antigenomic RNA of human hepatitis delta virus can undergo self-cleavage. J Virol 1988; 62:2674-2679.
29. Wu H.-N., Lin Y.-J., Lin F.-P., Makino S., Chang M.-F., Lai M.M.C. Human hepatitis δ virus RNA subfragments contain an autocleavage activity. Proc Natl Acad Sci USA 1989; 86:1831-1835.
30. Branch A.D., Robertson H.D. Efficient *trans* cleavage and a common structural motif for the ribozymes of the human hepatitis δ agent. Proc Natl Acad Sci USA 1991; 88:10163-10167.
31. Perrotta A.T., Been M.D. Cleavage of oligoribonucleotides by a ribozyme derived from the hepatitis delta virus RNA sequence. Biochemistry 1992: 31:16-21.
32. Perrotta A.T., Shih I., Been M.D. Imidazole rescue of a cytosine mutation in a self-cleaving ribozyme. Science 1999; 286:123-126.
33. Ferre-D'Amare A.R., Zhou K., Doudna J.A. Crystal structure of a hepatitis delta virus ribozyme. Nature 1998; 395:567-574.
34. Hampel A. Tritz R. RNA catalytic properties of the minimum (-)STRSV sequence. Biochemistry 1989; 28:4929-4933.
35. Forster A.C. Symons R.H. Self-cleavage of plus and minus RNAs of a virusoid and a structural model for the active sites. Cell 1987; 49:211-220.
36. Macnaughton T.B., Wang Y.-J., Lai M.M.C. Replication of hepatitis delta virus RNA: Effect of mutations of the autocatalytic cleavage sites. J Virol 1993: 67:2228-2234.
37. Jeng K.S., Daniel A., Lai M.M.C. A pseudoknot ribozyme structure is active in vivo and required for hepatitis delta virus RNA replication. J Virol 1996; 70:2403-2410.
38. Sharmeen L., Kuo M.Y.-P., Taylor J. Self-ligating RNA sequences on the antigenome of human hepatitis delta virus. J. Virol. 1989; 63:1428-1430.
39. Lazinski D.W. Taylor J.M. Intracellular cleavage and ligation of hepatitis delta virus genomic RNA: Regulation of ribozyme activity by *cis*-acting sequences and host factors. J. Virol 1995; 69:1190-1200.
40. Reid C.E. Lazinski D.W. A host-specific function is required for ligation of a wide variety of ribozyme-processed RNAs. Proc Natl Acad Sci USA 2000; 97:424-429.
41. Chang M.-F., Baker S.C., Soe L.H., Kamahora T., Keck J.G., Makino S., Govindarajan S., Lai M.M.C. Human hepatitis delta antigen is a nuclear phosphoprotein with RNA-binding activity. J Virol 1988; 62:2403-2410.
42. Hwang S.B., Lee C.Z., Lai M.M.C. Hepatitis delta antigen expressed by recombinant baculoviruses: Comparison of biochemical properties and post-translational modifications between the large and small forms. Virology 1992; 190:413-422.
43. Mu J.-J., Wu H.-L., Chiang B.-L., Chang R.-P., Chen D.-S., Chen P.-J. Characterization of the phosphorylated forms and the phosphorylated residues of hepatitis delta virus delta antigens. J Virol 1999; 73:10540-10545.
44. Yeh T.S., Lo S.J., Chen P.J., Lee Y.H.W. Casein kinase II and protein kinase C modulate hepatitis delta virus RNA replication but not empty viral particle assembly. J Virol. 1996; 70:6190-6198.

45. Chen P.-J., Chang F.-L., Wang C.-J., Lin C.-J., Sung S.-Y., Chen D.-S. Functional studies of hepatitis delta virus large antigen in packaging and replication inhibition: Role of the amino-terminal leucine zipper. J Virol 1992; 66:2853-2859.
46. Xia Y.-P. Lai M.M.C. Oligomerization of hepatitis delta antigen is required for both the trans-activating and trans-dominant inhibitory activities of the delta antigen. J Virol 1992; 66:6641-6648.
47. Xia Y.-P., Yeh C.-T., Ou J.-H., Lai M.M.C. Characterization of nuclear targeting signal of hepatitis delta antigen: nuclear transport as a protein complex. J Virol 1992; 66:914-921.
48. Chou H.-C., Hsieh T.-Y., Sheu G.-T., Lai M.M.C. Hepatitis delta antigen mediates the nuclear import of hepatitis delta virus RNA. J Virol 1998; 72:3684-3690.
49. Lee, C.-Z., Lin J.-H., McKnight K., Lai M.M.C. RNA-binding activity of hepatitis delta antigen involves two arginine-rich motifs and is required for hepatitis delta virus RNA replication. J Virol 1993; 67:2221-2229.
50. Poisson F., Roingeard P., Baillou A., Dubois F., Bonelli R., Calogero R.A., Goudeau A. Characterization of RNA-binding domains of hepatitis delta antigen. J Gen Virol (1993) 74:2473-2477.
51. Chang, M.-F., Sun C.-Y., Chen C.-J., Chang S.-C. Functional motifs of delta antigen essential for RNA binding and replication of hepatitis delta virus. J. Virol. 1993; 67:2529-2536.
52. Otto J.C. Casey P.J. The hepatitis delta virus large antigen is farnesylated both in vitro and in animal cells. J Biol Chem 1996; 271:4569-4572.
53. Glenn J.S., Watson J.A. Havel C.M. White J.M. Identification of a prenylation site in delta virus large antigen. Science 1992; 256:1331-1333.
54. Hwang S.B. Lai M.M.C. Isoprenylation mediates direct protein-protein interactions between hepatitis large delta antigen and hepatitis B virus surface antigen. J Virol 1993; 67:7659-7662.
55. Lee C.-Z., Chen P.-J., Lai M.M.C., Chen D.S.. Isoprenylation of large hepatitis delta antigen is necessary but not sufficient for hepatitis delta virus assembly. Virology 1994; 199:169-175.
56. Chang M.-F., Chen C.-J., Chang S.-C. Mutational analysis of delta antigen: effect on assembly and replication of hepatitis delta virus. J Virol 1994; 68:646-653.
57. de Bruin W., Leenders W., Kos T., Yap S.H. In vitro binding properties of the hepatitis delta antigens to the hepatitis B virus envelope proteins: potential significance for the formation of delta particles. Virus Res 1994;.31:27-37.
58. Kuo M.Y.-P., Chao M., Taylor J. Initiation of replication of the human hepatitis delta virus genome from cloned DNA: Role of delta antigen. J Virol 1989; 63:1945-1950.
59. Chao M., Hsieh S.-Y., Taylor J. Role of two forms of hepatitis delta virus antigen: Evidence for a mechanism of self-limiting genome replication. J Virol 1990; 64:5066-5069.
60. Glenn J.S. White J.M. *Trans*-dominant inhibition of human hepatitis delta virus genome replication. J Virol 1991; 65:2357-2361.
61. Chang F.-L., Chen P.-J., Tu S.-J., Wang C.-J., Chen D.-S. The large form of hepatitis δ antigen is crucial for assembly of hepatitis δ virus. Proc Natl Acad Sci USA 1991; 88:8490-8494.
62. Ryu W.-S., Bayer M., Taylor J. Assembly of hepatitis delta virus particles. J Virol 1992; 66:2310-2315.
63. Casey J.L., Gerin J.L. Genotype-specific complementation of hepatitis delta virus RNA replication by hepatitis delta antigen. J Virol 1998; 72:2806-14.

64. Lazinski D.W. Taylor J.M. Expression of hepatitis delta virus RNA deletions: *cis* and *trans* requirements for self-cleavage, ligation and RNA packaging. J Virol 1994; 68:2879-2888.
65. Jeng K.S., Su P.Y., Lai M.M.C. Hepatitis delta antigens enhance the ribozyme activities of hepatitis delta virus RNA in vivo. J Virol.1996; 70:4205-4209.
66. Huang Z.S., Wu H.N. Identification and characterization of the RNA chaperone activity of hepatitis delta antigen peptides. J Biol Chem 1998; 273:26455-26461.
67. Hsieh S.-Y. Taylor J. Regulation of polyadenylation of hepatitis delta virus antigenomic RNA. J Virol. 1991; 65:6438-6446.
68. Hsieh S.-Y., Yang P.-Y., Ou J.T., Chu C.M., Liaw Y.F. Polyadenylation of the mRNA of hepatitis delta virus is dependent upon the structure of the nascent RNA and regulated by the small or large delta antigen. Nucleic Acids Res 1994; 22:391-396.
69. Ponzetto A., Cote P.J., Popper H., Hoyer B.H., London W.T., Ford E.C., Bonino F., Purcell R.H., Gerin J.L. Transmission of hepatitis B virus-associated delta agent to the eastern woodchuck. Proc Natl Acad Sci USA 1984; 81:2208-2212.
70. Chang J., Morabda G., Taylor J. Limitations to replication of hepatitis delta virus in avian cells. J Virol 2000; 74:8861-8866.
71. Sureau C., Jacob J.R., Eichberg J.W., Lanford R.E. Tissue culture system for infection with human hepatitis delta virus. J Virol 1991; 65:3443-3450.
72. Choi S.-S., Rasshofer R., Roggendorf M.. Propagation of woodchuck hepatitis delta virus in primary woodchuck hepatocytes. Virology 1988; 167:451-457.
73. Taylor J., Mason W., Summers J., Goldberg J., Aldrich C., Coates L., Gerin J., Gowans E. Replication of human hepatitis delta virus in primary cultures of woodchuck hepatocytes. J Virol.1987; 61:2891-2895.
74. Macnaughton T.B., Beard M.R., Chao M., Gowans E.J., Lai M.M.C. Endogenous promoters can direct the transcription of hepatitis delta virus RNA from a recircularized cDNA template. Virology 1993; 196:629-636.
75. Tai F.-P., Chen P.-J., Chang F.-L., Chen D.-S. Hepatitis delta virus cDNA can be used in transfection experiments to initiate viral RNA replication. Virology 1993; 197:137-142.
76. Glenn J.S., Taylor J.M., White J.M. In vitro-synthesized hepatitis delta virus RNA initiates genome replication in cultured cells. J Virol.1990; 64:3104-3107.
77. Modahl L.E., Lai M.M.C. Transcription of hepatitis delta antigen mRNA continues throughout hepatitis delta virus (HDV) replication: A new model of HDV RNA transcription and replication. J Virol 1998;.72:5449-5456.
78. Bichko V., Netter H.J., Taylor J. Efficient introduction of hepatitis delta virus into animal cell lines via cationic liposomes. J Virol 1994; 68:5247-5252.
79. Dingle K., Bichko V., Zuccola H., Hogle J., Taylor J. Initiation of hepatitis delta virus genome replication. J Virol 1998; 72:4783-4788.
80. Sheu G.-T., Lai M.M.C. Recombinant hepatitis delta antigen from E. coli promotes hepatitis delta virus RNA replication only from the genomic strand but the antigenomic strand. Virology 2000; 278:578-586.
81. Hwang S.B., Jeng K.S., Lai M.M.C. "Studies of functional roles of hepatitis delta antigen in delta virus RNA replication." In *The unique hepatitis delta virus*, G. Dinter-Gottlieb, ed. Austin: R. G. Landes Company, 1995.
82. Ponzetto A., Negro F., Popper H., Bonino F., Engle R., Rizzetto M., Purcell R.H., Gerin J.L. Serial passage of hepatitis delta virus infection in chronic hepatitis B virus carrier chimpanzees. Hepatology 1988; 8:1655-1661.
83. Rizzetto M., Canese M.G., Gerin J.L., London W.T., Sly D.L., Purcell R.H. Transmission of the hepatitis B virus-associated delta antigen to chimpanzees. J Infect Dis 1980; 141:590-602.

84. Netter H.J., Kajino K., Taylor J.M. Experimental transmission of human hepatitis delta virus to the laboratory mouse. J Virol 1993; 67:3357-3362.
85. Polo J.M., Jeng K.S., Lim B., Govindarajan S., Hofman F., Sangiorgi F., Lai M.M.C. Transgenic mice support replication of hepatitis delta virus RNA in multiple tissues, particularly in skeletal muscle. J Virol. 1995; 69:4880-4887.
86. Polo J.M., Lim B., Govindarajan S., Lai M.M.C. Replication of hepatitis delta virus RNA in mice after intramuscular injection of plasmid DNA. J Virol. 1995; 69:5203-5207.
87. Gowans E.J., Baroudy B.M., Negro F., Ponzetto A., Purcell R.H., Gerin J.L. Evidence for replication of hepatitis delta virus RNA in hepatocyte nuclei after in vivo infection. Virology 1988; 167:274-278.
88. Macnaughton T.B., Gowans E.J., Jilbert A.R., Burrell C.J. Hepatitis delta virus RNA, protein synthesis and associated cytotoxicity in a stably transfected cell line. Virology 1990; 177:692-698.
89. Fu T.-B. Taylor J. The RNAs of hepatitis delta virus are copied by RNA polymerase II in nuclear homogenates. J Virol 1993; 67:6965-6972.
90. Filipovska J. Konarska M.M.. Specific HDV RNA-templated transcription by pol II in vitro. RNA 2000; 6: 41-54.
91. Modahl L.E., Macnaughton T.B., Zhu N., Johnson D.L., Lai M.M.C. RNA-dependent replication and transcription of hepatitis delta virus RNA involve distinct cellular RNA polymerases. Mol Cell Biol 2000; 20:6030-6039.
92. Branch A.D., Robertson H.D. A replication cycle for viroids and other small infectious RNAs. Science 1984; 223:450-455.
93. Lai M.M.C. The molecular biology of hepatitis delta virus. Annu Rev Biochem 1995; 64:259-286.
94. Beard M.R., Macnaughton T.B., Gowans E.J. Identification and characterization of a hepatitis delta virus RNA transcriptional promoter. J Virol 1996; 70: 4986-4995.
95. Modahl L.E., Lai M.M.C. The large delta antigen of hepatitis delta virus potently inhibits genomic but not antigenomic RNA synthesis: A mechanism enabling initiation of viral replication. J Virol 2000; 74:7375-7380.
96. Chen P.-J., Kalpana G., Goldberg J., Mason W., Werner B., Gerin J.L., Taylor J. Structure and replication of the genome of hepatitis delta virus. Proc Natl Acad Sci USA 1986; 83:8774-8778.
97. Hsieh S.-Y., Chao M., Coates L., Taylor J. Hepatitis delta virus genome replication: a polyadenylated mRNA for delta antigen. J Virol 1990; 64:3192-3198.
98. Lo K., Hwang S.B., Duncan R., Trousdale M., .Lai M.M.C. Characterization of mRNA for hepatitis delta antigen: Exclusion of the full-length antigenomic RNA as an mRNA..Virology 1998; 250:94-105.
99. Taylor J.M. Hepatitis delta virus. Intervirology 1999; 42:173-178.
100. Luo G., Chao M., Hsieh S.Y., Sureau C., Nishikura K., Taylor J. A specific base transition occurs on replicating hepatitis delta virus RNA. J Virol 1990; 64:1021-1027.
101. Polson A.G., Bass B.L., Casey J.L. RNA editing of hepatitis delta virus antigenome by dsRNA- adenosine deaminase. Nature 1996; 380:454-456.
102. Casey J.L., Gerin J.L. Hepatitis D virus RNA editing: Specific modification of adenosine in the antigenomic RNA. J Virol 1995; 69:7593-7600.
103. Casey J.L., Bergmann K.F., Brown T.L., Gerin J.L. Structural requirements for RNA editing in hepatitis delta virus: evidence for a uridine-to-cytidine editing mechanism. Proc Natl Acad Sci USA 1992; 89:7149-7153.
104. Zheng H., Fu T.-B., Lazinski D., Taylor J.. Post-transcriptional modification of genomic RNA of human hepatitis delta virus. J Virol. 1992; 66:4693-4697.

105. Polson A.G., Ley H.L., Bass B.L., Casey J.L. Hepatitis delta virus RNA editing is highly specific for the amber/W site and is suppressed by hepatitis delta antigen. Mol Cell Biol.1998; 18:1919-1926.
106. Xia Y.-P., Chang M.-F., Wei D., Govindarajan S., Lai M.M.C. Heterogeneity of hepatitis delta antigen. Virology 1990; 178:331-336.
107. Lazinski D.W. Taylor J.M. Relating structure to function in the hepatitis delta virus antigen. J Virol 1993; 67:2672-2680.
108. Bichko V.V., Taylor J.M. Redistribution of the delta antigens in cells replicating the genome of hepatitis delta virus. J Virol 1996; 70:8064-8070.
109. Bell P., Brazas R., Ganem D., Maul G.G. Hepatitis delta virus replication generates complexes of large hepatitis delta antigen and antigenomic RNA that affiliate with and alter nuclear domain 10. J Virol 2000; 74:5329-5336.
110. Cunha C., Monjardino J., Chang D., Krause S., Carmo-Fonseca M. Localization of hepatitis delta virus RNA in the nucleus of human cells. RNA 1998; 4:680-693.
111. Lee C.H., Chang S.C., Chen C.J., Chang M.F. The nucleolin binding activity of hepatitis delta antigen is associated with nucleolus targeting. J Biol Chem. 1998; 273:7650-7656.
112. Lo K., Sheu G.-W., Lai M.M.C. Inhibition of cellular RNA polymerase II transcription by delta antigen of hepatitis delta virus. Virology 1998; 247:178-188.
113. Wei Y., Ganem D. Activation of heterologous gene expression by the large isoform of hepatitis delta antigen. J Virol 1998; 72:2089-2096.
114. Wang C.-J., Chen P.-J., Wu J.-C., Patel D., Chen D.-S. Small-form hepatitis B surface antigen is sufficient to help in the assembly of hepatitis delta virus-like particles. J Virol.1991; 65:6630-6636.
115. Lee C.-Z., Chen P.J., Chen D.S. Large hepatitis delta antigen in packaging and replication inhibition: Role of the carboxyl-terminal 19 amino acids and aminoterminal sequences. J Virol 1995; 69:5332-5336.
116. Wang H.-W., Chen P.-J., Lee C.-Z., Wu H.-L., Chen D.-S. Packaging of hepatitis delta virus RNA via the RNA-binding domain of hepatitis delta antigens: Different roles for the small and large delta antigens. J Virol 1994; 68:6363-6371.
117. Lin J.-H., Chang M.-F., Baker S.C., Govindarajan S., Lai M.M.C. Characterization of hepatitis delta antigen: Specific binding to hepatitis delta virus RNA. J Virol 1990; 64:4051-4058.
118. Sureau C., Moriarty A.M., Thornton G.B., Lanford R.E. Production of infectious hepatitis delta virus *in vitro* and neutralization with antibodies directed against hepatitis B virus pre-S antigens. J Virol 1992; 66:1241-1245.

Chapter 6

THE MOLECULAR BIOLOGY OF HEPATITIS E VIRUS

Gregory R. Reyes

Infectious Diseases and Tumor Biology, Schering-Plough Research Institute, 2015 Galloping Hill Road, Kenilworth, NJ 07033

1. INTRODUCTION

There are two characterized forms of fecal-orally transmitted viral hepatitis: A and E (hepatitis A virus [HAV] and hepatitis E virus [HEV] respectively). The first *confirmed* epidemic of hepatitis E was in New Delhi, India in 1955 (1) and although the epidemiological features (e.g., fecal-oral transmission) suggested HAV as the etiologic agent, by 1980 there was clear evidence that a unique form of fecal-orally transmitted hepatitis was responsible for much of the epidemic hepatitis formerly ascribed to HAV. Indeed, HAV was excluded as the cause of the New Delhi epidemic by retrospective analysis of archived samples for evidence of HAV infection (2). The original observation of a common mode of transmission for hepatitis A and E is now eclipsed by our knowledge of the distinct epidemiology and molecular biology of HAV and HEV. Since the initial molecular cloning reported in 1990 (3) much has been learned about this virus. The unique features and our current understanding of HEV will be reviewed while highlighting issues requiring further investigation.

2. EPIDEMIOLOGY

2. 1 Transmission, Incubation and Time Course

The first human volunteer study established hepatitis E as a distinct form of orally transmitted viral hepatitis after acute hepatitis developed in a HAV immune volunteer upon ingestion of filtered pooled fecal specimens from patients with acute hepatitis (4). Typical 27-32 nm nonenveloped particles were detected by immunoelectron microscopy (IEM) in stool matching those seen in nonhuman primates infected with HEV (see below). In the non-epidemic setting the typical course is one of an acute self-limited disease with complete resolution and no evidence of chronicity. The second volunteer study, conducted after HEV had been cloned and characterized, used reverse-transcription-polymerase chain reaction (RT-PCR) (5) and an anti-HEV antibody ELISA to fully monitor the incubation and time course of hepatitis E. After an incubation period of 35 to 45 days, ALT becomes elevated preceded by overt clinical signs and symptoms (e.g., epigastric pain and discolored urine at day 30). HEV was detectable in serum by day 22 post-infection and about one week later in stool by RT-PCR. ALT peaked by day 46. Rhesus monkeys were successfully infected with human fecal filtrates collected on days 34 and 37. Anti-HEV IgG was first detected on day 41 and persisted 2 years. The disease course in this individual was unusually protracted with both biochemical and clinical signs of hepatitis lasting 3 months. Icterus, evident on day 38, lasted until day 120. The general parameters of viral infection as developed by this and other studies are schematically presented in Figure 1.

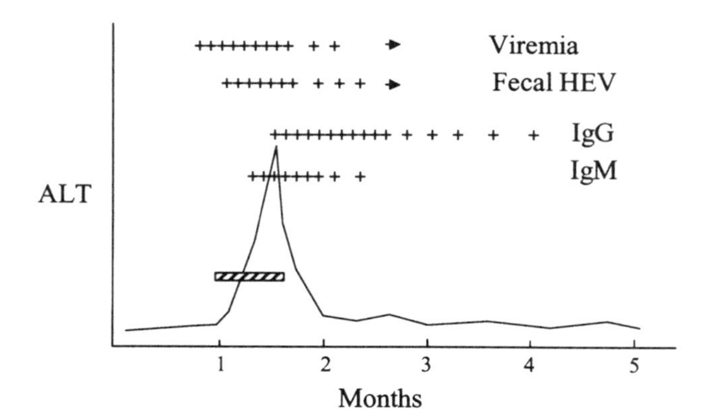

Figure 1. The general time course of acute hepatitis E is illustrated. The incubation period for HEV infection is reportedly 35 to 45 days with clinical signs and symptoms (hatched bar) preceding ALT elevation (5-10x normal), peaking by about day 45 post exposure. The acute resolving nature of the disease is reflected by the return of ALT to baseline. Viremia occurs prior to fecal shedding of virus, and although both generally resolve, prolonged states of viremia and viral excretion have been reported (arrows). The humoral immune response to virus infection appears typical at the early acute stages of infection with typical resolution, however, IgG antibody titers appear to wane over time.

Considerable variations can be seen in certain of the features of HEV infection including length of viremia, virus shedding and duration of ALT elevation. Protracted HEV viremia of 45-112 days duration was detected by PCR in approximately a quarter of patients presenting with acute sporadic hepatitis E (6). A prolonged period of virus shedding in feces (to day 52) was also seen in one patient for whom samples were available. This study suggested that protracted viremia in hepatitis E may be more common than previously thought and may contribute to a prolonged state of infectivity and as a point-source of virus in the community.

2.2 Endemic Hepatitis E

2.2.1 Sporadic and subclinical infection

As already noted, beyond similar fecal-oral transmission, the epidemiology of hepatitis E and hepatitis A are quite distinct. In endemic regions HAV seroconversion generally occurs in the first decade, approaching 100%, with resulting lifelong immunity. The same environmental conditions support endemic HEV infection. Exposure undoubtedly occurs early in life, however unlike hepatitis A, the highest attack rates are seen in young adults (age 15 - 40 years). Subclinical or anicteric hepatitis E is common in childhood (7) but detectable antibody titers are never present in greater than 40% of the population. High sporadic rates of pediatric infection (7) were also documented in Egypt, Somalia and the Sudan (8-10).

Antibody to HEV is believed to be protective but multiple studies indicate that antibody titers tend to decrease over time. The significance of antibody titers that develop and then decrease or disappear remains to be determined. High rates of subclinical infections are also seen outside of Africa. In Chile, sero-positivity rates in children were high (36%) when compared to prisoners (8%) and normal blood donors (7%) (11). A high incidence of subclinical infection was also seen in Hong Kong where anti-HEV IgG antibody was found in 16% of healthy subjects without any prior history of hepatitis (12). Hepatitis E would appear to be under reported and it is unclear what combination of host (e.g. age) or viral factors result in milder clinical disease but significant rates of seroconversion.

An Italian study found that volunteer blood donors and normal healthy individuals had similar seroprevalence rates (0.95 and 0.74% respectively) (13) whereas higher rates in intravenous drug users (1.94%) indicated the possibility of intravenous transmission as suggested elsewhere (14). Interestingly, 6.5% of patients with nonA, nonB, nonC hepatitis were found to be anti-HEV antibody positive, without any history of travel or contact with patients from endemic areas, implicating HEV as a cause of sporadic acute hepatitis in Italy.

2.2.2 Hepatitis E in industrialized nations

The severity of disease and the nature of the outbreak are largely dictated by the setting in which infection occurs. HEV in western industrialized countries was formerly viewed as an imported travelers disease as reflected in numerous case reports (15-18). However, this conflicted with data from studies reporting sporadic cases of hepatitis E in patients having no history of travel or exposure to individuals from endemic areas (13, 18) and from serosurveys (19) using recombinant HEV fusion proteins. A high seroprevalence was detected in normal blood donors in Europe (1.4%), South Africa (2.3%), Thailand (2.8%), Saudi Arabia (9.5) and Egypt (24.5%). The high rates in North Africa and the Middle East were not unsuspected given the reports of sporadic disease in these endemic areas. The findings in Europe, South Africa, and Thailand substantiate infection in developed and developing countries in the absence of reports of clinical hepatitis of unknown etiology. The HEV seropositivity rate in acute sporadic hepatitis was very high in samples tested from South Africa (8.2 %) and Japan (14.1 %) with Europe intermediate (9.5%). Out of the total 100 positive samples only 14% were IgM positive. HEV is clearly responsible for a substantial portion of sporadic hepatitis seen in these areas. It is also evident that the absence of reports of clinical hepatitis, in conjunction with high rates of seroconversion (1-2%), indicate substantial rates of subclinical HEV infection (19-21).

The zoonotic nature of the HEV is now recognized and the role played by animal hosts such as domesticated pigs (22-24) or rats (25) now requires consideration in order to develop a complete infection control plan for communities at risk. Indeed, experimental transmission of the US swine strain into nonhuman primates and the reciprocal experiment of transmitting the human US2 isolate into pathogen free pigs has been performed (26). It will also be important to determine the degree to which zoonotic maintenance of the virus contributes to the sporadic rate of infection seen in endemic communities.

2.3 Epidemic Hepatitis E

Fecal-oral transmission of HEV becomes most evident in the course of an epidemic outbreak. These outbreaks are point source and tend to be unimodal, although in a prolonged outbreak there may be multiple peaks of disease. Epidemics concentrate in the rainy or monsoon season when a higher probability exists of sewage contaminating drinking water. Inability to identify and interrupt the source contamination can lead to prolonged epidemics such as that seen in Xinjiang, China (1986-1988) (27). The other well-characterized fecal-oral transmitted hepatitis virus is HAV, but there are undoubtedly other *uncharacterized* hepatitis viruses that are similarly spread (28).

HEV outbreaks and sporadic cases have been reported on every continent with the exception of Antarctica (29-39). Epidemic hepatitis E strikes on a recurring, intermittent, basis in the developing world. Of the last 17 Indian epidemics of waterborne hepatitis spanning a period of nearly 40 years, 16 were attributable to hepatitis E (40). Isolates sequenced from three of these 16 epidemics, as well as 4 other epidemic or sporadic cases, showed that all seven isolates were highly related to one another. The interplay between predisposing environmental conditions, the endemic nature of HEV, or even the significance of prior subclinical hepatitis E infection, remains unclear.

The incidence of disease in the epidemic setting can be quite high. Contaminated water led to the Kanpur, India epidemic in 1992 with a 3.7% incidence of icteric hepatitis and an estimated 79,091 person afflicted (41). Attack rates were highest in adults versus children (4.3% vs 1.3%)) and higher for males versus females (5.3% vs. 3.3 %). Secondary intrafamilial spread is not a major factor in HEV disease transmission. In the 2-year epidemic in Xinjiang, nearly 120,000 people were infected with the highest attack rate in the 15-44 age range (5.2 %) compared to attack rates of 0.87 % for children < 14 years and 2.5% adults >45 years of age (27). Case fatality rates (~21 %) were again reportedly highest in pregnant women in their third trimester (see below).

All ages are susceptible to infection with specific attack rates highest in young adults (15- 40 years). Age may be considered a risk factor for the development of clinical disease. The overall clinical attack rate of 6.3% (1702/26920) in a Somali refugee camp in Kenya showed an increasing trend with each decade of life with greater than 50% of the cases reportedly anicteric (42). As already noted, children develop both acute sporadic hepatitis as well as anicteric hepatitis with IgG titers decreasing over the course of a year (7). The high attack rate in young adults may signify that initial infection as a child may not provide a protective antibody response, indeed, reinfection may predispose to a later bout of overt hepatitis on reinfection. A prospective study is needed to determine if a protective anamnestic immune response develops after primary infection.

Risk factors for infection and transmission focus on exposure and ingestion of contaminated drinking water. Boiling water prior to use or distance from point source contaminations are negative risk factors for disease development. The low incidence of person-to-person spread of hepatitis E perhaps indicates that the virus is less stable, has lower environmental titers or is less infective than HAV (42). Travel to an endemic areas is a recognized risk factor.

2.4 HEV Infection in Pregnancy

A key discerning feature of hepatitis E is the unexplained high mortality and morbidity in pregnant women with case fatality rates from fulminant hepatitis in the range of 20% (43, 44). The severity of hepatitis E in pregnancy cannot be simply explained by other host or environmental factors such as immunosuppression or malnutrition since the higher mortality / morbidity is not seen in other non-HEV viral infections. Viral hepatitis is reported as the most common cause of jaundice in pregnancy (45). There are however a number of conditions that cause jaundice including intrahepatic cholestasis of pregnancy, acute fatty liver of pregnancy, hyperemesis gravidarum, eclampsia and preeclampsia. Bile stasis is not uncommon in pregnant women and possibly indicates a common pathogenesis in the severe clinical course seen with hepatitis E. The physiological strain of pregnancy itself, superimposed on viral hepatitis, and in particular the peculiar pathology of hepatitis E (see below) may be sufficient to severely compromise hepatic function in pregnant women where the physiologic reserve may already be marginal.

In an attempt to model the high morbidity and mortality of HEV infection in pregnancy, four pregnant rhesus monkeys were intravenously infected with HEV in the third trimester (46). The average incubation period in nonpregnant monkeys who demonstrated elevated ALT was 36.4 ± 4.9 days but shortened to 9 and 13 days in two pregnant monkeys who delivered normal seronegative babies. A seropositive baby (tested at 11 months) delivered only two days after inoculation of a third animal (with a normal incubation period of 36 days) was not noted as a premature delivery, and the possibility of maternal antibody transfer was not excluded. The last animal delivered a "macerated" fetus on the 36th day post inoculation and the incubation period for this animal was 41 days. Unfortunately, there is no mention of any testing that was done on this fetus to detect the presence of HEV specific sequences. No mortalities were reported among the pregnant monkeys and indeed all were reportedly active without clinical signs of hepatitis. This study suggests that the hepatitis E incubation period in pregnant monkeys may be considerably shortened.

Vertical transmission of HEV was investigated in ten consecutive pregnant women presenting with hepatitis E (47). There were two deaths among the six women who developed fulminant hepatic failure and of the eight delivered infants, two were hypothermic / hypoglycemic and died; one with massive hepatic necrosis. The livers of these infants were not analyzed for viral sequence. PCR on cord blood or early birth samples showed 5 of 8 infants with detectable HEV together with elevated ALTs. This study clearly establishes that intrauterine infection with HEV is a cause of significant perinatal morbidity and mortality.

Outside of pregnancy, hepatitis E is implicated in fulminant hepatitis on a sporadic basis. The presence of the HEV genome was confirmed in 50% -60% of hepatocytes in 2 patients with fulminant hepatitis E using in situ hybridization with immunohistochemical detection (48). Viral transcript and antigens were detected in infected cells. The absence of an inflammatory infiltrate supports the hypothesis that hepatitis injury in HEV fulminant hepatitis is due to a direct cytopathic mechanism rather than immune mediated. In another report, seventeen patients with sporadic fulminant or subfulminant hepatitis of presumed non-A, non-B etiology did not have evidence of HEV infection (49).

The histopathologic changes in the liver after HEV infection are easily differentiated from other hepatitis viruses. These findings include the presence of cholestatic hepatitis with bile ductule proliferation (2, 50, 51). Pseudoglandular arrangement of hepatocytes surrounding distended bile caniliculi is evident with frank parenchymal necrosis. The complete pathologic picture includes mild portal and lobular inflammation with a predominantly polymorphonuclear leukocyte and macrophage infiltration.

3. HEV VIROLOGY

Ingestion of a putative NANBH inoculum satisfied Koch's postulates in 1983 (4). The pooled inoculum, derived from individuals who had developed hepatitis in the course of an outbreak, led to acute icteric hepatitis that resolved without evidence of any chronic infection (4). These same researchers were the first to passage what later became known as HEV into non human primates (4) with resulting hepatitis and excretion of viruslike particles in feces identical to those observed in man and distinct from HAV (52).

Development of an animal model for virus propagation was critical to the early recognition that HEV was distinct from hepatitis A. Primates provided homogeneous source material for the molecular characterization of HEV. The virus was first cloned from nonhuman derived samples using recombinant DNA procedures in 1990 (3). There quickly followed further molecular characterization and definitive evidence through its primary nucleic acid sequence that HEV was a unique positive strand RNA virus, distinct from HAV and other RNA viruses infecting humans (53, 54).

3.1 Virion Morphology, Structure and Size

Hepatitis E virus causes disease when introduced experimentally into a number of nonhuman primates (4, 55, 56). Passage of HEV in primates provided sufficient amounts of virus for biophysical characterization to establish it as a unique and distinct agent from HAV. The virus is a nonenveloped icosahedral particle ranging in size between 27 and 32 nm (55, 57). HEV resembles and was postulated to be a member of the caliciviridae (58); a family of small round structured viruses associated with nonbacterial gastroenteritis in man and animals (see below). Electron micrographs of the virus clearly indicate icosahedral symmetry (59).

3.2 Genome Length and Composition

Bile from infected cynomolgus macaques was found to contain large numbers of virus-like particles and therefore considered an excellent source material for molecular cloning (3). Differential hybridization using complementary DNA (cDNA) probes made from RNA extracted from bile of infected and uninfected cynos was used to identify putative HEV molecular clones (3). Confirmation of the clone as derived from an infectious pathogen was based on: 1) its being exogenous to both human and cynomolgus genomes, 2) its presence in fecal specimens from other unrelated outbreaks in Somalia, Tashkent, Borneo, Pakistan and Mexico, 3) a recognized sequence motif commonly seen in positive strand RNA viruses (see below) and 4) detection of a ~7,500 nt sequence present only in polyA RNA from infected cynomolgus macaque liver.

The first complete genome of HEV (Burma strain) were obtained and sequenced from a set of overlapping cDNA clones (60). The nucleic acid composition, genome sequence and genomic organization clearly indicated that HEV was distinct from HAV. Translation of the genomic sequence also indicated that unlike the single long open reading frame (ORF) seen in HAV, HEV contained at least three partially overlapping ORFs. HEV(Burma) ORF 1 was 5,079 nts in length and contained sequence motifs associated with nonstructural gene products utilized in viral replication (61) (Figure 2). These include the RNA-dependent, RNA polymerase, helicase, methyl transferase and several other regions having unknown function but related to sequences present in other positive strand RNA viruses such as the brome mosaic virus (62).

ORF 2 encodes the major structural gene product (60). The deduced protein sequence of ORF2 (1980 nt) contains an amino terminal hydrophobic segment (signal segment) that implies virion maturation in association with the endoplasmic reticulum (61). ORF 2 begins 37 nts downstream of ORF 1 and overlaps most of ORF 3. The ORF 2 translation product has a high pI (pI=10.35) at its amino terminus indicating a possible role in genomic RNA encapsidation (60). The highly immunogenic epitopes in the carboxy terminal portion of ORF 2 also suggests it encodes a structural gene (63). Two different consensus glyosylation sites were identified but it would appear that only the nonglycosylated form is stable in cells (64). An earlier report indicates that the glycosylated 88 KD product of ORF 2 is capable of forming noncovalent homodimers (65). Expression cloning of the complete ORF 2 in baculovirus leads to self-assembly into empty capsids (66) as also reported for certain human enteric *caliciviridae* (67, 68). The formation of virus-like particles will assist in defining the pathways of virion maturation and may ultimately prove critical for successful vaccine development.

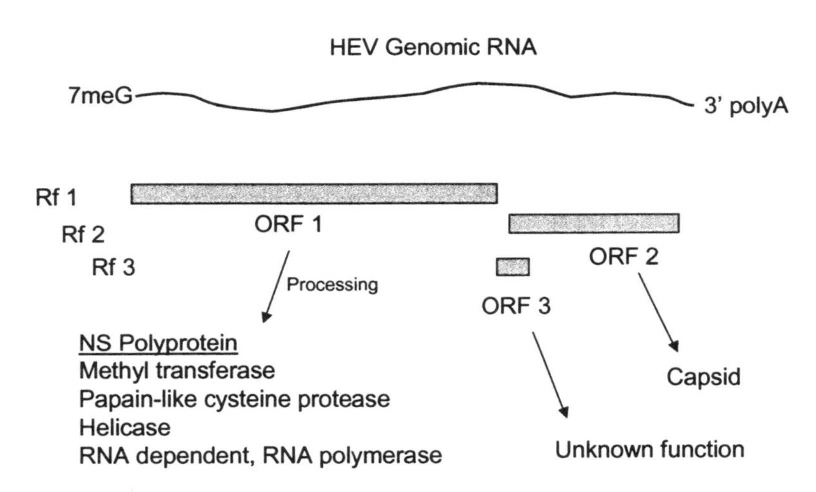

Figure 2. The general expression strategy for hepatitis E virus is illustrated. All forward reading frames are utilized (Rf1 to Rf3) for the three different open reading frames identified for viral protein expression (ORFs 1-3). Typical motifs and consensus sequences (e.g., RNA-dependent RNA polymerase) have been identified within ORF 1 for various nonstructural (NS) proteins involved in viral replication. These are initially expressed as a polyprotein that undergoes proteolytic processing after initial translation from the genomic plus strand transcript. The expression of ORFs 2 and 3 in the distal carboxy terminal third of the genome has yet to be clarified but presumably occurs through translation from subgenomic transcripts (see Figure 3). The utilization of ORFs 2 and 3 was originally confirmed by identification of antigenic epitopes recognized by sera from infected animals and humans. ORF 2 encodes the virus capsid and the function of ORF 3 has not been determined.

ORF 3 is only 369 nts in length and overlaps ORF 1 by a single nucleotide and ORF 2 by 328 nts. As with ORF 2, a highly immunoreactive epitope, localized to the carboxy terminus of ORF3, was identified by immunoscreening using lambda gt11 expression (63) and confirmed the utilization of ORF 3 by the virus. ORF 3 has two highly hydrophobic regions in its amino terminus that might constitute transmembrane segments or a signal sequence followed by a transmembrane region. The role of the ORF3 gene product is not known. The ~13.5 kD ORF 3 gene product has recently been found to homodimerize in vivo and there is the suggestion that the protein may be involved in intracellular signal transduction pathways based on the finding of a putative *src* homology domain (SH3) and a mitogen-activated protein kinase phosphorylation site (69).

The essential features of the HEV genome, as established by the cloning of the Burma isolate have been confirmed by reports of full-length genome cloning from a number of different isolates (70-73). One of the most notable efforts was the isolation and cloning of an isolate from a patient with acute hepatitis in the US with no known history of travel to endemic areas or other high risk exposures (74, 75). The phylogenetic analysis of this isolate shows it to be significantly divergent from previously characterized strains and in fact more highly related to a US swine isolate (26). Numerous other cloning efforts have uncovered novel variants of HEV the world-over requiring a consensus on defining genotypes (76-80). One of the most interesting reports would appear to indicate that an alternative strategy may exist for expressing ORFs 2 and 3 based on a single nucleotide insertion (81). These provocative results require further confirmation.

3.3 Replication Strategy

As noted above, the length of the HEV genome was first estimated by Northern blot using RNA extracted from infected cynomolgus liver. A single transcript of ~7,500 nts was detected using probes located in ORF 1. Probes from the extreme 3' end of the genome identified two smaller transcripts in addition to the full length genomic transcript. Two subgenomic polyadenylated transcripts of 3,700 nts and 2, 000 nts in length localized to the 3' end of HEV were detected in both the Burma (60) and Mexico isolates (63). There is recent evidence that the HEV genomic RNA is capped based on a well-controlled anti-cap (m^7G) antibody capture based PCR protocol (82).

A hypothetical scheme for viral replication was developed based on the essential characteristics of the viral genomic organization and expression strategy (Figure 3) (83). As in other positive strand RNA viruses, the replication strategy presumes expression of the NS gene products (i.e., ORF 1) from the full-length positive sense genome with co- or posttranslational processing of the individual NS gene products from the NS polyprotein. The NS gene products are involved at the earliest stages of viral replication to generate the negative strand RNA (anti-genomic strand). Recently it has been show that the HEV encoded RNA-dependent RNA polymerase (RdRp) will specifically bind to the 3' end of the viral genome in specific interactions with defined stem-loop structures (84). Both negative and positive strand viral RNAs are presumed generated by the RdRp as shown for other positive strand RNA viruses. The anti-genome RNA acts as template for production of full length genomic RNA. The latter is encapsidated or used for further expression and accumulation of nonstructural gene products. Host encoded proteins are probably involved in various stages of viral replication.

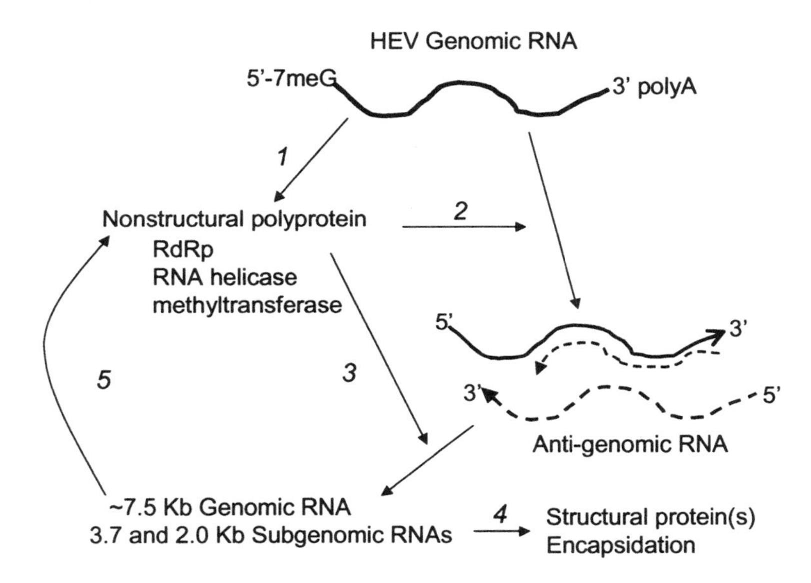

Figure 3. Based on available data, a hypothetical replication strategy for HEV is illustrated. After virion uncoating, the ~ 7.5 Kb capped polyadenylated genome is expressed by the host translational machinery to yield the ORF 1 nonstructural gene products responsible for viral replication (step *1*). After processing, the NS proteins then act to generate the antigenomic (negative strand [dashed line]) of the virus (step *2*) from the plus strand template which in turn serves as template for synthesis of both the full length genomic plus strand and possibly also the two subgenomic transcripts (step *3*). Subgenomic polyA transcripts co-terminal with the 3' end of the genome were detected from HEV infected cynomolgus livers from two disparate isolates (Mexico and Burma) and probably serve to express both ORF 2 and ORF 3. The expression of capsid from ORF 2 permits virion formation (self assembly) and encapsidation of the plus stranded genome (step *4*). Further rounds of NS protein production are possible from the newly synthesized plus strand genome (step *5*).

Production of viral structural protein(s) constitutes the second phase of the proposed replication model. Either full-length or 3' end anti-genomic RNA(s) serves as the template for the subgenomic messages. Translation of these subgenomic messages is presumed to result in capsid protein(s) production. The relative abundance of the individual subgenomic transcripts is not equimolar; the smaller 2.0 kb transcript more abundant than the 3.7 kb transcript (60). This may represent regulatory processing, initiation or a differential half-life of the individual RNA species. The variable abundance of the individual RNA species may be used to regulate expression of the individual gene products.

Genomic RNA encapsidation occurs by its association with the capsid protein. The high basic amino acid content (~10 % arginine) of the amino terminal half of the capsid protein(s) is presumably critical to its role in genomic RNA encapsidation as postulated for other viruses (85).

4. CLASSIFICATION: SEROTYPES AND GENOTYPES

Several lines of evidence indicate that classical neutralization serotypes do not exist. Antibody blocking experiments with convalescent antisera from disparate outbreaks effectively blocks the binding of reference antisera to liver sections from infected animals (86). Immunoelectron microscopy (IEM) also indicated that antibody from different outbreaks, collected the world over, efficiently aggregated virus-like particles from both humans and nonhuman primates infected with different geographic isolates. Finally, it was discovered that animals immunized with recombinant antigens from the HEV (Burma) isolate were protected from infection with HEV (Mexico) which is recognized as one of the most genotypically unrelated isolates (87).

The absence of HEV serotypes may be artefactual due to our use of diagnostic tests based on a limited repertoire of recombinant HEV antigens. Development of type-specific tests will aid in our understanding of HEV's complex epidemiology. The existence of type-specific antibody might be predicted given the genomic sequence divergence reported for the various HEV isolates. In fact it has now been shown that diagnostics based on the Burma and Mexico ORF 2 and 3 epitopes failed to detect anti-HEV antibodies from individuals infected with a Chinese sequence variant (79). An epitope survey is needed in order to determine whether type-specific antigens are able to differentiate divergent strains or properties linked to certain strains. Although at this time untested, it is possible that many of the epidemiologic characteristics of HEV can be attributed to phenotypes discernible at the antigenic level.

As already noted, similarities between HEV and members of the *caliciviridae* prompted a provisional grouping with this family (58). Characterization of the Norwalk virus indicates that the general genomic organization appears to be maintained with the NS genes at the 5' end and the S genes at the 3' end (88, 89). There are however significant differences with the *caliciviridae* and absence of significant nucleic or amino acid sequence homology. The relationship of HEV to the other small round structured viruses is distant when the viruses are compared using sequence algorithms and phylogenetic analyses. HEV is placed in a distinct group separate from the *caliciviridae, picornaviridae* and *togaviridae* and currently remains unclassified (90, 91).

5. LABORATORY DIAGNOSIS

Immunogenic regions of the viral genome expressed as antigens using recombinant DNA tools, have led to sensitive and reproducible diagnostic tests. These same antigens have been used to characterize the neutralizing antibody response to HEV and have shown preliminary utility in eliciting a protective immune response in animals as a first step in establishing the feasibility of a recombinant subunit vaccine for man.

Serologic assays. Antibody from HEV infected convalescent animals/humans can detect viral antigen in liver sections from infected non human primates (86). Antibody from putative HEV infected sources can then be tested for its ability to block binding of a qualified fluorescently labelled antibody. Using characterized specimens from infected animals, antiviral antibody or the presence of virus-like particles in clinical specimens can be detected by immunoelectron microscopy (4). Antibody blocking assays provided important information on the nature of the antibody response to infection as well as documenting the presence of viral specific antigen in the liver. Moreover, the ability of this test to be used in a competitive fashion provided important insight into the nature of the immune response to viral infection and indicated that there was considerable similarity among hepatitis E virus isolates. This latter finding strongly indicated that there existed only a single or predominant serotype for HEV.

Serodiagnostics based on recombinant antigens and recombinant peptides are available for HEV. HEV recombinant antigens have been used to detect and quantify IgG, IgM, and IgA antibodies (20, 92). The antibody response to the virus varies with age as well as the region of the genome utilized in the ELISA. Comprehensive surveys performed on the entire viral genome, dissecting the various ORFs using synthetic peptides revealed immunoreactive epitopes in the RdRp nonstructural region of the genome (93) but failed to identify other regions that were later identified as immunoreactive. In later experiments, peptides were synthesized based on the hydrophilic composition of the particular region and the presence of predicted secondary structure such as α-helices, ß-sheets, ß-turns or random coils. By these methods additional immunodominant peptides in ORF 2 and ORF 3 were identified (94, 95) as having diagnostic utility (96).

6. PREVENTION AND THERAPY

6.1 Immunoprophylaxis

Immunoglobulin prepared from nonendemic areas does not protect against disease. Although there is a high seroprevalence in developed countries the nature of the antibody that is being measured is problematic given the fact that disease or history of hepatitis has never been elicited. IgG prepared form endemic populations should be protective for hepatitis E based on decreasing disease prevalence and increasing antibody prevalence with age (97). When contacts to hospitalized patients were studied in a prospective manner those with pre-existing anti-HEV IgG did not become ill compared to an infection rate of at least 35% among contacts who were anti-HEV IgG negative suggesting that preexisting antibody to HEV ORF 2 is protective (97).

An analysis of passive immunization was also investigated using late convalescent antisera from an animal challenged with a pool of two Chinese isolates. Although infection occurred in animals receiving sufficient anti-HEV plasma to generate titers of 1:40 or 1:200, the amounts of virus in feces and serum (by PCR) were less than in the nonimmunized group and there was no significant ALT elevation in the passively immunized group (98). It is conceivable that immune globulin could have a benefit in pre-exposure prevention of hepatitis E but there have been no controlled clinical studies in patients.

6.2 Immunization

Two different studies have looked at active immunization with subunit recombinant antigens expressed from ORF 2. In the first study the carboxy two thirds of ORF 2 was used as the immunogen and protection against the homologous challenge inoculum (Burma) as gauged by virus in feces, HEV specific antigen in the liver and ALT elevation was complete (62). Interestingly protection against the most distantly related (by sequence analysis) genotype HEV Mexico was only partial in that HEV genome was detected in stool and HEV antigen in liver but there were no histopathologic changes in the liver and no ALT elevation. The second study utilized the full length ORF 2 (Pakistan strain) expressed in baculovirus (98). In this study, complete protection was achieved with a dose of 50µg i.m. injection either once or twice. The challenge inoculum was the homologous HEV Pakistan strain (SAR-55). There was no evidence by PCR of viremia after immunization or any evidence of hepatitis by biopsy or ALT elevation. Virus shedding in the feces was observed in three out of the four animals that received a single dose but in neither of the two animals that received two doses. The feasibility of a recombinant subunit vaccine was unequivocally demonstrated. A vaccine for hepatitis E is under development by GlaxoSmithkline Pharmaceutical.

6.3 Treatment

Hepatitis E is largely a self-limited disease without any evidence of chronic sequelae. Allowed to run its course, recovery will be complete and patients in all likelihood protected from future infection with HEV (see above). Supportive treatment is indicated for symptomatic relieve in the absence of any approved therapy.

The most serious and life threatening complication is the development of fulminant hepatic failure. Medical management in an intensive care setting is indicated in the case of progressive deterioration and development of fulminant hepatic failure as a result of acute viral hepatitis. Management of the patient who develops fulminant hepatic failure has changed little over the last decade. A variety of recommendations for supportive measures can result in effective symptomatic relief.

REFERENCES

1. Viswanathan, R. Infectious hepatitis in Delhi (1955-1956): A critical study; epidemiology. Ind J Med Res 1957; 45(Suppl):1-30.
2. Khuroo, M.S. Study of an epidemic of non-A, non-B hepatitis: possibility of another human hepatitis virus distinct from post-transfusion non-A, non-B type. Am J Med, 1980; 68:818-823.
3. Reyes G.R., Purdy M.A., Kim J.P., Luk K.-C., Young L.M., Fry K.E., Bradley D.W. Isolation of a cDNA from the virus responsible for enterically transmitted non-A, non-B hepatitis. Science 1990; 247:1335-1339.
4. Balayan M.S., Andzhaparidze A.G., Savinskaya S.S., Ketiladze E.S., Barginsky D.M., Savinov A.P., Poleschuk V.F. Evidence for a virus in non-A/non-B hepatitis transmitted via the fecal oral route. Intervirology 1983; 20:23-31.
5. Chauhan A., Jameel S., Dilawari J.B., Chawla Y.K., Kaur U., Ganguly N.K. Hepatitis E transmission to a volunteer. Lancet 1993; 341:149-150.
6. Nanda S.K., Ansari I.H., Acharya S.K., Jameel S., Panda S.K. Protracted viremia during acute sporadic hepatitis E infection. Gastroenterology ,1995; 108:225-230.
7. Goldsmith R., Yarbough P.O., Reyes G.R., Fry K.E., Gabor K.A., Kamel M., Zakaria S., Amer S., Ghaffar Y. Enzyme-linked immunosorbent assay for diagnosis

of acute sporadic hepatitis E infections in Egyptian children. Lancet 1992; 339:328-331.

8. Hyams K.C., Purdy M.A., Kaur M., McCarthy M.C., Hussain A.M., El-Tigani A., Krawczynski K., Bradley D.W., Carl M. Acute sporadic hepatitis E in Sudanese children: Analysis based on a new western blot assay. J Infect Dis 1992; 65:1001-1005
9. Hyams K.C., McCarthy M.C., Kaur M., Purdy M.A., Bradley D.W., Mansur M.M., Gray S., Watts D.M., Carl M. Acute sporadic hepatitis E in children living in Cairo, Egypt. J Med Virol 1992; 37:274-277.
10. El-Zimaity D.M.T., Hyams K.C., Imam I.Z.E., Watts D.M., Bassily S., Naffea E.M., Sultan Y., Emara K., Burans J., Purdy M.A., Bradley D.W., Carl M. Acute sporadic hepatitis E in an Egyptian pediatric population. Am J Trop Med Hyg 1993; 48:372-376.
11. Ibarra H.V., Riedemann S.G., Siegel F.G., Reinhardt G.V., Toledo C.A., Frösner G. Hepatitis E virus in Chile. Lancet 1994; 344:1501.
12. Lok A.S.F, Kwan W., Moeckli R., Yarbough P.O., Chan R.T., Reyes G.R., Lai C., Chung H., Lai T.S.T. A seroepidemiological survey of hepatitis E in Hong Kong using recombinant based enzyme immunoassays. Lancet 1992; 340:1205-1208.
13. Zanetti A.R., Dawson G.J. and the Study Group of Hepatiits E. Hepatitis type E in Italy: A seroepidemiological survey. J Med Virol 1994; 42:318-320.
14. Wang C.H., Flehmig B., Moeckli R. Transmission of hepatitis E virus by transmission. Lancet 1993; 341:825-826.
15. Skidmore S.J., Yarbough P.O., Gabor K.A., Tam A.W., Reyes G.R., Flower A.J.E. Imported hepatitis E in the U.K. Lancet 1991; 337:1541.
16. DeCock K.M., Bradley D.W., Sandford N.L., Govindarajan S., Maynard J.E., Redeker A.G. Epidemic non-A, non-B hepatitis in patients from Pakistan. Ann Intern Med 1987; 106:227-230.
17. Shidrawi R.G., Skidmore S.J., Coleman J.C., Dayton R., Murray-Lyon I.M. Hepatitis E-an important cause of imported non-A, non-B hepatitis among migrant workers in Qatar. J Med Virol 1994; 43: 412-414.
18. Zaajer H.L., Kok M., Lelie P.N., Timmerman R.J., Chau K., van der Pal H.J.H. Hepatitis E in the Netherland: imported and endemic. Lancet 1993; 341:826.
19. Paul D.A., Knigge M.F., Ritter A., Gtierrez R., Pilot-Matias T., Chau K.H., Dawson G.J. Determination of hepatitis E virus seroprevalence by using recombinant fusion proteins and synthetic peptides. J Infect Dis 1994; 169:801-806.
20. Dawson G.J., Chau K.H., Cabal C.M., Yarbough P.O., Reyes G.R., Mushawar I.K. Solid-phase enzyme-linked immunosorbent assay for hepatitis E virus IgG and IgM antibodies utilizing recombinant antigen and synthetic peptides. J Virol Methods 1992; 38:175-186.
21. Mast E.E., Kuramoto I.K., Favorov M.O., Sehocning V.R., Burkholder B.T., ShapiroC.N., Holland P.V. Prevalence of risk factors for antibodies to hepatitis E virus seroreactivity among blood donors in northern California. J Infect Dis 1997; 176:34-40.
22. Balayan M.S., Usmanov R.K., Zamyatina N.A., Djumalieva D.I., Karas F.R. Experimental hepatitis E infection in domestic pigs. J Med Virol 1990; 32:58-59.
23. Clayson E.T., Innis B.L., Myint K.S.A, Narupiti S., Vaughn D.W., Giri S., Ranabhat P., Shrestha M.P. Detection of hepatitis E virus infection among swine in the Kathmandu Valley of Nepal. Am J Trop Med Hyg 1995; 53:228-232.
24. Meng X.-J., Purcell R.H., Halbur P.G., Lehman J.R., Webb D.M., Tsareva T.S., Haynes J.S., Thacker B.J., Emerson S.U. A novel virus in swine is closely related to the human hepatitis E virus. Proc Natl Acad Sci USA 1997; 94:9860-9865.

25. Maneerat Y., Clayson E.T., Myint K.S.A, Young G.D., Innis B.L. Experimental infection of the laboratory rat with the hepatitis E virus. J Med Virol 1996; 48:121-128.
26. Meng X.-J., Halbur P.G., Shapiro M.S., Govindarajan S., Bruna J.D., Mushahwar I.K., Purcell R.H., Emerson S.U. Genetic and experimental evidence for cross-species infection by swine hepatitis E virus. J Virol 1998; 72: 9714-9721.
27. Cao X.Y., Ma X.Z., Liu Y.Z., Liu Z.E., Jin X.M., Gao Q., Dong H.J., Zhuang H., Liu C.B., Wang G.M. Studies of the epidemiology and aetiological agent of enterically transmitted non-A, non-B hepatitis in the south part of Xinjiang. Chinese J Exp Clin Virol 1989; 3:1-10.
28. Arankalle V.A., Chadha M.S., Tsarev S.A., Emerson S.U., Risbud A.R., Banerjee K., Purcell R.H. Seroepidemiology of water-borne hepatitis in India and evidence for a third enterically-tranmitted hepatitis agent. Proc Natl Acad Sci USA 1994; 91:3428-3432.
29. Myint H., Soe M.M., Khin T., Myint T.M., Tin K.M. A clinical and epidemiological study of an epidemic of non-A, non-B hepatitis in Rangoon. Am J Trop Med Hyg 1985; 34:1183-1189.
30. Sreenivasan M.A., Banerjee K., Pandya P.G., Kotak R.R., Pandya P.M., Desai N.J., Vaghela L.H. Epidemiological investigations of an outbreak of infectious hepatitis in Ahmadabad city during 1975-76. Ind J Med Res 1978; 67:197-206.
31. Tandon B.N., Joshi Y.K., Jain S.K., Gandhi B.M., Mathiesen L.R., Tandon H.D. An epidemic of non-A, non-B hepatitis in north India. Ind J Med Res, 1982; 75:739-744.
32. Belabbes E.H., Bourguermouh A., Benatallah A., Illoul G. Epidemic non-A, non-B hepatitis in Algeria: strong evidence for its spreading by water. J Med Virol 1985; 16:257-263.
33. Sarthou J.L., Budkowska A., Sharma M.D., Lhuillier M., Pillot J. Characterization of an antigen-antibody system associated with epidemic non-A, non-B hepatitis in West Africa and experimental transmission of an infectious agent to primates. Ann Inst Pasteur Microbiol 1986; 137E:225-232.
34. Zhuang, H., Cao, X.Y., Liu, C.B., Wang, G.M. "Enterically transmitted non-A, non-B hepatitis in China." In *Viral hepatitis C, D, E*, T. Shikata, R.H. Purcell, T. Uchida, eds. Amsterdam: Elsevier Science Publishers, 1991.
35. Morrow R.H., Smetana H.F., Sai F.T., Edgcomb J.H. Unusual features of viral hepatitis in Accra, Ghana. Ann Intern Med 1968; 68:1250-1264.
36. Molinie C., Saliou P., Roue R., Denee J.M., Farret O., Vergeau B., Vindrios J. "Acute epidemic non-A, non-B hepatitis: a clinical study of 38 cases in Chad." In *Viral hepatitis and liver disease*, A.J. Zuckerman, ed. New York: Alnan R. Liss 1988.
37. Centers for Disease Control. Enterically transmitted non-A, non-B hepatitis: Mexico. MMWR 1987; 36:597-602.
38. Centers for Disease Control. Enterically transmitted non-A, non-B hepatitis: East Africa. MMWR 1987; 36:241-244.
39. Huang R.T., Li D.R., Wei J., Huang X.R., Yuan X.T., Tian X. Isolation and identification of hepatitis E virus in Xinjiang China. J Gen Virol 1992; 73:1143-48.
40. Arankalle V.A., Choibe L.P., Jha J., Chadha M.S., Banerjee K., Favorov M.O., Kalinina T., Fields H. Aetiology of acute sporadic non-A, non-B viral hepatitis in India. J Med Virol 1993; 40:121-125.
41. Naik S.R., Aggarwal R., Salunke P.N., Mehrotra N.N. A large waterbortne viral hepaittis E epidemic in Kanpur, India. Bulletin of the World Health Organization 1992; 70:597-604.
42. Mast E.E., Polish L.B., Favorov M.O., Khudyakova N.S., Collins C., Tuket P.M., Koptich D., Khudyakov Y.E., Fields H.A., Margolis H.S. and the Somali refugee

Medical Team. Hepattis E among refugees in Kenya: minimal apparent person-to-person transmission, evidence for age-dependent disease expression, and new serologic assays. Viral hepatitis and Liver Disease 1994; pp.375-378.

43. Khuroo M.S., Teki M.R., Skidmore S., Sofi M.A., Khuroo M. Incidence and severity of viral hepatitis in pregnancy. Am J Med 1981; 70:252-255.

44. Tsega E., Hansson B.G., Krawczynski K., Nordenfelt E. Acute sporadic viral hepatitis in Ethiopia: causes, risk factors, and effects on pregnancy. Clin Infect Dis 1992; 14:961-965.

45. Mishra L., Seeff L.B. Viral hepatitis A through E complicating pregnancy. Gastroenterology 1992; 21:873-887.

46. Arankalle V.A., Chadha M.S., Manerjee K., Srinivasa M.A., Chobe L.P. Hepatitis E virus infection in pregnant rhesus monkeys. Ind J Med Res 1993; 97:4-8.

47. Khuroo M.S., Kamili S., Jameel S. Vertical transmission of hepatitis E virus. Lancet 1995; 345:1025-1026.

48. Lau J.Y.N., Sallie R., Fang J.W.S., Yarbough P.O., Reyes G.R., Portmann B.C., Mieli-Vergani G., Williams R. Detection of hepatitis E virus genome and gene products in two patients with fulminant hepatitis E. J Hepatol 1995; 22:605-610.

49. Liang T.J., Jeffers L., Reddy R.K., Silva M.O., Cheinquer H., Findor A., De Medina M., Yarbough P.O., Reyes G.R., Schiff E.R. Fulminant or sub-fulminant non-A, non-B viral hepatitis: the role of hepatitis C and E viruses. Gastroenterology 1993; 104: 556-562.

50. Gupta D.N., Smetana H.F. The histopathology of viral hepatitis as seen in Delhi epidemic (1955-1956), Ind J Med Res 1957; 1(suppl):145-151.

51. Dienes H.P., Hütteroth T.H., Bianci L., Grü M.M., Theones W. Hepatitis A-like non-A, non-B hepatitis: light and electron microscopic examination of three cases. Virchows Archiv[A]. Pathol Anat Histol, 1986; 409:657-667.

52. Bradley D.W., Andjaparidze A., Cook E.H., McCaustland K.A., Balayan M., Stetler H., Velazquez O., et al. Aetiologic agent of enterically transmitted non-A, non-B hepatitis. J Gen Virol 1988; 69:731-738.

53. Reyes G.R., Huang C.-C., Yarbough P.O., Tam A.W. Hepatitis E Virus: Comparison of "New and Old World" isolates. J Hepatol 1991; 13(Suppl):S155-S161.

54. Uchida T., Suzuki K., Hayashi N, Iida F, Hara T, Oo SS, Wang C.-K., Shikata T., Ichikawa M., Rikihisa T., Mizuno K., Win K.M. Hepatitis E virus: cDNA cloning and expression, Microbiol Immunol 1992; 36:67-79.

55. Bradley D.W., Krawczynski K., Cook E.H., McCaustland K.A., Humphrey C.D., Spelbring J., Myint H., Maynard J. Enterically transmitted non-A, non-B hepatitis: serial passage of disease in cynomolgus macaques and tamarins and recovery of disease-associated 27-34 nm viruslike particles. Proc Natl Acad Sci USA 1987; 84:6277-6281.

56. Kane M.A., Bradley D.W., Shrestha S.M., Maynard J.E., Cook E.H., Mishra P.P., Joshi D.D. Epidemic non-A, non-B hepatitis in Nepal: recovery of possible etiologic agent and transmission studies in marmosets. JAMA 1984; 252:3140-3145.

57. Arankalle V.A. Ticehurst J., Sreenivasan M.A., Kapikian A.Z., Popper H., Pavri K.M., Purcell R.H. Aetiological association of a virus like particle with enterically transmitted non-A, non-B hepatitis. Lancet 1988; 1:550-554.

58. Bradley D.W., Balayan M.S. Virus of enterically transmitted non-A, non-B hepatitis. Lancet 1988; 1:819.

59. Ticehurst J. "Identification and characterization of hepatitis E virus." In *Viral Hepatitis and Liver Disease*, B.F. Hollinger, S.M. Lemon, H.S. Margolis, eds. Baltimore: Williams and Wilkins, 1991.

60. Tam A.W., Smith M.W., Guerra M.E., Huang C.-C., Bradley D.W., Fry K.E., Reyes G.R. Hepatitis E Virus (HEV): molecular cloning & sequencing of the full-length viral genome. Virology 1991; 185:120-131.
61. Koonin E.V., Gorbalenya A.E., Purdy M.A., Rozanov M.N., Reyes G.R., Bradley D.W. Computer-assisted assignment of functional domains in the non-structural polyprotein of hepatitis E virus: delineation of a new group of animal and plant positive-strand RNA viruses. Proc Nat Acad Sci USA 1992; 89: 8259-8264.
62. Purdy M.A., Tam A.W., Huang C.-C., Yarbough P.O., Reyes G.R. Hepatitis E virus: a nonenveloped member of the "alpha-like" RNA virus supergroup? Seminars in Virology 1993; 4:319-326.
63. Yarbough P.O., Tam A.W., Fry K.E., Krawczynski K., McCaustland K.A., Bradley D.W., Reyes G.R. Hepatitis E Virus: identification of type-common epitopes J Virol 1991; 65: 5790-5797.
64. Torresi J., Li F., Locarnini S.A., Anderson D.A. Only the non-glycosylated fraction of hepatitis e virus capsid (open reading frame 2) protein is stable in mammalian cells. J Gen Virol 1999; 80:1185-1188.
65. Jameel S., Zafrullah M., Osdener M.H., Panda S.K. Expression in animal cells and characterization of the hepaitits E virus structural proteins. J Virol 1996; 70:207-216.
66. Tsarev S.A., Tsareva T.S., Emerson S.U., Kapikian A.Z., Ticehurst J., London W., Purcell R.H. ELISA for antibody to hepatitis E virus (HEV) based on complete open reading frame 2 protein expressed in insect cells: identification of HEV infection in primates. J Infect Dis 1993; 168:369-378.
67. Jiang X., Wang M., Graham D.Y., Estes M.K. Expression, self-assembly, and antigenicity of the Norwalk virus capsid protein. J Virol 1992; 66:6527-6532.
68. Dingle K.E., Lambden P.R., Caul E.O., Clarke I.N. Human enteric Caliciviridae: the complete genome sequence and expression of virus-like particles from a genetic group II small round structured virus. J Gen Virol 1995; 76:2349-2355.
69. Tyagi S., Jameel S., Lal S.K. Self-association and mapping of the interaction domain of hepatitis E virus ORF3 protein. J Virol 2001; 75:2493-2498.
70. Huang C.-C., Nguyen D., Fernandez J., Yun-Choe K., Tam A.W., Fry K.E., Bradley D.W., Reyes G.R. Molecular cloning and sequencing of the hepatitis E virus Mexico strain. Virology 1992; 191: 550-558.
71. Tsarev S.A., Emerson S.U., Reyes G.R., Tsareva T.S., Legters J.P., Malik I.A., Iqbal M., Purcell R.H. Characterization of a prototype strain of hepatitis E virus. Proc Natl Acad Sci USA 1992; 89: 559-563.
72. Bi S.-L., Purdy M.A., McCaustland K.A., Margolis H.S., Bradley D.W. The sequence of hepatitis E virus isolated directly from a single source during an outbreak in China. Virus Res 1993; 28:233-247.
73. Gouvea V., Snellings N., Popek M.J., Longer C.F., Innis B.L. Hepatitis E virus: complete genome sequence and phylogenetic analysis of a Nepali isolate. Virus Res. 1998; 57:21-26.
74. Schlauder G.G., Dawson G.J., Erker J.C., Kwo P.Y., Knigge M.F., Smalley D.L., Rosenblatt J.E., Desai S.M., Mushahawar I.K. The sequence and phylogenetic analysis of a novel hepatitis E virus isolated from a patient with acute hepatitis reported in the United States. J Gen Virol 1998; 79:447-456.
75. Erker, J.C., Desai S.M., Schlauder G.G., Dawson G.J., Mushahwar I.K. A hepatitis E virus variant from the United States: molecular characterization and transmission in cynomolgus macaques. J Gen Virol 1999; 80:681-90.
76. Chatterjee R., Tsarev S., Pillot J., Coursaget P., Emerson S.U., Purcell R.H. African strains of hepatitis E virus that are distinct from Asian strains. J Med Virol 1997; 53:139-144.

77. Schlauder G.G., Desai S.M., Zanetti A.R., Tassopoulos N.C., Mushahwar I.K. Novel hepatitis E virus (HEV) isolates from Europe: evidence for additional genotypes of HEV. J Med Virol 1999; 57:243-251.

78. Schlauder G.G., Frider B., Sookolan, S., Castano G.C., Mushahwar, I.K. Identification of 2 novel isolates of hepatitis E virus in Argentina. J Infect Dis 2000; 182:294-297.

79. Wang Y., Ling R., Erker J.C., Zhang H., Li H., Desai S., Mushahwar I.K, Harrison T.J. A divergent genotype of hepatitis E virus in Chinese patients with acute hepatitis. J Gen Virol 1999; 80:169-177.

80. Worm H.C., Schlauder G.G., Wurzer H., Mushahwar I.K. Identification of a novel variant of hepatitis E virus in Austria: sequence, phylogenetic and serological analysis. J Gen Virol 2000; 81:2885-2890.

81. Wang Y., Zhang H., Ling R., Li H., Harrison T.J. The complete sequence of hepatitis E virus genotype 4 reveals an alternative strategy for translation of open reading frames 2 and 3. J Gen Virol 2000; 81:1675-1686.

82. Kabrane-Lazizi Y., Meng X.J., Purcell R.H., Emerson S.U. Evidence that the Genomic RNA of hepatitis E virus is capped. J Virol 1999; 73:8848-8850.

83. Reyes G.R., Huang C.-C., Tam A.W., Purdy M.A. Molecular organization and replication of hepatitis E virus (HEV). Arch Virol 1993; 7(suppl);15-25.

84. Agrawal S., Gupta D., Panda S.K. The 3' end of hepatitis E virus (HEV) genome binds specifically to the viral RNA-dependent RNA polymerase (RdRp). Virology 2001; 282:87-101.

85. Dalgarno L., Rice C.M., Strauss J.H. Ross river virus 26S RNA: Complete nucleotide sequence and deduced sequence of the encoded structural proteins. Virology 1983; 129: 170-187.

86. Krawczynski K., Bradley D.W. Enterically transmitted non-A, non-B hepatitis: identification of virus associated antigen in experimentally infected cynomolgus macaques. J Infect Dis 1989; 159:1042-1047.

87. Purdy M.A., McCaustland K.A., Krawczynski K., Spelbring J., Reyes G.R., Bradley D.W. Preliminary evidence that a trpE-HEV fusion protein protects cynomolgus macaques against challenge with wild-type HEV. J Med Virol 1993; 41:90-94.

88. Jiang X., Wang M., Wang K., Estes M.K. Sequence and genomic organization of Norwalk virus. Virology 1993; 195:51-61.

89. Lambden P.R., Caul E.O., Ashley C.R., Clarke I.N. Sequence and genome organization of a human small round-structured (Norwalk-like) virus. Science 1993; 259:516-519.

90. Berke T., Matson D.O. Reclassification of the *caliciviridae* into distinct genera and exclusion of hepatitis E virus from the family on the basis of comparative phylogenetic analysis. Arch Virol 2000; 145:1421-1436.

91. Green K.Y., Ando T., Balayan M.S., Berke T., Clarke I.N., Estes M.K., Matson D.O., Nakata S., Neill J.D., Studdert M.J. Thiel H.-J. Taxonomy of the caliciviruses. J Infect Dis 2000; 181(Suppl 2):S322-330.

92. Chau K.H., Dawson G.J., Bile K.M., Magnius L.O., Sjogren M.H., Mushahwar I.K. Detection of IgA class antibody to hepatitis E Virus in serum samples from patients with hepatitis E virus infection. J Med Virol 1993; 40:334-338.

93. Kaur M., Hyams K.C., Purdy M.A., Krawczynski K., Ching W.M., Fry K.E., Reyes G.R., Bradley D.W., Carl M. Human linear B-cell epitopes encoded by the hepatitis E virus include determinants in the RNA-dependent RNA polymerase. Proc Natl Acad Sci USA 1992; 89:3855-3858.

94. Khudyakov Y.E., Khudyakova N.S., Fields H.A., Jue D., Starling C., Favorov M.O., Krawczynski K., Polish L., Mast E., Margolis H. Epitope mapping in proteins of hepatitis E virus. Virology 1993; 194:89-96.

95. Khudyakov Y.E., Favorov M.O., Jue D.L., Hine T.K., Fields H.A. Immunodominant antigenic regions in a structural protein of the hepatitis E virus,.Virology 1994; 198: 390-393.

96. Favorov M.O., Khudyakov Y.E., Fields H.A., Khudyakova N.S., Padhye N., Alter M.J., Mast L., Polish L., Yashina T.L., Yarasheva D.M., Onischenko G.G., Margolis H.S. Enzyme immunoassay for the detection of antibody to hepatitis E virus based on synthetic peptides, J Virol.Methods 1994; 46:237-250.

97. Bryan J.P., Tsarev S.A., Iqbal M., Ticehurst J., Emerson S., Ahmed A., Duncan J., Rafiqui A.R., Malik I.A., Purcell R.H., Legters L.J. Epidemic hepatitis E in Pakistan: patterns of serologic response and evidence that antibody to hepatitis E protects against disease. J Infect Dis 1994; 170:517-521.

98. Tsarev S.A., Tsareva T.S., Emerson S.U., Govindarajan S., Shapiro M., Gerin G.L., Purcell R.H. Successful passive and active immunization of cynomolgus monkeys against hepatitis. Proc Natl Acad Sci USA 1994; 91:10198-10202.

Chapter 7

THE MOLECULAR BIOLOGY OF GB VIRUSES

Thomas P. Leary and Isa K. Mushahwar

Abbott Laboratories, North Chicago, IL 60064, USA

1. INTRODUCTION

Following the development of specific and sensitive assays for the detection of the five recognized human hepatotropic viruses, it became apparent that additional human hepatitis viruses must exist as 5-20% of community-acquired and parenterally transmitted hepatitis cases could not be attributed to the known viruses. Further evidence for the existence of such viruses included varying incubation periods prior to disease onset (1), multiple disease episodes (2), chronic or fulminant hepatitis of cryptogentic origin (3, 4), and the visualization of virus-like particles (5-7). Finally, serial passage of the filterable GB agent was described in non-human primates (8, 9). This agent, originally obtained from a surgeon experiencing acute hepatitis, was inoculated into tamarins that later developed acute biochemical hepatitis. Although the disease could be further passaged in these animals, isolation of the putative agent remained elusive.

With the advent of modern molecular technologies, the GB agent was finally characterized in 1995, more than three decades after the initial description (10). Upon isolation, the agent was not a single virus as expected, but in fact two related viruses, termed GB virus A (GBV-A) and GB virus B (GBV-B). Additional studies searching for the presence of these two viruses in human sera resulted in the isolation of a third related virus, GB virus C (GBV-C) (11, 12), later described as Hepatitis G virus (13). Detailed studies to extend the three viruses to genome length revealed that each were related to one another, and to Hepatitis C virus (12, 14). Thus far, GBV-A and GBV-B have been detected only in non-human primates. While it has been possible to passage each of the viruses independently in these animals, the observed hepatitis can only be attributed to GBV-B (15).

Through a mechanism that is not understood, GBV-A appears to persist for extended periods of time in the absence of disease (15, 16), as do a number of species-specific variants of GBV-A (16, 17). On the contrary, GBV-C has only been naturally detected in humans, however, passage in non-human primates has been reported (18), and a variant of the virus has been isolated from chimpanzees (19, 20). While GBV-C is found at very low levels in normal blood donors, greatly increased rates of infection are found in those at risk for exposure to parentally transmitted viruses (21).

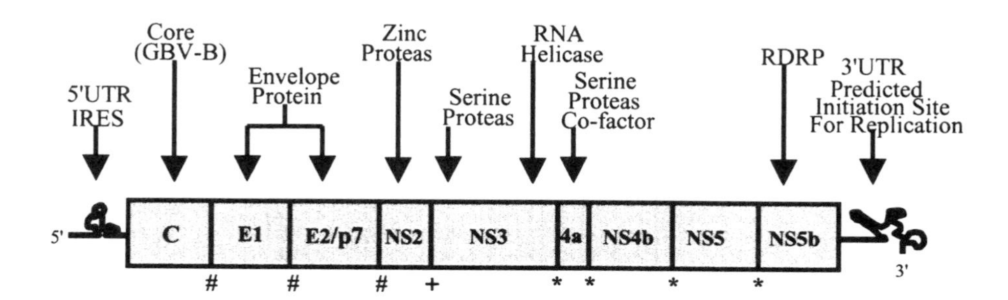

Figure 1. Organization of the GB virus genomes. Coding regions are shaded while non-coding regions are designated by thick lines. # host protease cleavage sites; + zinc protease cleavage site; * serine protease cleavage site.

2. GENOMIC STRUCTURE AND ORGANIZATION OF THE GB VIRUSES

Studies into the nature of the GB viruses have been hampered by the inability to obtain large quantities of virus by way of cell culture or purification from infectious sources, though recent efforts in this regard should facilitate this effort (22, 23). The virions are 50-100 nm, enveloped particles that in all likelihood contain a single copy of the RNA genome as do other closely related viruses. The particles are associated with a significant amount of lipoprotein and also contain varying levels of carbohydrate moieties (24). Each of the three genomes are greater than 9000 nucleotides in length (Figure 1 and Table 1), are single-stranded and of positive polarity, and contain a long open reading frame (ORF). In addition, non-coding sequences are found both upstream and downstream of the ORFs. The ORFs of the GB viruses encode large polyprotein precursors that are post-translationally cleaved into the individual proteins responsible for virus replication and packaging (Figure 1). Based on the above criteria, as well as similarities with other known viruses, the GB viruses have been classified as three distinct genera within the *Flaviviridae* (12, 14). Other genera also classified within this family include the Hepatitis C virus (HCV), the pestiviruses (Hog Cholera, Bovine Viral Diarrhea, Border Disease, etc.) and the flaviviruses (Yellow Fever, Dengue, West Nile, etc.).

Table 1. Properties of the GB Virus Genome

	GBV-A	GBV-B	GBV-C	HCV-1
GenBank Accession	U22303	U22304	U36380	M62321
Genome (nt)*	9653	9143	9377	9401
Polyprotein (aa)°	2954	2864	2843	3011
5' NTR (nt)	593	445	533	341
3' NTR (nt)	195	103	312	72

*nucleotides, °amino acids

2.1 Non-translated Virus Sequences

Each of the three GB viruses possess long, non-translated sequences that flank the single large ORF (Table 1). These sequences are thought to be of importance in genome replication and packaging, as well as in the regulation of gene expression. Indicative of their importance, they are highly conserved between distinct isolates of the virus (25), and to some degree between the three GB viruses and other members of the *Flaviviridae* (14). While it is possible that several very small proteins are encoded by ORFs present within the 5' NTR, no such proteins have been characterized at this time. Of great importance is the presence of multiple, inverted repeat sequences that are present within the 5' NTR. These sequences are indicative of a highly ordered structure within this region and consistent with the presence of an internal ribosome entry site (IRES) that has been attributed to these sequences (26, 27). In this manner, translation of the large ORF occurs in a cap-independent fashion in which the IRES and other cellular proteins direct the 40S ribosomal subunit into position at the initiating methionine codon positioned somewhat distant from the 5'-end of the genome. This is as opposed to the cap-dependent translation that is utilized by eucaryotic cell proteins and many other viruses. In this case, the ribosome binds at a 7-methylguanylate residue located at the 5'-end of the mRNA, then proceeds to scan the genome in a 5' to 3' direction until an initiating methionine codon is located. In addition to the IRES activity that has been ascribed to the 5'NTR of HCV, the substantial secondary structure of the region has also been demonstrated to be a translational inhibitor (28). While a similar characteristic has not been formally demonstrated for the GB viruses, it can likely be inferred based on the similar folding pattern that occurs between these viruses and HCV (26, 27).

Within GBV-A, GBV-B and GBV-C, the 5' non-translated region (NTR) sequences are 594, 445 and 524 nucleotides in length, respectively. Similar sequences have been described in other members of the *Flaviviridae*, including HCV and the pestiviruses (341 to 385 nucleotides). In contrast, these sequences are absent in the flaviviruses (approximately 100 nucleotides). Though the 5' NTR of GBV-B is significantly larger than that of HCV, nucleic acid sequence alignments of GBV-B, HCV and the pestiviruses reveal significant levels of identity clustered in distinct regions of the 5' NTR, with the greatest similarity occurring between GBV-B and HCV (14).

Like other members of the *Flaviviridae*, the GB viruses also contain long sequences downstream of the large ORF in the 3' NTR (Table 1). These sequences are thought to be important in viral packaging and replication. Like HCV, it has been speculated that substantial structure exists in these regions much like that found in the 5' NTR (29, 30). Because these viruses do not utilize a DNA intermediate during genome replication, by default, an antigenomic intermediate is necessary for the RNA-dependent RNA-polymerase to produce genomic length RNA molecules for polyprotein synthesis and the production of progeny virus. The structure that has been described within 3' NTR is thought to confer a docking site for the viral replicase encoded by the NS5B gene. Indicative of the importance of these sequences to virus replication, the 3' sequences have been shown to be indispensable in the construction of an infectious molecular clone of GBV-B (30). An additional point of interest is that GBV-B contains a poly-U tract just downstream of the translational stop codon, identical to that found in some isolates of HCV (14). This is as opposed to host mRNAs that contain a similar poly-A tract. While the significance of this finding is unclear, it is possible that some unknown advantage is conferred upon the genome in terms of stability, replication or enhanced protein synthesis.

2.2 The GB Virus Structural Proteins

Each of the three GB viruses encode a single large polyprotein that is post-translationally cleaved into the respective viral structural and non-structural proteins. Typical of the *Flaviviridae*, the GB structural proteins are located at the N-terminus of the polyprotein and appear to be processed into the individual virion components within the lumen of the ER by host-encoded proteases (12). Hydropathy plots of the GB viruses within the structural region, as compared to other members of the *Flaviviridae*, demonstrate a significant degree of structural similarity between the various viral agents. It is also apparent from this analysis that GBV-A and GBV-C are distinct viruses(14). Despite these differences, four potential eucaryotic signal sequence cleavage sites are predicted to occur within the structural region of the GB viruses, resulting in four distinct proteins or peptides encoded within the structural region of the genome by these viruses (Table 2).

Table 2. Properties of the GB Virus Structural Genes

	GBV-A	GBV-B	GBV-C	HCV-1
Core (aa)	None	156	None	191
PI of core	-	11.1	-	11.9
E1				
Cys Residues	9	11	10	8
Glycosylation Sites*	1	3	1	4
E2				
Cys Residues	23	16	8	18
Glycosylation Sites*	2	6	3	11

*Predicted N-linked Glycosylation Consensus Sites

2.2.1 Core

A core protein that associates with the genomic RNA prior to encapsidation is traditionally the first protein encoded by members of the *Flaviviridae*. In HCV, this protein is highly basic (pI>11) and functions as a nucleocapsid-like protein in the virion. The mature protein is associated with the cytosolic side of the ER and has been shown to interact with the E1 protein. Similar to HCV, the GBV-B core protein is 156 amino acids in length with a calculated pI of 11.1 (14). Despite the fact that it is significantly smaller than the core proteins of HCV, BVBV and YFV, GBV-B core appears to serve an analogous role. It is possible that this protein undergoes further processing into several smaller fragments as is the case with the previously mentioned viruses, however, further proteolytic processing has not yet been demonstrated.

It is within the expected core region of GBV-A and GBV-C that distinguishes these viruses from GBV-B and HCV (Table 2), and all other members of the *Flaviviridae* (12, 14, 31-33). In both viruses, no apparent core protein is encoded in this region, or in any other region of the genome.

While a potential eucaryotic signal cleavage site is present just upstream of the putative E1 protein in both GBV-A and GBV-C, only a very short peptide is synthesized in this region. Apparently, this short peptide is translated to allow translocation of the E1 protein into the endoplasmic reticulum of the host cell, and is then cleaved from the mature E1 protein by the host cell machinery. In this manner, E1 is allowed access to the glycosylation and secretory pathways of the cell, a necessity for membrane-exposed proteins. Extensive studies have been performed to elucidate core coding regions within both GBV-A and GBV-C genomes, and to this point, none have convincingly established the existence of a core protein encoded by either virus. One hypothesis is clear, core-like proteins are not encoded upstream of the E1 protein as is demonstrated by studies establishing the presence of IRES elements in the 5' NTR of these viruses (26). While it is likely that a core-like protein is present in the virions of both GBV-A and GBV-C, the source of that protein is currently unclear. It is possible that the protein is encoded elsewhere on the viral genome, though it is more likely that a host cell protein is performing this function. Such studies are currently pending.

2.2.2 E1

Each of the GB viruses encodes at least two distinct envelope proteins (E1 and E2), and strong evidence suggests further post-translational cleavage of E2. Such a structural organization would be directly comparable to HCV. Both proteins contain sugar moieties that substantially increase their molecular mass, and a large number of cysteine residues present suggest that they possess extensive tertiary structure (Table 2). The E1 protein of GBV-C is highly conserved between different isolates of the virus (31), and a conserved Asn-Cys-Cys motif is absolutely conserved between each of the GB viruses at the E1 N-terminus (14, 31). The high level of conservation between isolates may suggest a level of functional significance, or perhaps the lack of significant host immune selective pressures. It may be that the E1 proteins are not directly exposed to the external environment as may be suggested from the net hydrophobicity of the E1 protein (14), but instead buried on the virion surface. Regardless of the reason, the conservation observed in E1 is in direct contrast to that of HCV where the level of diversity is as great as that found elsewhere in the polyprotein (31).

2.2.3 E2

The GB virus E2 proteins are similar to E1 in terms of proposed structure. Multiple cysteine residues are conserved among the three viruses as well as HCV, and the size of each appears to be dramatically enhanced by the addition of sugar moieties (Table 2). Variability within the E2 protein is significantly greater than at any other region of the polyprotein, nevertheless, the observed variability in this region is not nearly as extensive as that found within the E2 region of various HCV isolates (31). This could reflect a lack of host immune selective pressures or a lack of tolerance for substitutions in this protein. In HCV, it is well established that the host immune response is a major influence behind variability within a small region at the N-terminus of E2 (34, 35). In a patient lacking immunoglobulin, the E2 protein sequence remained unchanged for several years in the absence of an anti-E2 immune response (36), and an infected chimpanzee maintained a highly conserved E2 for years in the absence of a detectable immune response (37). Of significance, seven years after the chimpanzee infection was initiated, the animal developed an anti-E2 antibody response that resulted in variability occurring within the E2 protein. This observation is noteworthy because patients infected with GBV-C can remain viremic for extended periods of time in the absence of a detectable anti-E2 response, though the virus is eliminated upon development of such a response (38, 39). This would support the hypothesis that GBV-C has the inability to escape immune clearance and persist by mutation within the envelope region as has been speculated for HCV. Further studies are necessary to detail the exact nature of GBV-C clearance as well as HCV persistence, perhaps leading to new anti-viral therapies for HCV.

Current evidence suggests that the E2 proteins of the GB viruses are further reduced into two distinct proteins, E2 and p7. This is directly comparable to HCV E2 where p7 appears to be associated with the virion. In this case, the E2/p7 cleavage is rather inefficient, resulting in the generation of two distinct species, E2 and E2/p7. Although the significance of p7 and E2/p7 is not understood at this time, they may play a role in virion assembly. Of note, the E2 proteins of GBV-A, and several variants of GBV-A possess much larger E2 proteins than GBV-B, GBV-C or HCV. Interestingly, amino acid sequence comparisons demonstrate that the additional length is isolated to the carboxyl-terminus of the GBV-A E2 proteins, downstream of the last potential signal sequence cleavage site. Therefore, the additional sequence in E2 is contained within p7. This finding may be of value in deciphering the exact role of p7 in virus replication and/or packaging for these and other member of the *Flaviviridae*.

2.3 The GB Virus Nonstructural Proteins

The nonstructural proteins of the GB viruses are encoded on the 3' two-thirds of the genome in a manner similar of other members of the *Flaviviridae*. These proteins are liberated from the carboxyl-terminus of the polyprotein into the individual viral components by way of viral proteases encoded within the nonstructural genes. Additionally, a helicase, a replicase and a protease co-factor are encoded in this region of the polyprotein. These proteins are involved in genome replication, protein synthesis and the production of proteins for the nascent virions.

2.3.1 NS2

The first of the proteins encoded within the nonstructural genes is a zinc-dependent metalloproteinase within NS2. This protease is responsible for the liberation of NS2 from the remainder of the polyprotein. Although little identity exists between the GB viruses and HCV within the NS2 region, the histidine and cysteine residues shown to be essential to the function of the HCV protease (40) are spacially conserved in the GB viruses (12, 14, 32). Additionally, a point mutation introduced at either position in the GBV-C protease abrogates cleavage of NS2 from NS3 in a recombinant system (41). This would seem to suggest that despite the observed differences in the amino acid sequences of the GB viruses, each process the NS2/NS3 cleavage in an identical fashion. This suggestion is affirmed by comparison of the NS2/NS3 cleavage sites in HCV and GBV-C where absolute conservation is maintained in the P5, P1' and P2' positions. Further, when the proposed cleavage sites for GBV-A and GBV-B are considered in this analysis, these same positions are conserved, as is the P1 position in each of the GB viruses (41). Therefore, conservation in this region of the GB viruses extends to the point that the amino acids flanking the scissile bond are maintained.

2.3.2 NS3

The remainder of the polyprotein is processed into the individual viral proteins by way of a serine protease found at the N-terminus of NS3. As was the case with the zinc protease in NS2, very little amino acid sequence conservation is found between the GB viruses or HCV in this region of the polyprotein. Despite these differences, the amino acid residues that have been demonstrated to constitute the active site of the HCV serine protease are conserved and appropriately spaced in each of the GB viruses (12, 14, 32). Like other serine proteases, the functional site of the HCV protease is composed of a triad involving a histidine residue located 24 amino acids upstream of an aspartic acid residue, followed by a serine residue 58 amino acids downstream of the aspartic acid. As was the case for the HCV serine protease (42), alteration of the serine residue in the catalytic site abolishes processing of NS3 cleavage sites in GBV-C (41). Additionally, the GBV-B serine protease expressed in *E. coli* had the ability to process regions of the HCV polyprotein normally processed by the HCV serine protease (43). In total, these studies would seem to suggest that the GB viruses process the downstream polyprotein cleavages by way of a serine protease located within the NS3 gene, in much the same fashion as HCV.

Studies to understand the precise cleavage sites of the GB virus NS3 protease also underscore the similarity to HCV. In HCV, the NS3 protease cleaves the HCV polyprotein at specific sites in which a threonine or cysteine residue is located in the P1 position of the scissile bond, and a serine or alanine residue is located in the P1' position. Early studies seemed to suggest a requirement for an acidic reside in the P6 position, though it has since been demonstrated that this residue is not an absolute requirement for the protease to be functional (44, 45). By way of amino acid sequence comparisons with HCV, protease cleavage sites have been identified in the GB viruses (12, 14, 32). Despite the fact that each of these sites has yet to be empirically determined, they are unique to the individual viral proteases and differ from those found in HCV and amongst one another. From the few studies that have been performed, it is clear that the GBV-B NS3 cleavage sites are very similar to HCV (43), and that they differ from the NS4B/NS5A site determined for GBV-C (41).

In the three GB viruses, the carboxy-terminus of the NS3 protein possesses an RNA helicase that is located in this same position of the polyprotein for all members of the *Flaviviridae*. These helicases are readily identified as members of the supergroup II RNA helicases based on amino acid residues conserved in these regions of the polyprotein. As is also the case in HCV (46), amino acid sequence motifs found in nucleic acid helicases (DECHXXD) and NTPases (GXGKS) are well conserved in each of the GB viruses. Based on the identity shared between the GB viruses and other well characterized RNA helicases, it is clear that the carboxyl-terminus of these NS3 proteins possesses helicase activity. In the cases of GBV-B and GBV-C, the helicase regions of the NS3 protein have been expressed in *E. coli* based systems. (47, 48). These enzymes appear to be similar to one another and to HCV in that the RNA helicase proceeds in a 3' to 5' direction and is dependent on ATP and divalent metal ions. Both enzymes possessed NTPase activity, and in each case the active site was mapped to the amino acid residues predicted based on other known enzymes.

2.3.3 NS4

The NS4 protein of HCV virus is further subdivided into NS4A and NS4B. While the function of the NS4B domain is unknown at this time, the NS4A subunit serves as a co-factor for the enzymatic activity of the NS3 protease. The HCV NS4A subunit is required for processing the NS3/NS4A and NS4B/NS5A cleavage sites, and it greatly facilitates cleavage of the NS5A/NS5B junction. Although the GBV-B NS3 serine protease has the ability to process the HCV polyprotein in an NS4A independent fashion (43), cleavage of the GBV-B polyprotein by this same enzyme requires the presence of the NS4A co-factor (49, 50). Of significance, the regions of co-factor activity within NS4A overlap between HCV and GBV-B, and two spacially conserved amino acids in each have been identified as critical to the functioning of this co-factor (49). This analysis underscores the idea that despite the lack of amino acid sequence identity that exists between HCV and the GB viruses within this region of the polyprotein, as indicated by the hydropathy plots, these proteins seem to have a similar structure, indicative of a functional similarity (14).

2.3.4 NS5

The final proteins encoded at the carboxyl-terminus of the polyprotein are found within the NS5 region. As was the case with the NS4 region, the NS5 region is also processed into two distinct proteins, NS5A and NS5B. In HCV, the function of the protein encoded within NS5A is not known at this time, though the protein clearly undergoes post-translational modification that significantly alters the molecular mass. Further, this protein seems to be responsible for the ineffectiveness of interferon therapy observed in specific genotypes of HCV. At this time, no studies have been performed on the GB viruses to decipher the functional role of these proteins. Hydropathy plots would seem to indicate that functional analogies exist between these proteins and that encoded by HCV.

The NS5B protein of HCV has been clearly identified as an RNA-dependent RNA polymerase responsible for the replication of the HCV genome. Characteristic of replicases, this protein contains the GDD protein motif thought to be at the active site of the enzyme. In addition to the structural similarities suggested by hydropathy plots, clear amino acid sequence identities exist between the NS5B proteins encoded by the GB viruses and those of other members of the *Flaviviridae*. Sequences motifs conserved by members of the supergroup II replicases are found in each of the GB viruses, as is the GDD signature sequence. At this time, the only replicase of the GB viruses that has been extensively studied is the one encoded by GBV-B. Expression studies in *E. coli* have demonstrated that the GBV-B NS5B protein does possess RNA-dependent RNA polymerase activity that requires Mn^{2+} as a co-factor (51). In these studies, mutational analyses confirmed that specific amino acid sequence motifs of know importance to the replicases of other similar viruses were also essential to the functioning of the GBV-B replicase. While detailed studies on the replicases of GBV-A and GBV-C are pending, based on the extensive sequence identity that exists with other members of the *Flaviviridae*, it is expected that they will function similarly.

3. PHYLOGENETIC ANALYSIS

Comparisons of amino acid sequence identity within the GB virus polyproteins demonstrate the relationship between these viruses, as well as how they differ from one another and HCV (Table 3). Despite the fact that each shares only marginal identity with HCV (26-32%), GBV-A and GBV-C are the most closely related sharing 48% identity. The identity between GBV-B and the other GB viruses is significantly less (27-28%), and GBV-B is actually more closely related to HCV (32%) than to the other GB viruses. Comparisons of conserved amino acid substitutions yields a much higher level of similarity between the GB viruses, as well as with HCV. Despite the relatively low level of overall identity between these viruses and HCV, significant regional identity occurs between these viruses and other members of the *flaviviridae*. Between viruses as disparate as HCV and GBV-C, identity achieving 50% is found within the NS3 helicase and the NS5B replicase genes (12), and additional levels of significant identity occur within E2, the NS3 protease, NS4A and NS5A. Similar comparisons between GBV-A and GBV-C, two viruses that are much more closely related, demonstrate much greater localized identity. Finally, similar levels of regional identity can be found between GBV-A, GBV-B and HCV (14).

Table 3. Sequence Identity Amongst GB Virus Polyproteins*

	GBV-A	GBV-B	GBV-C
GBV-A	100%	27%	48%
GBV-B	27%	100%	28%
HCV-1	26%	32%	29%

*Sequence alignments were performed using the Wisconsin Package from the Genetics Computer Group.

Although localized identity can provide insights into the functional role of the individual viral proteins, phylogenetic analysis of these regions is necessary to determine the exact evolutionary relationship of these proteins to one another, as well as the relatedness of the viruses as a whole. Phylogenetic studies using the helicase and replicase proteins of the GB viruses have unequivocally demonstrated the relationship of these viruses to other members of the *Flaviviridae* (14). Analyses of the helicase and replicase regions establish that these proteins in the GB viruses are very closely related to HCV, branching on the same limb of the respective trees. The same proteins of the pesti- and flaviviruses, while clearly related evolutionarily, branch distantly on additional limbs, as do similar proteins found in several different plant viruses.

Phylogenetic analysis encompassing the entire polyprotein has been used to classify the GB viruses into the *Flaviviridae*. As demonstrated in Figure 2, each of the three GB viruses branch onto a separate limb of the evolutionary tree. GBV-B is most closely related to HCV, forming a single branch of which it is the only isolate. The diversity of this virus is not currently understood as only a single isolate has been described at this time. Though GBV-A and GBV-C localize to the same major branch of the tree, it is further subdivided on two smaller branches. Based on the breath of the analysis presented in this figure, the individual isolates of GBV-C, even the most distant, cannot be distinguished from one another. This is as compared to HCV in which the individual genotypes of the virus are clearly apparent. The inability to delineate these distinct isolates speaks to the overall conservation that has been reported for GBV-C (31). The virus most closely related to GBV-C is a variant (GBV-C_{tro}) isolated from chimpanzees (19). Finally, GBV-A occurs on the last branch of the tree, as do several distinct isolates of the virus (32, 33). Interestingly, as compared to the GBV-C variants, these isolates also are apparent on the present tree, further demonstrating the evolutionary conservation of GBV-C.

Phylogenetic analysis has also been performed within the non-coding region of GBV-C, upstream from the large polyprotein. Such analyses have been useful to delineate specific genotypes of the virus that seem to correlate with the geographic region in which the isolate was recovered (52, 53). Of interest is the fact that genotypes are readily deciphered from the non-coding coding region of GBV-C, though these same genotypes cannot be identified upon examination of coding region sequences. This is directly opposed to HCV, in which only coding region sequences have been successfully used to determine virus genotype. The significance of this observation is unclear; it may provide some type of selective advantage to GBV-C, or it may simply suggest that nucleotide substitutions in the coding regions are not well tolerated. To the contrary, genotypic differentiation of GBV-A isolates can be performed utilizing sequences obtained from either the non-coding region or the polyprotein (16, 33). Analyses from non-coding region sequences can further be used to determine the species origin of a specific isolate (16). The functional significance of this observation, if any, is unclear at this time, However, it more likely this reflects the independent selection of GBV-A in distinct hosts over an extended period of time.

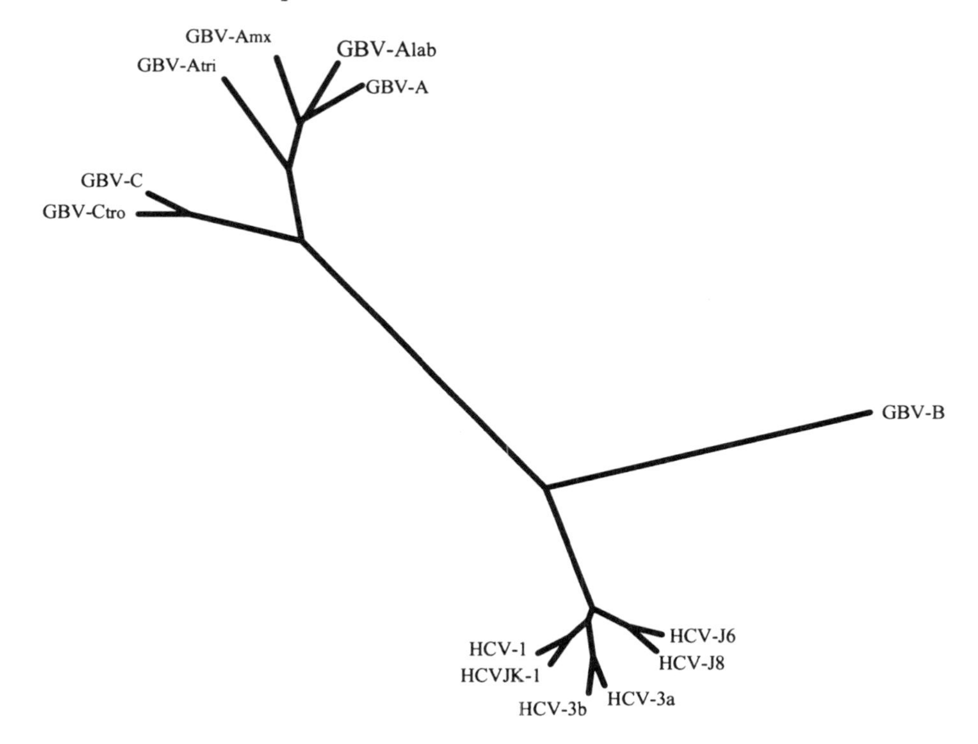

Figure 2. Phylogenetic analysis of the *Flaviviridae* polyproteins. Phylogenetic distances were determined using the PRODIST program of the PHYLIP package, and the distances obtained were used to construct the tree using the program FITCH. TREEVIEW produced the final output.

4. DETECTION OF THE GB VIRUSES

Detection systems for the GB viruses have been hampered by the relative poor antigenecity of the nonstructural genes for each of the three viruses (54). These findings have resulted in a proliferation of probe-based assays for the detection of viral nucleic acids. At this time, most assays have been developed for the detection of GBV-C, emphasizing the human aspect of the virus. The first probe-based assay to be described was designed within conserved regions of the NS3 helicase (55). Degenerate primers were utilized in this assay system in an attempt to alleviate potential detection problems with regard to genetic variation in the virus population. Additionally, cycling conditions were adjusted as to further address this problem. Though this assay system is quite reliable, it is also cumbersome as nucleic acid blotting is required due to background resulting from the permissive cycling conditions. Though positive samples usually yield definitive results, negative samples require blotting due to background amplification.

More recent probe-based assays for the detection of GBV-C are designed to detect sequences contained within the 5' NTR (56). As is found in HCV, sequences within the 5' NTR of GBV-C are much more conserved than in other regions of the genome. Comparisons between isolates of GBV-C reveal several regions that are absolutely conserved between all virus isolates known. Such sequence conservation provides excellent sites for oligonucleotide primer design, allowing target amplification to be performed under more stringent conditions. Additionally, amplified products can be readily sequenced to provide the genotype identity with the same product. The greatest problem associated with the probe-based assay systems is the difficulty associated with reliably testing large sample numbers. This has resulted in the development of an automated system that as detects sequences within the GBV-C 5'NTR (57). Because this system utilizes an enzymatic-based detection system, confirmation of the target identity is integral to the assay. Further, due to the detection system, the assay appears to be more sensitive than assays utilizing gel electrophoresis.

Finally, the remaining system for the detection of GBV-C is an ELISA assay that detects antibodies directed against a recombinant E2 antigen (38, 39). This antigen is truncated as compared to the native protein and is produced in eucaryotic cells transfected with a plasmid (58). Utilizing this assay, seroprevalence rates range from 3-8% in volunteer blood donors, with much higher rates in individuals exposed to parenterally transmitted viruses. It is important to note that this assay can not be used independently of the above mentioned probe-based assays as the production of anti-E2 antibodies is associated with the clearance of GBV-C. Therefore, both probe- and antibody-based assays are required to develop the complete picture of GBV-C prevalence. The probe-based assays are utilized for the identification of the currently infected, while antibody-based assays identify those who have been previously exposed to the virus but have since cleared the infection immunologically.

REFERENCES

1. Hollinger F.B., Mosley J.W., Szmuness W., Aach R.D., Peters, R.L., Stevens C. Transfusion-transmitted viruses study: Experimental evidence for two non-A, non-B hepatitis agents. J Infect Dis 1980; 142:400-407.
2. Mosley J.W., Redeker A.G., Feinstone S.M., Purcell R.H. Multiple hepatitis viruses in multiple attacks of acute viral hepatitis. New England J Med 1977; 296:75-80.
3. Alter M.J., Margolis H.S., Krawczynski K., Judson F.N., Mares A., Alexander W.J., Hu P.Y., Miller J.K., Gerber M.A., Sampliner R.E., Meeks E.L., Beach M.J. The natural history of community-acquired hepatitis C in the United States. New England J Med 1992; 327:1899-1905.
4. Wu J.C., Chen C.L., Hou M.C., Chen T.Z., Lee S.D., Lo K.L. Multiple viral infection as the most common cause of fulminant and subfulminant viral hepatitis in an area endemic for hepatitis B: application and limitations of the polymerase chain reaction. Hepatology 1994; 19:833-840.
5. Fagan E.A., Ellis D.S., Tovey G.M. Toga-like virus as a cause of fulminant hepatitis attributed to sporatic non-A, non-B hepatitis. J Med Virol 1989; 28:150-155.
6. Fagan E.A., Ellis D.S., Tovey G.M., Lloyd G., Smith H.M., Portmann B., Tan K.-C., Zuckerman A.J., Williams, R. Toga virus-like particles in acute liver failure attributed to sporatic non-A, non-B hepatitis and recurrence after liver transplantation. J Med Virol 1992; 38:71-77.
7. Phillips M.J., Blendis L.M., Poucell S., Patterson J., Petric M., Roberts E., Levy G.A., Superina R.A., Greig P.D., Cameron R., Langer B., Purcell R.H. Syncytial giant-cell hepatitis: sporadic hepatitis with distinctive pathological features, a severe clinical course, and paramyxoviral features. New England J Med 1991; 324:455-460.
8. Deinhardt F., Holmes A.W., Capps R.B., Popper H. Studies on the transmission of disease of human viral hepatitis to marmoset monkeys. I. Transmission of disease, serial passage and description of liver lesions. J Exp Med 1967; 125:673-687.
9. Deinhardt F., Holmes A.W. "Transmission of Human Viral Hepatitis to Non-Human Primates." In *Prospectives in Virology,* New York, New York: Academic Press Inc., 1968.
10. Simons J.N., Pilot-Matias T.J., Leary T.P., Dawson G.J., Desai S.M., Schlauder G.G., Muerhoff A.S., Erker J.C., Buijk S.L., Chalmers M.L., van Sant C.L., Mushahwar I.K. Identification of two flavivirus-like genomes in the GB hepatitis agent. Proc Natl Acad Sci USA 1995; 92:3401-3405.
11. Simons J.N., Leary T.P., Dawson G.J., Pilot-Matias T.J., Muerhoff A.S., Schlauder G.G., Desai S.M., Mushahwar I.K. Isolation of novel virus-like sequences associated with human hepatitis. Nature Med 1995; 1:564-569.
12. Leary T.P., Muerhoff A.S., Simons J.N., Pilot-Matias T.J., Erker J.C., Chalmers M.C., Schlauder G.G., Dawson G.J., Desai S.M., Mushahwar I.K. The sequence and genomic orginization of GBV-C: a novel member of the *Flaviviridae* associated with human non A-E hepatitis. J Med Virol 1996; 48:60-67.
13. Linnen J., Wages J., Zhang-Keck Z.-Y., Fry K.E., Krawczynski K.Z., Alter H., Koonin E., Gallagher M., Alter M., Hadziyannis S., Karayiannis P., Fung K., Nakatsuji Y., Shih J.W.-K., Young L., Jr., M.P., Hoover C., Fernandez J., Chen S., Zou J.-C., Morris T., Hyams K.C., Ismay S., Lifson J.D., Hess G., Foung S.K.H., Thomas H., Bradley D., Margolis H., Kim J.P. Molecular cloning and disease

association of hepatitis G virus: a transfusion-transmissible agent. Science 1996; 271:505-508.

14. Muerhoff A.S., Leary T.P., Simons J.N., Pilot-Matias T.J., Erker J.C., Chalmers M.C., Schlauder G.G., Dawson G.J., Desai S.M., Mushahwar I.K. Genomic organization of GBV-A and GBV-B: two new members of the flaviviridae associated with GB-agent hepatitis. J Virol 1995; 69:5621-5630.
15. Schaluder G.G., Dawson G.J., Simons J.N., Pilot-Matias T.J., Gutierrez R.A., Heynen C.A., Knigge M.F., Kurpiewski G.S., Buijk S.L., Leary T.P., Muerhoff A.S., Desai S.M., Mushahwar I.K. Molecular and serologic analysis in the transmission of the GB hepatitis agents. J Med Virol 1995; 46:81-90.
16. Leary T.P., Desai S.M., Yamaguchi J., Chalmers M.L., Schlauder G.G., Dawson G.J., Mushahwar I.K. Species-specific variants of GB virus A in captive monkeys. J Virol 1996; 70:9028-9030.
17. Bukh J., Apgar C.L. Five new or recently discovered (GBV-A) virus species are indigenous to New World monkeys and may consitute a separate genus of the Flaviviridae. Virol 1997; 229:429-436.
18. Bukh J., Kim J.P., Govindarajan S., Apgar C.L., Foung S.K.H., Wages J., Yun A.J., Shapiro M., Emerson S.U., Purcell R.H. Experimental infection of chimpanzees with hepatitis G virus and genetic analysis of the virus. J Infect Dis 1998; 177:855-862.
19. Birkenmeyer L.G., Desai S.M., Muerhoff A.S., Leary T.P., Simons J.N., Montes C.C., Mushahwar I.K. Isolation of a GB virus-related genome from a chimpanzee. J Med Virol 1998; 56:44-51.
20. Adams N.J., Prescott L.E., Jarvis L.M., Lewis J.C., McClure M.O., Smith D.B., Simmonds, P. Detection in chimpanzees of a novel flavivirus related to GB virus-C/hepatitis G virus. J Gen Virol 1998; 79:1871-7.
21. Dawson G.J., Schlauder G.G., Pilot-Matias T.J., Thiele D., Leary T.P., Murphy P., Rosenblatt J. E., Simons J. N., Martinson F.E.A., Gutierrez R.A., Lentino J.R., Pachucki C., Muerhoff A.S., Widell A., Tegtmeier G., Desai S., Mushahwar I.K. Prevalence studies of GB virus-C using reverse-transcriptase-polymerase chain reaction. J Med Virol 1996; 50:97-103.
22. Xiang J., Wunschmann S., Schmidt W., Shao J., Stapleton J.T. Full-length GB virus C (Hepatitis G virus) RNA transcripts are infectious in primary CD4-positive T cells J Virol 2000; 74:9125-33.
23. Beames B., Chavez D., Guerra B., Notvall L., Brasky K.M., Lanford R.E. Development of a primary tamarin hepatocyte culture system for GB virus-B: a surrogate model for hepatitis C virus. J Virol 2000; 74:11764-72.
24. Melvin S.L., Dawson G.J., Carrick R.J., Schlauder G.G., Heynen C.A., Mushahwar I.K. Biophysical characterization og GB virus C from human plasma. J Virol Meth 1998; 71:147-157.
25. Muerhoff A.S., Simons J.N., Erker J.C., Desai S.M., Mushahwar I.K. Conserved nucleotide sequences within the GB Virus C 5' untranslated region: Design of PCR primers for detection of viral RNA. J Virol Meth 1996; 62:55-62.
26. Simons J.N., Desai S.M., Schultz D.E., Lemon S.M., Mushahwar I.K. Translation initiation in GB viruses A and C: evidence for internal ribosome entry and implications on genome organization. J Virol 1996; 70:6126-6135.
27. Rijnbrand R., Abell G., Lemon S.M. Mutational analysis of the GB virus B internal ribosome entry site. J Virol 2000; 74:773-83.
28. Yoo B.J., Spaete R.R., Geballe A.P., Selby M., Houghton M., Han J.H. 5' end-dependent translation initiation of hepatitis C viral RNA and the presence of

putative positive and negative translational control elements within the 5' untranslated region. Virol 1992; 191:889-99.

29. Sbardellati A., Scarselli E., Tomei L., Kekule A.S., Traboni C. Identification of a novel sequence at the 3' end of the GB virus B genome. J Virol 1999; 73:10546-50.
30. Bukh J., Apgar C.L., Yanagi M. Toward a surrogate model for hepatitis C virus: An infectious molecular clone of the GB virus-B hepatitis agent. Virol 1999; 262:470-8.
31. Erker J.C., Simons J.N., Muerhoff A.S., Leary T.P., Chalmers M.L., Desai S.M., Mushahwar I.K. Molecular cloning and characterization of a GB virus C isolates from a patient with non-A-E hepatitis. J Gen Virol 1996; 77:2713-2720.
32. Leary T.P., Desai S.M., Erker J.C., Mushahwar I.K. The sequence and genomic organization of a GB virus A variant isolated from captive tamarins. J Gen Virol 1997; 78:2307-2313.
33. Erker J.C., Desai S.M., Leary T.P., Chalmers M.C., Montes C.C., Mushahwar I.K. Genomic analysis of two GB virus A variants isolated from captive monkeys. J. Gen. Virol. 1998; 79:41-45.
34. Weiner A.J., Brauer M.J., Rosenblatt J., Richman K.H., Tung J., Crawford K., Bonino F., Saracco G., Choo Q.L., Houghton M., Han J.H. Variable and hypervariable domains are found in the regions of HCV corresponding to the flavivirus envelope and NS1 proteins and the pestivirus envelope glycoproteins. Virol 1991; 180:842-848.
35. Hijikata M., Kato N., Ootsuyama Y., Nakagawa M., Ohkoshi S., Shimotohno K. Hypervariable regions in the putative glycoprotein of hepatitis C virus. Biochem Biophys Res Commun 1991; 175:220-8.
36. Kumar U., Monjardino J., Thomas H.C. Hypervariable region of hepatitis C virus envelope glycoprotein (E2/NS1) in an agammaglobulinemic patient. Gastroenterol 1994; 106:1072-5.
37. van Doorn L.J., Capriles I., Maertens G., DeLeys R., Murray K., Kos T., Schellekens H., Quint W. Sequence evolution of the hypervariable region in the putative envelope region E2/NS1 of hepatitis C virus is correlated with specific humoral immune responses. J Virol 1995; 69:773-8.
38. Pilot-Matias T.J., Carrick R.J., Coleman P.F., Leary T.P., Surowy T.K., Simons J.N., Muerhoff A.S., Buijk S.L., Chalmers M.L., Dawson G.J., Desai S.M., Mushahwar I. K. Expression of the GB virus C E2 glycoprotein using the Semliki forest virus vector system and its utility as a serologic marker. Virology 1996; 225:282-292.
39. Dille B.J., Surowy T.K., Gutierrez R.A., Coleman P.F., Knigge M.F., Carrick R.J., Aach R.D., Hollinger F.B., Stevens C.E., Barbosa L.H., Nemo G.J., Mosley J.W., Dawson G.J., Mushahwar I.K. An ELISA for detection of antibodies to the E2 protein of GB virus C. J Infect Dis 1997; 175:458-461.
40. Grakoui A., McCourt D.W., Wychowski C., Feinstone S.M., Rice C.M. A second hepatitis C virus-encoded proteinase. Proc Natl Acad Sci USA 1993; 90:10583-10587.
41. Belyaev A., Chong S., Novikov A., Kongpachith A., Masiarz F.R., Lim M., Kim J.P. Hepatitis G virus encodes protease activities which can effect processing of the virus putative nonstructural proteins. J Virol 1998; 72:868-872.
42. Grakoui A., McCourt D.W., Wychowski C., Feinstone S.M., Rice C.M. Characterization of the hepatitis C virus-encoded serine proteinase: determination of proteinase-dependent polyprotein cleavage sites. J Virol 1993; 67:2832-43.

43. Scarselli E., Urbani A., Sbardellati A., Tomei L., Francesco R.D., Traboni C. GB virus B and hepatitis C virus NS3 serine proteases share substrate specificity. J Virol 1997; 71:4985-4989.
44. Kolykhalov A.A., Agapov E.V., Rice C.M. Specificity of the hepatitis C virus NS3 serine protease: Effects of substitutions at the 3/4A, 4A/4B, 4B/5A, and 5A/5B cleavage sites on polprotein processing. J Virol 1994; 68:7525-7533.
45. Bartenschlager R., Ahlborn-Laake L., Yasargil K., Mous J., Jacobsen H. Substrate determinants for cleavage in *cis* and in *trans* by the hepatitisC virusNS3 proteinase. J Virol 1995; 69:198-205.
46. Suzich J.A., Tamura J.K., Palmer-Hill F., Warrener P., Grakoui A., Rice C.M., Feinstone S.M., Collett M.S. Hepatitis C virus NS3 protein polynucleotide-stimulated nucleoside triphosphatase and comparison with the related pestivirus and flavivirus enzymes. J Virol 1993; 67:6152-8.
47. Zhong W., Ingravallo P., Wright-Minogue J., Skelton A., Uss A.S., Chase R., Yao N., Lau J.Y., Hong Z. Nucleoside triphosphatase and RNA helicase activities associated with GB virus B nonstructural protein 3. Virol 1999; 261:216-26.
48. Gwack Y., Yoo H., Song I., Choe J., Han J.H. RNA-Stimulated ATPase and RNA helicase activities and RNA binding domain of hepatitis G virus nonstructural protein 3. J Virol 1999; 73:2909-15.
49. Butkiewicz N., Yao N., Zhong W., Wright-Minogue J., Ingravallo P., Zhang R., Durkin J., Standring D.N., Baroudy B.M., Sangar D.V., Lemon S.M., Lau J.Y., Hong Z. Virus-specific cofactor requirement and chimeric hepatitis C virus/GB virus B nonstructural protein 3. J Virol 2000; 74:4291-301.
50. Sbardellati A., Scarselli E., Amati V., Falcinelli S., Kekule A.S., Traboni C. Processing of GB virus B non-structural proteins in cultured cells requires both NS3 protease and NS4A cofactor. J Gen Virol 2000; 81:2183-2188.
51. Zhong W., Ingravallo P., Wright-Minogue J., Uss A.S., Skelton A., Ferrari E., Lau J.Y., Hong Z. RNA-dependent RNA polymerase activity encoded by GB virus-B non-structural protein 5B. J Viral Hepat 2000; 7:335-342.
52. Muerhoff A.S., Simons J.N., Leary T.P., Erker J.C., Chalmers M.L., Pilot-Matias T.J., Dawson G.J., Desai S.M., Mushahwar I.K. Sequence heterogeneity within the 5'-terminal region of the hepatitis GB virus C genome and evidence for genotypes. J Hepatology 1996; 25:379-384.
53. Tucker T.J., Smuts H.E. GBV-C/HGV genotypes: proposed nomenclature for genotypes 1-5. J Med Virol 2000; 62:82-3.
54. Pilot-Matias T.J., Muerhoff A.S., Simons J.N., Leary T.P., Buijk S.L., Chalmers M.L., Erker J.C., Dawson G.J., Desai S.M., Mushahwar I.K. Identification of antigenic regions in the GB hepatitis viruses GBV-A, GBV-B and GBV-C. J Med Virol 1996; 48:329-338.
55. Leary T.P., Muerhoff A.S., Simons J.N., Pilot-Matias T.J., Erker, J. C., Chalmers M.L., Schlauder G.G., Dawson G.J., Desai S.M., Mushahwar I.K. Consensus oligonucleotide primers for the detection of GB virus C in human cryptogenic hepatitis. J Virol Meth 1996; 56:119-121.
56. Muerhoff A.S., Simons J.N., Erker J.C., Desai S.M., Mushahwar I.K. Identification of conserved nucleotide sequences within the GB virus C 5'-untranslated region: design of PCR primers for detection of viral RNA. J Virol Meth 1996; 62:55-62.
57. Marshall R.L., Cockerill J., Friedman P., Hayden M., Hodges S., Holas C., Jennings C., Jou C.K., Kratochvil J., Laffler T., Lewis N., Scheffel C., Traylor D., Wang L., Solomon N. Detection of GB virus C by the RT-PCR LCx system. J Virol Meth 1998; 73:99-107.

58. Surowy T.K., Leary T.P., Carrick R.J., Knigge M.F., Pilot-Matias T.J., Heynen C., Gutierrez R.A., Desai S.M., Dawson G.J., Mushahwar I.K. GB virus C E2 glycoprotein: expression in CHO cells, purification and characterization. J Gen Virol 1997; 78:1851-1859.

Chapter 8

IMMUNOBIOLOGY OF HEPATITIS VIRUSES

Michael P. Curry[1] and Margaret James Koziel[2]

[1]Hepatology Fellow, Department of Medicine, Beth Israel Deaconess Medical Center and Harvard Medical School, Boston, MA 02215, USA
[2]Staff Physician, Department of Infectious Diseases, Beth Israel Deaconess Medical Center, Associate Professor of Medicine, Harvard Medical School, Boston, MA 02215, USA

1. INTRODUCTION

Hippocrates described the existence of agents capable of causing epidemic jaundice as early as 400BC. Further outbreaks of catarrhal jaundice during World War II, the distinction of "infectious" and "serum" hepatitis by Krugman, the discovery of the Australia antigen by Blumberg and colleagues and the identification of hepatitis C virus (HCV) as the major cause of post-transfusion hepatitis by Houghton have resulted in vast increases in our knowledge of viral hepatitis. Intense research into the molecular structures, biological life cycles, and mechanisms of disease has occurred over the past 30 years (1, 2) . Researchers have made full use of epidemiological studies, clinical research trials and technologies such as animal models, tissue culture and human studies to develop sensitive and specific diagnostics tests, effective antiviral therapies and in the cases of hepatitis A and B viruses, (HAV, HBV) safe and effective vaccines. Despite the wealth of knowledge regarding the molecular biology of the viruses and the medical importance of the disease, the interactions between the immune system and hepatitis viruses, and mechanisms of viral persistence are not completely understood, largely because of in-efficient or non-existent culture systems for viral propagation and absence of reagents for analysis of the immune responses in existing animal models.

The development of percutaneous needle liver biopsy techniques in the 1930s resulted in the recognition of a necroinflammatory lesion that characterizes all forms of viral hepatitis. While certain histological features may suggest a particular infectious agent, it may not possible to distinguish between different forms of chronic viral hepatitis based on biopsy alone. The histological hallmark of chronic hepatitis is the inflammatory destruction of hepatocytes resulting in progressive fibrosis. The inflammatory infiltrate of chronic viral hepatitis is composed of 3 components: 1) a dense portal inflammatory infiltrate possibly with lymphoid aggregates of B- and T-cells and small number of macrophages and dendritic cells; 2) interface hepatitis at the limiting plate which is dominated by $CD4^+$ T lymphocytes accompanied by smaller numbers of $CD8^+$ T-cells, plasma cells and macrophages; and 3) lobular hepatitis consisting of small foci of lymphocytes and macrophages associated with focal hepatocyte apoptosis. This infiltrate is composed of virus specific and non-specific cells of both the innate and adaptive immune system. These viral specific-cells which are present in the peripheral blood and the liver are responsible for viral control and with the recruitment of other non-antigen specific cells possibly, liver injury. At present, the balance between the factors which determine whether the immune system can eliminate the virus or whether the virus persists despite a continuous antiviral response are not fully known.

2. MECHANISM OF LIVER INJURY

Until recently, it was generally thought that the normal human liver was devoid of lymphoid cells and that the presence of these cells in the liver was a reflection of disease processes. The unique position of the liver placed “between” the portal and systemic circulations provides it with a role of distinguishing between foreign antigens, some of which are harmful and some of which are necessary dietary components. The liver must have an immune system capable of making theses distinctions so that protective immune responses and tolerance can be maintained (See refs [3, 4] for review). Recently, isolation and study of lymphocytes from the normal adult human liver has revealed significant differences from those in the peripheral blood (5, 6) . Significant numbers of natural killer (NK) cells and γδ T cell $receptor^+$ (TCR) T-cells are found in the normal adult liver. In addition to these innate immune cells, both conventional $CD4^+$ T-cells, $CD8^+$ T cells and non-conventional T- cells are also found. The non conventional T-cells which are either $CD4^-/CD8^-$ T-cells (double negative) or $CD4^+/CD8^+$ T-cells (double positive) are also found in the murine liver and usually co-express the NK1.1 (human equivalents are CD56 or CD169) marker usually expressed on NK cells and are thus designated NK T-cells (6, 7) . These NK T-cells are capable of recognizing antigen through either the NK receptors or the T cell receptor when presented in the context of the monomorphic major histocompatibility complex (MHC) class I molecule CD1 (8) . The origins of these lymphoid cells remains controversial. The conventional T-cells within the liver appear to arise from the periphery and in fact during HCV infection, most intra-hepatic lymphocytes are found in the G0/G1 state of the cell cycle, suggesting that T-cell activation and expansion occurs outside the liver, followed by migration into the liver (9). However, lymphoid progenitors and the mechanism necessary for their development are found in the normal human liver suggesting that extra-thymic T-cell development may occur in the liver (10, 11) . The purpose of these lymphoid cells in both health and disease remains unclear. In health, it is proposed that the intra-hepatic lymphocytes are important in maintaining the balance between tolerance and an immune response generated by circulating lymphocytes.

A diffuse inflammatory process resulting in either acute and/or chronic liver injury characterizes viral hepatitis. The relative contributions of resident intra-hepatic lymphocytes and circulating lymphocytes remains unknown. The precise pathogenesis of hepatocyte cell injury is not known, although both antigen non-specific and antigen specific cells of the innate and adaptive immune systems are capable of inducing hepatocyte injury. The innate immune responses which constitute the first line of defence are generally antigen non-specific and include the production of interferons (IFN) and activation of Natural killer (NK) cells, macrophages and possibly γδ T-cells and Natural killer T-cells (NKT) which are capable of recognizing viral antigens in the absence of antigen presenting cells (APCs) (12, 13) . The importance of these cells in the immune response to hepatitis viruses is emphasized by the fact that large numbers of NK-cells, γδ T-cells and NKT-cells are demonstrated in both the normal and inflamed adult human liver (3, 7, 9) . The innate immune system provides time for the uptake and processing of viral antigens by professional APCs and presentation to the cells of the adaptive immune response (Figure 1). Double stranded RNA production during viral replication results in the formation of IFN-α and IFN-β, which further induces 2' 5'-oligoadenylate synthetase and double stranded RNA-dependent protein kinase, both of which inhibit protein synthesis in infected cells and thus viral replication. These interferons also increase the expression of major histocompatibility (MHC) class I antigens, thereby enhancing the presentation of viral antigens to $CD8^+$ cytotoxic T-cells (CTLs).

NK-cells are activated by the production of IFN-α and IFN-β and are capable of killing virally infected cells because down regulation of MHC class I expression induced by most viral infections results in loss of the inhibitory signals delivered to the NK Ly49 family of receptors. Unfortunately, many viruses have evolved ways of interfering with innate immune responses. Up regulation of MHC class I expression on target cells might render those target immune from NK mediated destruction (14, 15) . NK-cells can also kill targets through stimulation of their cell surface receptors by immunoglobulin, in a process known as antibody-dependent cell-mediated cytotoxicity (ADCC). Their mechanism of killing involves the release of preformed granules containing perforin and is similar to that of CTLs described below.

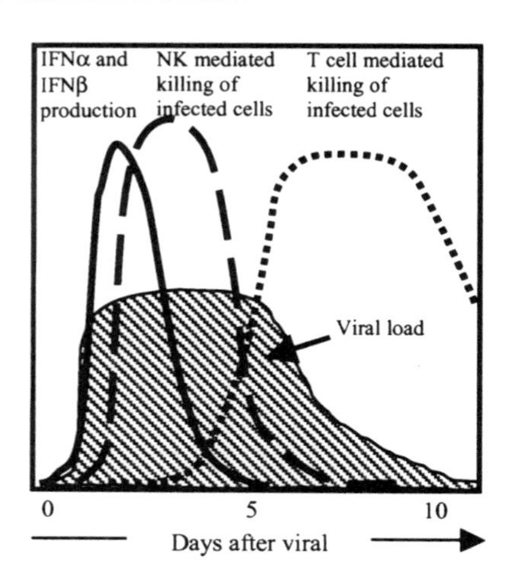

Figure 1. Time course development of innate and adaptive immune responses after viral infection.

The adaptive or acquired immune system is comprised of $CD8^+$ cytotoxic T-cell responses, $CD4^+$ Helper T-cell responses, and antibody responses, which typically become armed several days after a viral infection. Generally, $CD8^+$ cytotoxic T-cells characteristically recognize endogenous antigen presented in the context of the MHC class I antigen and β-2 microglobulin by the target cells. $CD4^+$ T-cells recognise exogenous antigen taken up, processed by antigen presenting cells and presented in the context of MHC class II molecules with co-stimulatory molecules required for activation (Figure 2). A number of exceptions to these general rules exist as "cross presentation". Dendritic cells are capable of engulfing exogenous apoptotic cells and processing exogenous antigen and presenting it on MHC class I molecules for recognition by $CD8^+$ T-cells (16, 17) ; and $CD4^+$ T-cells can recognize endogenously produced antigens presented to them by class II molecules. For example, exogenous hepatitis B surface antigen (HBsAg) particles have been shown to be capable of entering the class I pathway by a novel processing route (18) . A number of viruses interfere with processing of antigens and expression of the class I/antigen/ β2 microglobulin complex on the cell surface (19) . Although none of the hepatitis viruses are clearly known to interfere with antigen presentation to CTLs, HBV X protein has been demonstrated to bind to the proteosome, which is a key molecule

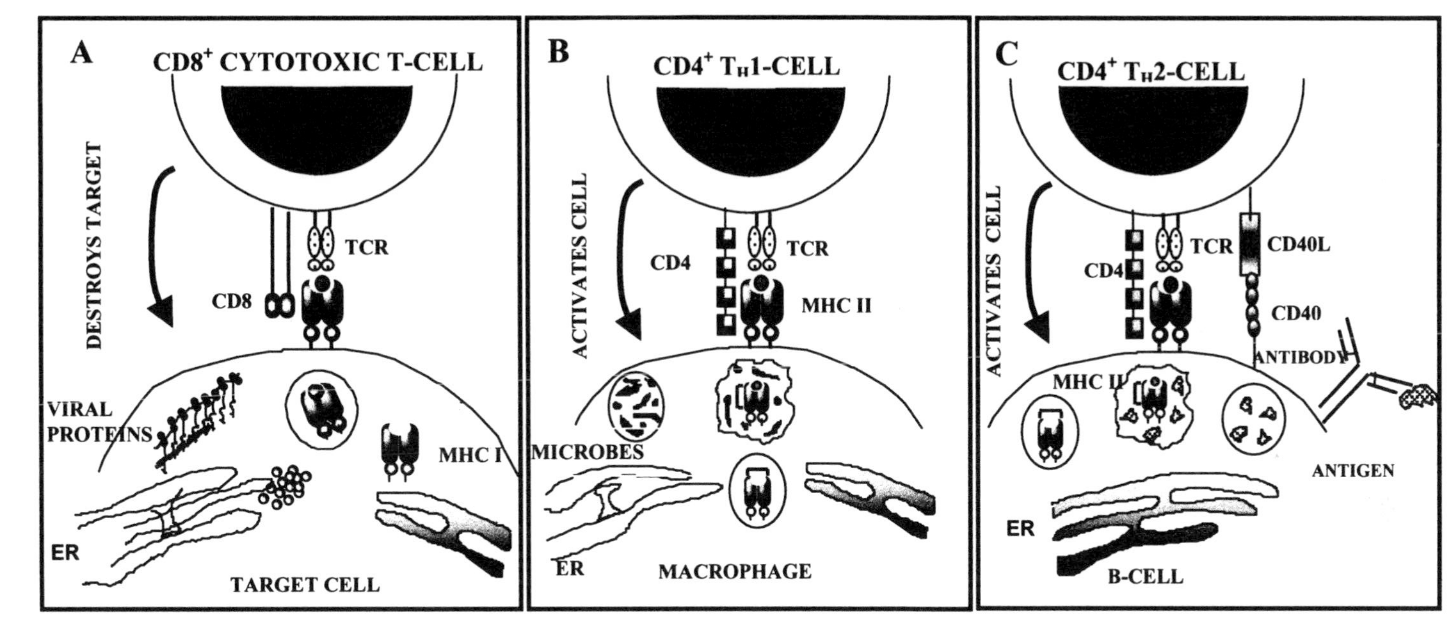

Figure 2. Classical antigen presentation in the context of MHC class I and II molecules. (A) Presentation of endogenously synthesized antigen by the infected target cell to antigen specific-CD8$^+$ T-cells in the context of class I molecules results in cytolytic activity mediated through Fas/Fas ligand of TNF-α mediated apoptosis. Hepatocytes which lack the expression of s-stimulatory molecules may not be able to activate T-cells and promote anergy. (B and C) Presentation of antigen by professional antigen presenting cells and B-cells in the context of MHC class II molecule results in either activation of the PAC to kill the antigen or the production of neutralizing antibodies.

involved in antigen processing (20) . Specific recognition of peptide antigen in the groove of the MHC molecule of the target cell by $CD8^+$ cytotoxic T-cells results in destruction of the target cell usually by a process known as apoptosis or programmed cell death (21) . The mechanisms by which the cytotoxic T-cells kill target cells are by release of preformed secretory granules, which contain perforin and granzymes, or by Fas or TNF-α mediated induction of apoptosis. The perforin molecules released from the granules has the capability to insert itself into the lipid membrane of the infected cell, thereby resulting in the formation of pores in the membrane of the target cell resulting in a lytic mechanism of killing. These pores also allow the entry of granzymes or fragmentins which results in activation of the apoptosis cascade with resultant DNA fragmentation into oligomers of 200 base pairs. Fas ligand (FasL) expression is induced in cytotoxic T cells and is capable of ligating the molecule Fas expressed on the surface of target cells. Fas is expressed on many cells, including hepatocytes, whereas FasL is expressed on activated T lymphocytes (22) . This ligation brings about activation of the apoptotic cascade resulting in cell suicide. In chronic HCV infection, areas of active inflammation are composed of cytotoxic T-cells surrounding areas of Fas expressing hepatocytes with acidophilic bodies, indicating that CTLs may be inducing the hepatocyte apoptosis. A third potential effector molecule used by CTL to induce cell death is tumor necrosis facto-α (TNF-α) TNF-α belongs to a family of nine ligands known as the TNF receptor superfamily produced by activated macrophages and lymphocytes and is an important mediator of apoptosis. TNF-α is induced by hepatitis B and C viruses in human liver tissue, in hepatoma cell lines transfected with HBV; and in PBMCs from patients with chronic HBV (23, 24) . TNF-α has been implicated in the process of bystander killing, in which neighbouring uninfected cells are damaged by antigen-specific CTLs (25) . Hepatocytes in particular are known to be sensitive to the effects of TNF-α.

The role of antigen-specific CTLs in mediating hepatocyte toxicity is well established in murine models of viral hepatitis. A hepatotropic strain of lymphocytic choriomeningitis virus (LCMV) is capable of developing MHC class I-restricted CTL mediated liver injury. The classic model of liver injury in viral hepatitis is the adoptive transfer of syngeneic HBV-specific CTL clones into HBV transgenic mice, which results in the development of acute hepatitis (26, 27) . Both of these experiments suggest that the liver injury occurs as a direct consequence of the CTLs. CTL cells can also contribute to host defence by the production of cytokines that are capable of non-cytopathic viral elimination. CTLs are capable of producing multiple cytokines, including IFN-γ, TNF-α, IL-8 and multiple chemokines (13, 28, 29) . Of these, the role of IFN-γ in inhibiting viral replication has been characterized the best. In HBV-transgenic mice, IFN-γ has been shown to contribute to the non-cytopathic reduction in viral gene expression and replication; super-infection of these mice with another pathogen such as LCMV, which results in IFN-γ leads to a reduction in viral replication (28, 30, 31) .

$CD4^+$ T-cells recognize peptide antigens when presented in the context of the MHC class II molecules. These antigens have been endocytosed, proteolytically processed from larger soluble proteins, and presented by professional APCs and are of two functional types: inflammatory $CD4^+$ T-cells (T_H1) and helper $CD4^+$ T-cells (T_H2). These two groups of activated $CD4^+$ T-cells can stimulate the antigen presenting cells, the antigen specific $CD8^+$ T-cells, non-antigen specific cells and antibody-producing B-cells by a host of cytokines. Type 1 or T_H1 responses are typically important in priming and maintenance of the cellular immune response, while type 2 of T_H2 responses promote humoral immunity. T_H1-cells are specialized for macrophage activation following specific recognition of peptide antigen-MHC II complex. These cells, through their production of IFN-γ, stimulate phagocytosis and killing of intracellular viruses, and up regulate class I and II MHC molecules on a variety of cells, thereby increasing antigen presentation to T-cells. There is growing evidence that clearance of hepatitis viruses is associated with activation of T_H1-cell responses. Proinflammatory cytokines such as Interleukin-12 (IL-12), have been shown to activate natural killer cells and promote the differentiation of T_H1 $CD4^+$ T-cells. IL-12 is secreted by antigen presenting cells including dendritic cells and macrophages, both in tissues and in secondary lymphoid organs. Experimental studies have shown that administration of this cytokine rapidly activates both innate and adaptive immune responses, resulting in enhanced cellular responses and clearance of virus infection. Among patients who clear hepatitis B and C infections, there are higher levels of IL-12 than in chronic carriers and their $CD4^+$ T-cell responses preferentially produce IFN-γ and IL-2 (32-34) .

However, production of cytokines by antigen-specific T cells may contribute to liver injury as well as mediate viral clearance. In animal models of Concanavalin A induced hepatitis, the absence of $CD4^+$ T-cells or IFN-γ is associated with failure to develop liver injury (35) . A novel mouse model of HBV liver injury established by Ohta *et al.* failed to develop liver injury in the absence of TNF-α and IFN-γ indicating that these cytokines produced by HBsAg-specific T_H1-cells are important in the effector phase of immune mediated liver injury (36) . Similarly, transgenic mice constitutively producing IFN-γ develop liver injury independent of any pathogen (37) . This illustrates a key paradox in understanding viral hepatitis, as one of the same molecules, IFN-γ, involved in recovery is also implicated in liver injury; the result may be a matter of local kinetics of release.

Recognition of an appropriate antigen in the MHC class II grove triggers the $CD4^+$ T_H2-cell to synthesize membrane bound and secreted CD40 ligand which synergizes with B cell CD40 molecules resulting in B cell proliferation and clonal expansion. Secretion of interleukin-4 (IL-4), IL-5, and IL-6 by the T_H2-cells drives the proliferation and differentiation of B-cells into antibody secreting plasma cells. In addition to their stimulatory effects, T_H1- and T_H2-cells cross regulate each other. The IFN-γ secreted by T_H1-cells suppresses the IL-4 production and therefore inhibits differentiation of T_H0-cell to T_H2-cells. Conversely, production of IL-4 and IL-10 inhibits IL-12 and IFN-γ induction of T_H1-cells.

3. ENTERIC VIRUSES

3.1 Hepatitis A

Documented outbreaks of jaundice have occurred as early as the 8th century BC, however, while a viral etiology was suspected, it was not until crucial experiments were performed in volunteers and mentally handicapped children that a distinct "infectious" agent named MS-1 (subsequently HAV) was clinically recognised. In 1973, HAV was isolated in human stool of prisoners who had been infected with the MS-1 strain and subsequently, the development of sensitive and specific serologic techniques were important advances in understanding the clinical course of HAV (38) .

HAV is a heat and acid resistant non-enveloped single stranded RNA virus measuring 27-28 nm in diameter of the *Hepatovirus* genus, which causes an acute resolving inflammation of the liver without progression to chronicity. The clinical manifestations of HAV infection are variable, from asymptomatic infection to relapsing HAV (3-20%) and occasionally fulminant hepatic failure (1-5%). Young children (< 2 years) are rarely symptomatic whereas children 5 year and older and adults usually develop jaundice.

The mechanism by which HAV infects the hepatocyte remains poorly characterized (39) . Viral uptake by hepatocytes is thought to occur by cell surface-receptors. Recently glycoproteins (HAVcr-1) on the surface of African green monkey kidney cells have been identified as receptors for HAV, although it is possible there are multiple receptors. After oral ingestion, viral replication within the intestinal epithelial cells may occur. Studies with animal models have provided conflicting evidence for replication in the intestinal epithelium (40) . The most recent data suggests that viral replication can take place within the crypt cells of the small intestine, which is in keeping with studies indicating significant histological changes in this organ during acute HAV infection. During acute infection, HAV antigen has been demonstrated within the hepatocyte cytoplasm and following replication, HAV is secreted across the apical canalicular surface of the hepatocyte into the bile. In spite of significant progress achieved in understanding the molecular structure of HAV and the processing of the polyprotein, little is known about the HAV RNA replication within hepatocytes.

3.1.1 Cellular immunity

The pathogenesis of HAV-associated liver injury is not completely understood, although the presence of HAV within hepatocytes prior to an increase in transaminases and the absence of direct cytopathic injury in cell culture models support a non-cytopathic mechanism of injury. Only limited data are available on the immunology of HAV infection and these studies typically have analysed only the acute symptomatic phase of the infection in mild icteric cases. There are no data available on the immunology of fulminant HAV infection. Immunohistochemical studies of liver tissue from acutely infected HAV individuals have demonstrated the presence of $CD45RO^+$ memory T-cells, $CD8^+$ T-cells and B-cells in the portal areas and T-cells, and natural killer (NK) cells in the necrotic areas of liver biopsies from patients with acute HAV infection (41) . Phenotypic analysis of peripheral blood lymphocytes from individuals with acute HAV infection show an increase in the percentage of activated T-cells during the 2 week period after the onset of jaundice. In keeping with the antiviral effects of NK-cells, both absolute and relative numbers of peripheral blood NK-cells are significantly increased in acute HAV infection when compared with normal controls (42) .

Standard radiolabeled-chromium (^{51}Cr) release cytotoxicity assays demonstrate that peripheral NK-cells and lymphokine activated killer (LAK) cells from HAV infected individuals are capable of lysing cell lines infected with HAV more efficiently than NK-cells from HAV antibody negative individuals. In these same experiments, peripheral T-cells and recombinant interleukin-2 (rIL-2) treated (LAK) T-cells did not demonstrate significant lysis of HAV infected target cells. However, Vallbracht *et al.* demonstrated that peripheral blood HLA class I dependent $CD8^+$ T cells from individual with acute HAV are capable of lysing autologous HAV infected skin fibroblasts and can produce IFN-γ (43, 44) . In patients with acute resolving HAV infection the CTL response as measured by ^{51}Cr release assay peaked 2-3 weeks after the inset of jaundice, compared to a peak 8-12 weeks after symptoms in those patients who developed protracted HAV infection, indicating that the kinetics of the immune response dictate the clinical course of the infection. $CD8^+$ T-cells clones isolated from liver biopsies following the clinical onset of symptomatic HAV infection are likewise capable of mediating HAV specific cytolysis of HAV infected fibroblasts and predominate over $CD4^+$ T-cell clones in the early symptomatic period. Later in the clinical course of the disease, more $CD4^+$ T-cell clones can be isolated form the liver than $CD8^+$ clones, which may be associated with the development of recovery and antibody production.

While chronic HAV infection does not develop, the clinical entity of relapsing HAV infection can occur in up to 3-20% of individuals following acute HAV infection (45) . The characteristic pattern is one of biochemical and clinical relapse in an individual who has had a clinical recovery from the initial HAV infection. No differences in the clinical presentation of the initial infection are noted in these individuals, however in one individual with relapsing clinical course, 20% of the T-cell clones isolated from the liver demonstrated an unusual phenotype with T-cell and NK-cell characteristics, possibly suggesting that this recently described NK T-cell is in some respects involved in the pathogenesis of relapsing HAV infection. Further studies are necessary to clarify this phenomenon. This relapsing HAV infection may also result because of re-infection of the hepatocytes with HAV infection. Dotzauer et al. have demonstrated *in vitro* infection of the HepG2 cell line and primary hepatocytes via uptake of HAV-anti-HAV IgA antibody complexes by the asialoglycoprotein receptor (39) .These experiments suggest that liver injury in HAV infection is not caused by a viral cytopathoic effect but is due to an immunopathological reaction of sensitised cytotoxic T cells directed against infected hepatocytes.

3.1.2 Humoral immunity

The early development of immune electron microscopic methods for the detection of HAV in acute-phase stool suspensions and serological test for the detection of antibody to HAV in serum made it possible to differentiate recent infections with hepatitis A using paired acute and convalescent phase sera. Introduction of less cumbersome and time-consuming serologic test methods, including complement fixation and immune adherence hemagglutination, made it feasible to rapidly assay larger numbers of specimens for HAV or anti-HAV. Subsequent development of sensitive immunofluorescence assays, solid-phase radioimmunoassays (RIA), and enzyme immunoassays (EIA) for HAV and anti-HAV heralded intensive laboratory studies of the biophysical and biochemical properties of the virus as well as efforts to define the pathogenesis and clinical course of disease. HAV infection is diagnosed by the detection of immunoglobulin M (IgM) or IgG antibodies to the four major viral capsid proteins (VP1, VP2, VP3 and VP4). Examination of IgG and IgM antibody fractions of acute-phase sera from infected individuals can differentiate acute from resolved HAV infection. A diagnosis of acute HAV requires the demonstration of IgM anti-HAV in serum. The test is positive from the very onset of symptoms and remains positive for approximately four months (46, 47) . In general, antibodies to these capsid proteins are thought to be protective against a viral challenge with early experimental work demonstrating the development of neutralizing antibodies in rabbits in response to immunization with purified VP1, VP2 or VP3 viral proteins (48) . While some investigators have purported that VP1 induces the strongest response, others have argued that the immunodominant neutralization site is located on capsid protein, VP3. Similarly, studies in humans are discordant in the findings of the immunodominant epitopes with Gauss-Müller et al reporting that antibodies isolated from acute and convalescent sera from HAV infected humans showing that IgM preferentially recognised VP0 (VP2 plus VP4) and VP3 capsid antigens and IgG and IgA antibodies reacted more strongly with VP1 (49) . This contrasts with the findings of Wang *et al.* who demonstrated a preferential recognition of VP1 antigens by IgM antibodies in acute sera. Convalescent serum taken 10 year after acute infection from one individual demonstrated IgG reactivity against all antigens (50) . These contrasting results may be due to differences in the denaturation procedure of HAV antigens resulting in minor differences in epitope recognition. While the denatured capsid protein antigen used in these experiments represents a fraction of the antigen sequence of B cell epitopes and denaturing techniques can alter the conformational structures, these studies show that the immune response to naturally acquired HAV infection is directed against all three viral capsid antigens and can persist for at least up to 10 years after acute infection.

IgA antibodies are also produced during the acute infection. HAV–specific IgA antibodies produced by the lymphoid tissue has recently been suspected as a possible viral carrier and mediator of hepatocyte re-infection in relapsing HAV infection.

3.2 Hepatitis E

The existence of Hepatitis E Virus (HEV) was first suspected following studies of water-borne hepatitis in India in 1980, but it was not cloned until 1990. It is now recognized as a cause of epidemic and endemic hepatitis principally in Asia, Africa, the Middle East and Mexico. Hepatitis E virus infection spreads by the faecal-oral route, usually through contaminated water, and presents after an incubation period of 8-10 weeks with a clinical illness resembling other forms of acute viral hepatitis. Clinical attack rates are highest among young adults. Asymptomatic and anicteric infections are known to occur but chronic HEV infection is not observed. HEV RNA and viral excretion in the feces stops after resolution of biochemical hepatitis (51) . Although the mortality rate is usually low (0.07-0.6%), the illness may be particularly severe among pregnant women, with mortality rates reaching as high as 25%. It is a non-enveloped spherical virus approximately 27-34 nm in diameter with a single strand RNA 7.5 kb genome consisting of a 5' non-coding region, 3 over lapping reading reading frames (ORF_1, ORF_2, ORF_3) and a polyadenylated 3' non-coding region. The ORF_1 encodes at least seven defined non-structural proteins involved in RNA replication. The ORF_2 encodes the structural capsid protein. The function of the ORF_3 is unknown. The virus enters the host primarily through the oral route, but the mechanisms by which it reaches the liver from the intestine are as yet unknown.

3.2.1 Humoral immunity

The exact time course antibody response to HEV is not known but anti-HEV IgM begins to develop just before the ALT reaches a peak and reaches a maximum levels as the time of maximum liver damage as reflected by the height of the ALT. Animal models have characterized the sequence of events related to HEV infection. Cynomologus macaques inoculated with HEV develop histopathological changes of acute hepatitis in the third week associated with the presence of HEV antigen in the liver, HEV in the bile and elevations in ALT. By the fourth week, there is increased liver damage, with the development of anti-HEV IgM, which peaks in parallel with the ALT during the fifth week after infection. During the sixth week, there is loss of HEV antigen form the liver but persistent pathological changes suggesting that clearance of the viral infection occurs in association with the development of antibody responses. However, recovery from liver injury may lag behind clearance of infection. The IgM anti-HEV produced during acute infection persists for approximately 4 to 5 months. IgG anti-HEV gradually increases during the convalescent period and can remain positive up to 4.5 years and possibly 14 years after acute infection (52) . IgA antibodies specific to HEV can also be demonstrated in the serum during the acute phase but disappear during convalescence (53) . Antibodies against the ORF_2 and ORF_3 are highest during the early acute phase. Antibody to the capsid (ORF_2) proteins is sufficient to confer immunity and the specific epitopes identified by B-cells are the $ORF_{2.1}$ epitope and a linear epitope within the 434 to 457 amino acid sequence as they blocked convalescent patient antibody reactivity in VLP ELISAs (54) .

3.2.2 Cellular immunity

The pathogenesis of hepatitis E has not been thoroughly investigated and very little is known about the role of the cellular immune response in the pathogenesis of HEV infection. The variable clinical spectrum of disease associated with the infection suggests that the host immune response determines outcome. Immunohistochemical examination of the liver during infection demonstrates that HEV-infected hepatocytes are proximally related to a lymphocytic infiltrate. Histological features of the HEV infection suggest two types of illness. More than half of those infected develop a cholestatic illness with liver biopsy findings of bile stasis, bile duct proliferation and pseudo-glandular arrangements of hepatocytes, with ballooning and spotty necrosis of parenchymal cells. In the remainder, the process resembles the classical features of acute hepatitis, with focal and confluent hepatocyte necrosis, acidophilic bodies representing apoptosis, lobular inflammatory infiltrate consisting of lymphocytes and neutrophils and a portal infiltrate consisting of lymphocytes, eosinophils, and neutrophils. There are no reports published which have investigated the HEV-specific $CD4^+$ or $CD8^+$ T-cell responses in HEV infection.

4. PARENTERAL VIRUSES

4.1 Hepatitis B

Among the viral hepatitides, the immunopathogenesis of hepatitis B has been studied most extensively. HBV is an enveloped, double stranded DNA virus that causes both acute and chronic hepatitis (55). Before the identification of HCV infection, HBV was deemed the major cause of chronic liver disease cirrhosis and hepatocellular carcinoma (HCC) worldwide. It is spread through parenteral and sexual exposure but vertical transmission from mother to baby is the predominant route of transmission, particularly in developing countries. The rate of progression form acute HBV to chronic infection depends on age of acquisition. Ninety percent of perinatal infections become persistent while less than 5% of adults who acquire the HBV develop chronic infection (56) . Currently 300 million people worldwide have chronic HBV infection and these individuals have a 100-fold increased risk of developing HCC (57) . The currently available vaccine against HBV is the only true cancer vaccine available, as rates of HCC in communities which historically had high rates of HBV-associated HCC declined after introduction of mass vaccination against HBV (58) . In the United States, nearly 1 million people are chronically infected, of whom 5000 die each year from cirrhosis and HCC.

4.1.1 Cellular immunity

4.1.1.1 Acute HBV

The exact mechanisms by which chronic liver injury occurs in HBV infection are not known, although most studies suggest that the hepatitis virus is not directly cytopathic to the hepatocyte (59) . There is no robust tissue culture system but the existence of aymptomatic hepatitis B carriers with normal liver histology and function suggests that the virus is not directly cytopathic. Extensive human and animal studies have now shown that the liver injury mediated by HBV is initiated by a viral specific cellular and humoral immune response. In more than 95% of immunocompetent adults, this immune response is co-ordinated, vigorous, polyclonal and multi-specific and results in acute self-limited hepatitis with reduction of viral load, and the development of long lasting humoral and cellular immunity. In the remaining 5% of HBV infected individuals, persistent infection occurs with resultant chronic necroinflammatory activity, which eventually leads to cirrhosis. The mechanism for this viral persistence and immune mediated liver injury is not known.

The innate immune response to HBV infection is characterised by an increase in the non-MHC restricted NK-cell subset early in the infectious process. Webster *et al.* demonstrated an increase in circulating NK-cells occurs in the early asymptomatic period consistent with their role in early phases of viral infection, and a later reduction in NK-cell numbers, which ran parallel with the reduction in HBV DNA levels (60) . Guidotti has reported clearance of HBV DNA from the liver in infected chimpanzees before the peak of T-cell infiltration and the development of liver disease, implying that the innate immune response may mediate this early viral clearance through the production of IFN-γ and TNF-α (61) . The role of the innate immune response in mediating early HBV clearance from the liver is further supported by a mouse model, which demonstrates that activation of NK-cells and NKT by a glycolipid (α-GalCer) results in the production of IFN-γ and IFN-α/β. NK-cells are recruited into the liver and there is activation induced death of NKT cells with the subsequent clearance of HBV from the liver, which is probably mediated non-cytopathically by IFN-α/β (12).

Acute HBV infection is accompanied by a strong and transient expansion of $CD4^+$ T_H1- and T_H0-cells directed against multiple epitopes within the HBV. Hepatitis B core (HBc) is the dominant antigen recognised by $CD4^+$ T-cells in most case of acute resolving HBV infection (62, 63) . These HBc-specific $CD4^+$ cells also contribute to the induction of virus specific CTLs and provide help for the production of antibody to hepatitis B surface antigen (HBsAg) (64, 65) . The vigorous response to HBc peptides elicited from patients with acute self-resolving HBV infection is particularly associated with HLA class II molecules DR 13 (66) . When acute HBV infection resolves, a dramatic decline in the HBc-specific $CD4^+$ T-cell response is frequently observed. A strong memory response may persist for years after the elimination of virus. This reduction in $CD4^+$ T_H1-cells may result from loss of viral markers and so loss of antigenic stimulus, T-cell exhaustion, or the development of anergy of the HBc-specific $CD4^+$ T-cells (67). Whatever the method of reduction, it may serve as an immunoregulatory mechanism required to balance the development of fulminant hepatic failure against chronic hepatitis.

The polyclonal HBV specific-CD4 T cell response is associated with the development of a vigorous and polyclonal cytotoxic T-cell responses (CTL) capable of recognizing different eptitopes within HBV genome (67, 68) . Individuals who successfully clear HBV infection, either spontaneously or after interferon therapy maintain these broad and strong peripheral CTL responses against all HBV epitopes over time (69) . CTL memory in the presence of low levels of persisting HBV DNA has been shown to persist up to 23 years after infection despite markers of serological recovery. These CTL express the activation markers CD69 and HLA-DR indicating recent contact with antigen. (70) . This study suggests that sterilizing immunity fails to occur as despite an effective immune response, HBV can persistently replicate at low levels, probably stimulating CTL responses keep the virus in check.

The assays used to deduce these findings about the immune responses in the acute and chronic liver injuries have their limitations, as many studies have relied upon repeated stimulation in culture and measurement of killing using chromium release assays. Repeated stimulations of these highly activated PBMCs results in apoptosis and thus they may not be detected in the final analysis, resulting in an underestimation of the viral specific immune responses (71) . Furthermore, the nature of the assays prevents study of the kinetics of the immune response. Recent development of tetrameric human HLA class I/peptide complexes have helped identify antigen specific $CD8^+$ T cells (72) . These multimeric HLA class I/peptide complexes have a high avidity for T cells displaying the appropriate T-cell receptor and binding to specific cells can be detected by flow cytometric analysis after labelling with a fluorochrome. Application of these technologies to the study of the viral immune responses has shown clearly that previous methods of CTL analysis have underestimated the number of virus-specific CTLs. Further study of patients with acute and chronic HBV with this new technology has enhanced our understanding of the kinetics and vigour of the immune responses to this infection. HLA tetramers have documented that the immune response is characterized by a vigorous peripheral blood $CD8^+$ CTL responses directed against multiple epitopes within the HBV envelope, nucleocapsid or core and polymerase regions with approximately 1 HBV specific T cell per 150 $CD8^+$-cells. This represents about 30-45 fold increase over estimates made in older studies (73) . Tetramer positive cells specific for the core 18-27 epitope are found at higher frequency than those specific for the polymerase 575-583 and envelope 335-343 epitopes in most patients with acute HBV. These T-cell responses are associated with the maximum elevation in serum alanine aminotransferase (ALT) and precede the clearance of hepatitis B envelope and surface antigens and the development of neutralizing antibodies. A complete understanding the dynamics of the immune response in acute HBV has been furthered by the study of events in the days to several weeks immediately following infection by Webster *et al* (60) . Analysis of the NK, CTL and $CD4^+$ T-cell responses in the incubation phase has demonstrated that CTL response increases in parallel with ALT, consistent with the previously observed notions that CTL activity is responsible for liver injury, and to occur after the HBV DNA titers had begun to fall, which again is consistent with the data of Guidotti *et al.* (61) .

4.1.1.2 Chronic HBV

Persistent HBV infection may result because of the failure of initial innate and adaptive immune responses. In newborns who become infected at birth it appears that T-cell tolerance to HBeAg and HBcAg contributes to viral persistence. Cord lymphocytes from infants born to $HBeAg^+$ and $HBeAg^-$ mothers do not show proliferative responses to HBcAg initially, but after development of an acute flare of HBV, significant proliferative responses can be seen to HBcAg suggesting that HBeAg induces a CD4 T-cell tolerance (74) . For HBeAg to induce T-cell tolerance in the newborn it must cross the placenta and while animal studies have argued against this, recent human data supports that it can (78, 79) . Because HBeAg and HBcAg are cross-reactive at the T-cell level, deletion of the T_H-cell responses to HBeAg result in ineffective CTL responses to HBcAg. In addition to the host immunocompetence and level of viremia in the mother, infection with mixed populations of HBV appears to be associated with eradication while infection with wild type virus alone may contribute to chronicity (76) . In addition to the presence of tolerant T-cells, HBsAb production may also be impaired in these neonates because of defective antigen presentation by dendritic cells due to reductions in MHC class II and CD86 molecules. Treatment with IFN-γ restores the expression of these molecules and APC function (77) .

Adult immunocompetent individuals who do not successfully clear acute HBV infection also have less vigorous $CD4^+$ T-cell and CTL responses. In contrast to acute resolving infection, peripheral $CD4^+$ T-cell and CTL responses in those individuals who develop chronic infection are weak and more narrowly focused, at least as measured with ^{51}Cr release (78) . Despite the weak and narrowly focused response of the peripheral blood CTL, HBV-specific $CD4^+$ and $CD8^+$ T-cell clones can be isolated from the livers of chronic HBV patients which are capable of ex-vivo class I-restricted cytolytic activity in response to envelope and core peptides. The cytotoxicity mediated by these cells appears to be sufficiently strong enough to cause liver injury but not strong enough to eradicate virus from all hepatocytes (79, 80) . The reasons for the development of persistent infection are not clearly understood, because of the inability of HBV to infect cells *in vitro* and by the difficulty in studying the lymphocytes in the intra-hepatic compartment. In addition to the virus-specific cells, the inflamed liver contains other recruited cells, the function of which is not entirely clear, but which may participate in the hepatocyte damage by-stander activation (81) . Chronic HBV infection results in two patterns of disease. The classical "chronic carrier" status is characterized by a low level viral replication, with evidence of antibody to the envelope antigen (HBeAg) and no evidence of liver damage, while the patient with "chronic active hepatitis" has evidence of high viral replication, the presence of HBeAg and liver inflammation. Using HLA-A2 restricted tetramers specific for the envelope, core and polymerase peptides of HBV, Maini *et al.* have demonstrated that the median level of peripheral tetramer positive cells was significantly higher in the patients with low level HBV replication and without liver injury compared with those with high HBV DNA levels and abnormal ALT. On the other hand, analysis of the intra-hepatic compartment demonstrated that the actual number of HBV core tetramer positive $CD8^+$ T-cells was the same in both groups, but that those with low level replication and no liver damage had fewer antigen non-specific cells recruited to the liver and no staining for hepatitis B core antigen (HBcAg) in the hepatocytes. In addition, core-specific PBMCs from those with low level viral replication and no liver damage vigorously expanded in vitro in response to core 18-27 synthetic peptides and were capable of specific cytolysis and IFN-γ production, which was not the case in those with high levels of HBV DNA and liver damage. Contrary to previous notions that peripheral T-cell tolerance and cellular exhaustion may result in the absence of an active cellular immune response in HBV chronic carriers, these individuals appear to possess a functionally active HBV-specific immune response which controls HBV replication in the absence of liver damage (82).

It would appear that the essential process for the resolution of viral infection is the recognition and elimination of intracellular viral by a well orchestrated $CD4^+T_H1$ and CTL response. In chronic carriers of HBV infection there is a defect in the immune process which leads to viral persistence. Characterization of the proliferative $CD4^+$ and cytotoxic $CD8^+$

T-cell responses reveals a "hyporesponsiveness". The precise mechanism of this hyporesponsiveness has not been defined, although there may be a viral protein, which can interfere with antigen presentation or cellular activation. This hyporesponsiveness is illustrated by the observation that HBV-specific $CD4^+$ and CTL responses are re-established in patients who respond to either interferon or lamivudine treatment with declines in viral replication. $CD4^+$ T-cell responses become positive within 1-2 weeks of therapy and this is followed by a reduction in HBV DNA levels, after which CTL responses against peptides antigen recognised in acute HBV return. Tetramer staining of HBV specific CTLs shows a lower frequency than that observed during acute HCV infection (64, 83, 84) .

4.1.2 Humoral immunity

HBcAg is extremely immunogenic during HBV infection and after immunization. The mechanism by which this occur is not known, however murine HBcAg binds specifically to membrane bound immunoglobulins (mIg) antigen receptors on HBcAg-specific B-cells with resultant induction of co-stimulatory molecules. This B-cells activation and processing of core antigen allows presentation to naïve T_H-cells in a more efficient manner than antigen presentation by macrophages or dendritic cells. This probably accounts for the vigour of the antibody response to HBcAg that can be detected throughout the course of infection. IgM anti-HBc is the first antibody to be detected, usually appearing within 1 month of the appearance of HBsAg and 1-2 weeks before the rise in ALT. During convalescence the titer of IgM anti-HBc declines while the titers of IgG anti-HBc increases. While IgM anti-HBc is frequently considered to be associated with acute infection, it can persist for up to 2 years in 20% of individuals and low titers may be found in chronically infected individuals, which rise during acute flares in HBV. The titer of IgM anti-HBc has been reported to correlate with the titer of ALT and HBV DNA levels in patients with chronic infection (85).

The development of surface antibody (anti-HBs) follows the disappearance of surface antigen and marks recovery from HBV infection conferring immunity for life. Anti-HBs is sufficient for protection against HBV infection, as demonstrated by the success of HBV vaccines, even if it is not the sole operative mechanism clearing acute infection The development of HBV vaccines is considered one of the major achievements of modern medicine (86) . Successful vaccination is not only effective in preventing HBV infection but it also prevents the sequelae of chronic HBV infection and so is the first example that cancer can be prevented by vaccination (87) . Currently available vaccines are extremely safe and have an efficacy of greater than 90%. The "a" determinant is the predominant B-cell epitope common to all six HBV serotypes. Antibodies against this epitope confer immunity to all HBV subtypes. Co-existence of HBsAg and anti-HBs is reported in up to 24% of chronically infected individuals, in which case the anti-HBs is directed against one of the subtypic determinants and is therefore not able to neutralize the virus. Other surface antigens that stimulate antibody responses include the pre-S1 and pre-S2 antigens. Antibody to these develops during recovery and can be detected before anti-HBs, however, routine serological assays are not readily available. Mutation in the S gene usually occurs in the "a" determinant, which allows the virus to escape antibody neutralization. This may occur in response to the use of hepatitis B immunoglobulin for passive immunization of newborns and transplant recipients.

4.1.3 Mechanism of HBV persistence

All limbs of the immune system must co-operate to eliminate viral infection. Even in clinical recovery from HBV infection, with evidence of serological recovery and normalization of liver function tests, HBV DNA remains detectable for prolonged periods after what appears to be a successful immune response (70) . The mechanisms by which HBV evades the immune response and results in chronic infection in approximately 5% of exposed adults remain obscure. It has been suggested that the size of initial viral infection and viral kinetics maybe such that the immune system is overwhelmed by the virus and even at the height of the innate and adaptive T-cell responses, the virus may still persist (59) . However, as the virus specific CTL response is much more vigorous in acute than chronic infection, other mechanism must play a role.

Infection of immunologically privileged sites has been proposed as a possible mechanism to promote the development of persistent viral infection. There is evidence that HBV can infect extra-hepatic tissues and therefore may evade immune recognition. In addition, apoptosis of activated $CD8^+$ T-cells from the periphery in the liver may afford some protection against immune recognition as well. These activated T-cells may be induced to undergo apoptosis by hepatocytes acting as antigen presenting cells[8] . However, the human liver is also known to contain significant number of resident lymphoid cells and their contribution to the homeostasis of immune response during viral infections is not yet known. In addition to these mechanisms of evasion, hepatocyte can be induced to express Fas ligand during inflammatory response and therefore can induce Fas mediated death of CTLs that they encounter protecting themselves from CTL mediated killing (88) . Alternatively, down regulation of Fas expression on hepatocytes would protect against CTL mediated death. However, it has not yet been established, if HBV can result in Fas ligand expression by hepatocytes or down regulation of Fas expression by these cells. The development of CTL escape mutants has been proposed as a mechanism of HBV persistence, but has not been substantiated experimentally (59) .

4.2 Hepatitis D virus (deltavirus)

Hepatitis D virus (HDV) has tentatively been classified in a separate genus (*Deltavirus*) of the Deltaviridae family. It is a defective single-strand RNA virus that depends on the HBV for propagation but not replication (89) . The HDV is a 1.7-kb single stranded RNA genome, which contains highly conserved regions encoding the HDV antigens. The host range of HDV include man, chimpanzees and ducks carrying the HBV related hepadnavirus. It is a major cause of either acute or chronic hepatitis in humans, with outbreaks of fatal epidemics reported from several parts of the world. HBV is required for successful HDV infection as it provides the coat and allows virus assembly and infectivity. HDV infection can occur simultaneously with HBV or as a super-infection in a HBV carrier and can be eradicated along with HBV.

4.2.1 Cellular immunity

The mechanisms of liver damage are unclear and available study is controversial. It has been suggested, that the hepatocyte injury resulting from HDV infection is caused by a direct viral cytopathic effect rather than the immune mediated damage seen in HBV and HCV. Hepatocyte cell lines infected with plasmid containing the hepatitis D antigen (HDAg) gene resulted in impairment of RNA synthesis and cell death suggesting that HDV might be directly cytopathic to hepatocytes (90, 91) . In addition, microvesicular steatosis of hepatocytes has been observed in association with HDV infection and recurrence after liver transplantation again suggesting a direct cytopathic effect (92-94) . As evidence against this, transgenic mice that express HDAg do not develop liver disease and recurrent HDV in the absence of recurrent HBV can occur after transplantation without evidence of liver disease (95-97). Histological assessment demonstrates that the degree of cellular infiltration in the portal tracts and lobules correlates with the degree of staining for HDAg in the liver, suggesting immune mediated damage of infected hepatocytes. In addition, HBV chronic carriers who develop HDV super-infection undergo a more severe chronic liver injury associated with the presence of a dense infiltration of lymphocytes in the liver (98) . Nisini *et al.* have also demonstrated that PBMCs from individuals with "inactive" HDV super-infection (as defined by normal ALT and absence of IgM anti-HDV) proliferated *in vitro* in response to HDAg and produced IFN-γ, while those from individuals with elevated ALT and high IgM anti-HDV did not. This would suggest that the presence of HDV-specific responses are associated with control of HDV infection similar to that demonstrated for HBV (82, 99) . Woodchucks infected with HDV demonstrate an increased level of viremia when treated with cyclosporin A, and hepatitis D viremia is enhanced in HIV positive individuals indicating that T cell immunity is important in the control of HDV infection.

4.2.2 Humoral immunity

HDAg is the only viral protein known to be expressed during HDV infection. Detection of antibody is the usual method for diagnosis of acute infection. During the HDV infection antibodies of both IgM and IgG classes can be detected in the serum of infected individuals. A high titer of IgM anti-HDV is strongly associated both to elevated hepatitis D viremia and to the severity of liver injury, whereas a more favourable course to HDV infection is found in individuals with IgG-anti HDV. While these antibody responses are present during acute and chronic infection, there is no convincing evidence of a protective role of anti-HDV antibodies (99) . Woodchucks infected with woodchuck hepadnavirus and immunized with recombinant HDAg are only partially protected from subsequent challenge with HDV in the absence of humoral responses, which suggests that other mechanisms are responsible for immunity (100) .

4.3 HEPATITIS C

Cloned in 1989, hepatitis C virus (HCV) is now recognized as a common cause of chronic liver disease, cirrhosis, and hepatocellular carcinoma in the world (101) . More than 170 million people are estimated to be infected worldwide with an estimated 2.8 million persistently infected in the US. The natural history of the infection is such that only about 15% of infected individuals successfully clear the infection and the remainder develop chronic disease of varying severity. At this present time, it is the leading cause for liver transplantation in the US. The host and viral factors that are responsible for viral persistence and those that are associated with successful spontaneous eradication have not been clearly elucidated as of yet. This is mainly because the vast majority of acute HCV infections are sub-clinical with only 20% presenting with jaundice, therefore making it difficult to identify and study individuals during the critical period that is thought to dictate eventual disease outcome (102-104) . For the most part, patients with HCV infection present with chronic infection rather than the acute illness and because of this, the vast majority of studies on HCV specific immune responses have focused on chronic rather than acute/resolving infection. Furthermore, our understanding of the immune responses during the acute phase of HCV infection is also limited by the absence of a small suitable animal model. The only available animal model of HCV infection is the chimpanzee, which, because of limited number of animals, reagents, and expense has hampered our understanding of acute HCV (105) .

4.3.1 Cellular immunity

4.3.1.1 Acute HCV

The characteristics of HCV-specific T cell responses have been the focus of much research because they are believed to play critical roles in viral elimination and disease pathogenesis. Again, the absence of a robust tissue culture system limits our understanding of the effect of HCV upon cell survival, but at least two lines of evidence support the fact that HCV is not directly cytopathic in the classic sense. First, some transgenic animals expressing HCV proteins do not develop cytopathic changes or liver inflammation (106, 107) . Second, cell lines expressing HCV under the control of inducible promoters do not develop cytopathic changes once HCV expression is turned on, although the cells do have delayed growth kinetics (108). Some other lineages of transgenic animals do develop hepatic steatosis and HCC, presumably in the setting of T cell tolerance, which suggests that HCV proteins or RNA may interfere with host cell function (109, 110) .

Clinical observations suggest that the interaction between the virus and the immune response in the first few weeks after the infection may determine the outcome of HCV infection. Proliferative responses of $CD4^+$ T_H-cells from individuals who recover are vigorous and multi-specific. These $CD4^+$ T-cell subsets are crucial in helping B cell production of antibody, priming of $CD8^+$-CTLs and in maintaining CTL memory (111) . Diepolder *et al.* were the first to show that individuals with a vigorous peripheral blood proliferative response to HCV antigens, particularly the helicase domain (NS3), appear to have self-limited disease (112) . Since then, this and other groups have shown that these vigorous and polyclonal responses are associated with self-limited disease and response to interferon therapy (103, 113-117) . Analysis of the cytokine profile of bulk cultures and peripheral $CD4^+$ T_H-clones from individuals with acute self-limiting HCV infection reveals that viral clearance is more common in cases with secretion of a T_H1 cytokine profile (IFNγ and IL-2) than in individuals with predominantly a T_H2 cytokine profile and a less vigorous proliferative response (34) . Furthermore, these cellular responses can persist for long periods as suggested by the strong T_H1 responses directed against both structural and non-structural HCV antigens persisting in individuals up to 17 years after spontaneous clearance (118) . Of further interest is a group of patients who develop transient and low-level proliferative $CD4^+$ T_H responses to HCV antigens lasting up to 10 months following an acute infection, all of whom have a relapse of viremia coincident with loss of HCV-specific $CD4^+$ responses. The reasons for, or the consequences of, a partial cellular immune response in these individuals, are not known (104). Therefore, it would appear that the vigour, epitope specificity, and cytokine profile of the initial $CD4^+$ response have an important bearing on the outcome of acute HCV infection.

CTL-mediated lysis of virally infected cells may result in clearance of the infection or, if incomplete, persistence of the infection and eventual tissue damage. Based on this, it would appear that a successful CTL response is crucial in preventing liver damage. Analysis of CTL responses is limited because of the small numbers of patients presenting with acute HCV and because of the inherent technical difficulties in working with these cells. Moreover, it is not ethical to perform liver biopsies on patients with acute viral hepatitis and so all human studies examining HCV specific-CTL responses have been performed on peripheral blood samples. Many of the studies on HCV specific CTLs have used predicted epitopes based on known HLA class I binding motifs, such as HLA A2.1, which limits ones scope to those predicted peptides and HLA class I responses on which sufficient data is available. Similarly, the use of techniques such as chromium release cytotoxicity assays to enumerate antigen specific cells may result in underestimation of these cells because of apoptosis during repeated rounds of stimulation. Despite this we know from chimpanzee studies that elimination of HCV infection in the acute phase is associated with the development of a multispecific intra-hepatic CTL response, while the persistence of infection is associated with a delayed and weaker intra-hepatic response against fewer epitopes (119) . In humans, a similar early and polyclonal response is seen in those with self-limiting disease (120) . The development of newer techniques such as Class I tetramers and ELISPOT assays, which can detect antigen specific cells by virtue of their secreted cytokines in response to specific antigen, have provided further tools with which we can more efficiently measure the effectiveness of the early CTL responses (121, 122) . Gruner *et al.* have demonstrated using ELISPOT techniques, a significant $CD8^+$ T-cell response directed against multiple class I-restricted HCV epitopes of structural and non-structural regions of the HCV polyprotein, which correlates with elimination of the virus. The reported frequency reported by Gruner in the peripheral blood of individuals with acute HCV was 0.2% of total $CD8^+$-cells. Significant proliferative responses of $CD4^+$ T-cells to NS3 and NS4 proteins accompanied the CTL responses and were maintained in those patients with resolving infection (123) . Lechner *et al.* using a combination of class I tetramer staining and ELISPOT techniques have also demonstrated that the successful eradication of HCV is associated with a peripheral blood CTL frequency of greater than 7% at the peak of clinical illness, which is similar to that noted for infectious mononucleosis, and that this CTL response was simultaneously directed against 8 different epitopes. This frequency of CTL contrasts with the low levels (0.07%) found in the peripheral blood of those individuals with chronic infection (124) . In addition, a strong activated CTL response (7.4% of total CTLs) directed against the one particular epitope (NS5 2594) was detected when the individual was jaundiced and corresponded with maximum levels of alanine aminotransaminase (ALT) reflecting hepatocyte destruction. This reduction in ALT levels was mirrored by a reduction in the number and activation status of the NS5 2594-specific CTLs. Additionally, Lechner *et al.* also calculated the frequency of $CD4^{+}T_{H}$-cells during the acute infection. They found that antigen specific $CD4^{+}$ $T_{H}1$, IFN-γ producing cells were maximal at the first time point and comprised about 3% of the total $CD4^{+}$ lymphocytes, preceding the peak magnitude of the IFN-γ CTLs. The differences observed between the frequency observed by Gruner *et al.* and those of Lechner *et al.* may be related to the timing of the blood draw. In the latter study, CTLs were isolated at the time of acute hepatitis while Gruner *et al.* isolated CTL sometime within the 6 months after onset of infection. Several studies have demonstrated the persistence of HCV specific CTL responses in the absence of serological or molecular evidence of HCV infection indicating the presence of CTL memory, possible persisting undetectable virus, or cross reactive responses against heterologous proteins (121, 125) . Whether these "memory" CTL responses can protect against future exposure remains to be seen. Therefore, acute self-limiting HCV infection is characterized by vigorous multispecific $CD4^{+}$ $T_{H}1$ and CTL response (112, 119, 126) .

4.3.1.2 Chronic HCV

Failure to eradicate HCV infection results in the development of chronic HCV infection with resultant chronic hepatitis from which a significant proportion of individuals will develop cirrhosis and hepatocellular carcinoma. The mechanism by which chronic HCV infection develops in the majority of infected persons remains unclear, but it does so despite the presence of virus-specific $CD4^{+}$ and $CD8^{+}$ T-cell responses in the peripheral blood and the liver, suggesting that these responses are ineffective for the most part. In chronic HCV infection, MHC class I-restricted HCV specific

$CD8^+$ T-cells, detected in the liver and less frequently in the peripheral blood, are capable of recognizing conserved and variable regions of the HCV polyprotein in the context of different MHC molecules (127-129) . There is considerable heterogeneity in the CTL responses with no immunodominant response to a particular protein. While these CTL responses fail to eliminate virus, there is evidence to suggest that the CTL response in the peripheral blood and the liver is able to exert some control over viral replication (130-132) . The presence of these CTL in the liver in chronic infection suggests that these CTLs contribute to the pathogenesis of chronic liver disease through Fas/Fas-ligand and TNF-α mediated apoptosis of virally infected cells and "innocent bystanders" (133) .

In addition, peripheral $CD4^+$ T-cell proliferative responses, which are vigorous and persist indefinitely in the acute setting, respond weakly or not at all in the periphery of the chronically infected individuals (134) . There is an association between strong $CD4^+$ T-cell responses and the absence of histological progression in immunocompetent host and in determining disease progression in post liver transplant recurrence (134-136). Detailed animal studies surrounding the earliest events in HCV replication that facilitate the development of chronic infection, have not yet been published, but HCV has a number of potential mechanisms to evade host immune responses which will be discussed later.

By inference, cellular immune responses are implicated in liver damage, and there is some direct evidence for this. First, transgenic mice expressing HCV proteins do not develop the typical histologic pattern of liver inflammation characteristic of chronic HCV infection (106, 137). Second, cells isolated from liver tissue are capable of lysing autologous hepatocytes (138-140). Third, treatment of an HCV-infected patient with anti-CD8 monoclonal antibodies resulted in improvement of serum transaminases, suggesting that the $CD8^+$ T-cells were the cause if liver damage. The vigour of the inflammatory response as measured by cytokine production within the liver is also correlated with the extent of liver injury (141). However, HCV-specific cellular immune responses may not be the sole explanation, as evidence by multiple clinical observations that describe an accelerated natural history of HCV-related liver damage in patients with immunosuppression, whether that is exogenously induced after liver transplantation or subsequent to human immunodeficiency virus (HIV)-induced loss of $CD4^+$ T-cells (142, 143). As discussed previously, this suggests that there may be some direct effect of HCV expressed at high levels on cell function or some cytokine mediated liver injury. Understanding to what extent cellular immune responses limit viral replication and liver injury and to what extent, they induce liver injury remains a paradox that awaits resolution.

4.3.2 Humoral immunity

After HCV was cloned in 1989, screening of overlapping cDNA clones against sera from individuals with post-transfusion NANB hepatitis resulted in the identification of reactive peptides, which encoded for B-cell epitopes. Subsequently the development of multiple generations of antibody testing to detect HCV antigens provided clear evidence that HCV infection is associated with the development of antibodies against both structural and non-structural antigens. After exposure to HCV, the majority of immunocompetent individuals will seroconvert within a 4-week period (144). Both the order and pattern of antibody response is variable but the first antibody responses to develop are usually those against the NS3 (anti-c33) and core (anti-c22) with anti-NS4 and anti-E1/E2 developing later. Unlike the development of antibodies in HBV infection, the development of anti-HCV antibodies does not herald resolution of the disease process and lasting immunity. Similarly, IgM responses to core have been described in HCV infection, but they do not necessarily represent acute infection as the appearance of these antibodies may post-date the development of IgG response, are usually associated with the presence of viremia, and do not occur in all individuals (145). In the face of continuing viremia, antibody responses are directed against multiple epitopes, while acute self-limiting infection is generally associated with a transient antibody response that gradually declines or one that does not develop at all (119, 146, 147). Numerous studies have proposed associations between the pattern of antibody response and disease outcome, but have given conflicting results about the nature of this response and outcome (148, 149). Recently, animal and human studies have consistently shown that low titers of anti-E_2 antibodies are seen in acute infection when compared with chronic disease and the presence of anti-E_2 antibodies is not associated with viral clearance. This suggests that E_2 antibodies do not have neutralizing capacity or that the envelope is not relevant to viral persistence (119, 150-153). These HCV-specific antibody responses are present in the vast majority of individuals with persistent viremia, suggesting that they appear to be inefficient in the control of infection.

Studies in the chimpanzee have demonstrated that repetitive re-challenge of convalescent animals with homologous or heterologous strains of HCV results in reappearance of viremia in spite of the presence of antibodies to HCV indicating that they are no protective value (154). However, more recent animal studies have shown that neutralizing antibodies can be raised by repeated immunization that are protective against infection with homologous virus; and plasma infectivity can be neutralized by serum from the same animal after 2 years but not after 11 years indicating development of escape mutants over time (155, 156). Thus, it appears that while weak neutralizing antibody responses are formed, escape mutation allow evasion of the humoral immune response. Mutations within the N-Terminus of HVR1, which has been shown to occur rapidly in, infected individuals with rapid disappearance of the anti-HVR1 antibodies (167). Secondly, antibody responses to the envelope proteins develop slowly, achieve only modest titers and tend to be short lived, suggesting that neutralizing antibodies may emerge too late or may not be strong enough to prevent chronic infection (152).

Although there is little evidence to implicate the humoral immune response in the pathogenesis of HCV liver disease, there is accumulating evidence to suggest that B-cells are important mediators of some of the extrahepatic manifestations of chronic HCV disease. Numerous investigators have proposed that HCV infection is associated with the development of essential mixed cryoglobulinemia (EMC), (see ref [158] for review) autoantibody production and with the development of lymphomas. The findings of expanded $CD5^+$ B-cell subsets in the peripheral blood and liver of HCV infected individuals and the presence of functionally active clonal B-cell lymphoid aggregates in the liver of HCV infected individuals support the fact that B cells are important in HCV mediated EMC and B-cell dyscrasias (159-162). More recently, Chan *et al.* have provided strong evidence to suggest that these HCV associated B-cell proliferations are antigen driven. The B-cells from HCV associated EMC and lymphomas used the same restricted V_H gene sequence (V_H1-*69*) as B-cells that produce antibody against the HCV E_2 glycoprotein (163).

The immune response to HCV is polyclonal and multispecific, both in terms of antibody production and cellular immunity. Resolution of infection is associated with a vigourous $CD4^+$ T_H1 response directed against one or more antigens. CTL responses are directed against multiple non-dominant epitopes and appear to control viral replication but may contribute to chronic infection in those individuals who do not clear infection. Vaccine development for HCV faces a number of obstacles. Firstly, the absence of a reliable tissue culture system or small animal model makes it difficult to study the immunological correlates of viral eradication or disease progression; and secondly, there is a substantial amount of genetic heterogeneity among different genotypes and subtypes, which would make a global vaccine difficult. A better understanding of the humoral and cellular immune responses to HCV will enhance our understanding and aid future vaccine development.

4.3.3 Mechanisms of HCV persistence

In contrast to chronic HBV infection, almost 85% of patients infected with HCV will develop persistent infection with chronic hepatitis.

The reasons for this high frequency of chronicity are unknown but the evasion of the immune responses by HCV result in significant morbidity and mortality. One of these is the relative immune privilege of the liver, which may participate in the removal of activated $CD8^+$ T-cells (4). This is supported by lines of evidence demonstrating "compartmentalization" of virus specific $CD4^+$ T-cells in chronic HCV. Minutello *et al.* demonstrated that there is sequestration of virus-specific T cells in the liver during chronic HCV infection, as indicated by a different pattern of epitope specificity and T cell receptor usage (164). Schirren *et al.* have also demonstrated that the $CD4^+$ T-cell responses are compartmentalized to the liver as opposed to the peripheral blood, where they secrete a T_H1 cytokine profile (IFN-γ) which theoretically could continually stimulate the CTL and resulting in chronic liver disease (165). A number of other hypotheses have been proposed for the development of persistent HCV infection.

HCV is known to infect and replicate within peripheral blood mononuclear cells, which might contribute to evasion of immune responses (166). In addition to this, it has been suggested that HCV infection fails to induce an effective innate and adaptive immune response during the early infection, which may explain the low frequency of clinically evident acute infection (167). The intensity of the immune response is generally dependent on the cytopathic effects of virus, viral load, co-stimulatory signals, and type of APC. In HCV infection, dendritic cells may have impaired allostimulatory activity and this, coupled with antigen presentation by hepatocytes in the absence of co-stimulation, may result in an impaired immune response (168).

The genetic variability of HCV is a major contributing factor to the development of chronicity. The genetic diversity of HCV is manifest in six major genotypes, multiple subtypes, and a host of quasispecies that can be isolated from any chronically infected individual. The absence of proofreading activity by the RNA-dependent polymerase results in the continuous development of viral variants. Given the high degree of variability in the viral quasispecies within infected individuals, the emergence of viral escape mutants in antibody recognizing sites and CTL epitopes is another potential mechanism. Farci *et al.* have demonstrated during the acute infection that there is relative stasis of the quasispecies population in those who have acute self-limited infection, but significant variation in the quasispecies of those that develop progressive infection (169). Furthermore, complex quasispecies with sequence change variation in the HVR1 have been shown to correlate with disease persistence (170, 171). Mutations in the hypervariable region 1 (HVR1) of the envelope are less numerous in immunocompromised individuals than they are in immunocompetent individuals suggesting that antibody pressure may result in viral escape mutants, however, it has also been noted that persistent infection in chimpanzees can occur without variation in the HVR1 (150, 172, 173). Similarly, CTL escape mutants have been reported in chimpanzees by Weiner and colleagues. CTL escape mutants have also been reported by Kaneko *et al.* and Chang *et al.* in human studies (171, 174, 175). Tsai *et al.* have reported the development of mutant epitopes in HVR1 which function as TCR antagonists capable of inhibiting HVR1-specific CTL activity in two patients who developed chronic infection, consistent with the notion that early CTL escape promotes chronicity (34). Recently Christie *et al.* in a longitudinal study of virus mutations have challenged the notion of immune pressure inducing viral escape mutants. They have described mutations in the envelope region of HCV over a 2-year period, which are not associated with known B-cell or CTL epitopes. In addition, they noted no mutations in the core region despite CTL recognition (176). Other mechanism for viral persistence, which are as yet unproven in HCV persistence include alterations in cytokine secretion pattern induced by variation within cellular epitopes (177, 178); the virus mediated induction of Fas ligand expression by hepatocytes thereby destroying CTLs or virus mediated down regulation of Fas expression, thereby protecting itself from Fas-mediated apoptosis (179, 180).

5. CONCLUSIONS

Hepatitis viruses result in both acute and chronic infections with resultant liver injury, which for the most part is mediated by non-antigen-specific innate and antigen-specific adaptive immune responses. The clinical outcome of the infection is determined by the quality and vigour of the antiviral cellular and humoral immune responses produced by the host (59, 111). The cellular immune response, which dictates disease outcome are best, described for HBV infection due to the burden of disease, clinically apparent acute disease, and the availability of a small animal model, although much work is being done in HCV as well. Very little is known about the immune responses to HAV, HEV and HDV infections and the mechanisms which fail to control the development of fulminant liver failure still remain to be elucidated.

In acute resolving infections, a co-ordinated and well-orchestrated innate immune response resulting in non-cythopathic viral elimination mediated predominantly by the production of T_H1 cytokines followed by vigourous $CD4^+$ T_H1 responses which serve a critically important regulatory function by secreting cytokines that can facilitate B-cell maturation, expansion and antibody secretion or that foster the development of a strong CTL response (60, 181).

The precise mechanisms by which such a coordinated response fails are not well defined. However, as shown for HBV infection even the best outcome of a well co-ordinated immune response, which results in efficient cellular responses and serological recovery, does not result in sterilizing immunity leaving the host open to recurrent infection in the event of immunosuppression (70). In spite of all the progress in determining the mechanism of cell injury and many pieces of experimental evidence reported on viral chronicity, we still lack the knowledge to explain adequately why and how viral persistence occur in the case of HBV and HCV. It may results because of an inefficient immune response, evasion by the virus, escape mutations, or virus induced apoptosis of antigen specific-CTLs. It would appear that some or all of these mechanisms play some role to a greater or lesser degree depending on the virus involved. Understanding the kinetics of the immune response and the balance between effective control and liver damage remains an important challenge for the future.

REFERENCES

1. Blumberg B.S., Sutnick A.I., London W.T., Millman I. Australia antigen and hepatitis. N Engl J Med 1970; 283:349-354.
2. Kubo Y., Takeuchi K., Boonmar S., Katayama T., Choo Q.L., Kuo G., Weiner A.J., Bradley D.W., Houghton M., Saito I., et al. A cDNA fragment of hepatitis C virus isolated from an implicated donor of post-transfusion non-A, non-B hepatitis in Japan. Nucleic Acids Res 1989; 17:10367-10372.
3. Doherty D.G., O'Farrelly C. Innate and adaptive lymphoid cells in the human liver. Immunol Rev 2000; 174:5-20.
4. Mehal W.Z., Azzaroli F., Crispe I.N. Immunology of the healthy liver: old questions and new insights. Gastroenterology 2001; 120:250-60.
5. Curry M.P., Norris S., Golden-Mason L., Doherty D.G., Deignan T., Collins C., Traynor O., McEntee G.P., Hegarty J.E., O'Farrelly C. Isolation of lymphocytes from normal adult human liver suitable for phenotypic and functional characterization. J Immunol Methods 2000; 242:21-31.
6. Norris S., Collins C., Doherty D.G., Smith F., McEntee G., Traynor O., Nolan N., Hegarty J., O'Farrelly C. Resident human hepatic lymphocytes are phenotypically different from circulating lymphocytes. J Hepatol 1998; 28:84-90.
7. Doherty D.G., Norris S., Madrigal-Estebas L., McEntee G., Traynor O., Hegarty J.E., O'Farrelly C. The human liver contains multiple populations of NK cells, T cells, and CD3+CD56+ natural T cells with distinct cytotoxic activities and Th1, Th2, and Th0 cytokine secretion patterns. J Immunol 1999; 163:2314-2321.
8. Bertolino P., Klimpel G., Lemon S.M. Hepatic inflammation and immunity: a summary of a conference on the function of the immune system within the liver. Hepatology 2000; 31:1374-1378.
9. Nuti S., Rosa D., Valiante N.M., Saletti G., Caratozzolo M., Dellabona P., Barnaba V., Abrignani S. Dynamics of intra-hepatic lymphocytes in chronic hepatitis C: enrichment for Valpha24+ T cells and rapid elimination of effector cells by apoptosis. Eur J Immunol 1998; 28:3448-3455.

10. Crosbie O.M., Reynolds M., McEntee G., Traynor O., Hegarty J.E., O'Farrelly C. In vitro evidence for the presence of hematopoietic stem cells in the adult human liver. Hepatology 1999; 29:1193-1198.
11. Collins C., Norris S., McEntee G., Traynor O., Bruno L., von Boehmer H., Hegarty J., O'Farrelly C. RAG1, RAG2 and pre-T cell receptor alpha chain expression by adult human hepatic T cells: evidence for extrathymic T cell maturation. Eur J Immunol 1996; 26:3114-3118.
12. Kakimi K., Guidotti L.G., Koezuka Y., Chisari F.V. Natural killer T cell activation inhibits hepatitis B virus replication in vivo. J Exp Med 2000; 192:921-930.
13. Koziel M.J. Cytokines in viral hepatitis. Semin Liver Dis 1999; 19:157-169
14. Ballardini G., Groff P., Pontisso P., Giostra F., Francesconi R., Lenzi M., Zauli D., Alberti A., Bianchi F.B. Hepatitis C virus (HCV) genotype, tissue HCV antigens, hepatocellular expression of HLA-A,B,C, and intercellular adhesion-1 molecules. Clues to pathogenesis of hepatocellular damage and response to interferon treatment in patients with chronic hepatitis C. J Clin Invest 1995; 95:2067-2075.
15. Barbatis C., Woods J., Morton J.A., Fleming K.A., McMichael A., McGee J.O. Immunohistochemical analysis of HLA (A, B, C) antigens in liver disease using a monoclonal antibody. Gut 1981; 22:985-991.
16. Albert M.L., Sauter B., Bhardwaj N. Dendritic cells acquire antigen from apoptotic cells and induce class I-restricted CTLs. Nature 1998; 392:86-89.
17. Albert M.L., Pearce S.F., Francisco L.M., Sauter B., Roy P., Silverstein R.L., Bhardwaj N. Immature dendritic cells phagocytose apoptotic cells via alphavbeta5 and CD36, and cross-present antigens to cytotoxic T lymphocytes. J Exp Med 1998; 188:1359-1368.
18. Schirmbeck R., Melber K., Kuhrober A., Janowicz Z.A., Reimann J. Immunization with soluble hepatitis B virus surface protein elicits murine H-2 class I-restricted CD8+ cytotoxic T lymphocyte responses in vivo. J Immunol 1994; 152:1110-1119.
19. Tortorella D., Gewurz B., Schust D., Furman M., Ploegh H. Down-regulation of MHC class I antigen presentation by HCMV; lessons for tumor immunology. Immunol Invest 2000; 29:97-100.
20. Huang J., Kwong J., Sun E.C., Liang T.J. Proteasome complex as a potential cellular target of hepatitis B virus X protein. J Virol 1996; 70:5582-5591.
21. Lau J.Y., Xie X., Lai M.M., Wu P.C. Apoptosis and viral hepatitis. Semin Liver Dis 1998; 18:169-176
22. Berke G. Killing mechanisms of cytotoxic lymphocytes. Curr Opin Hematol 1997; 4:32-40.
23. Gonzalez-Amaro R., Garcia-Monzon C., Garcia-Buey L., Moreno-Otero R., Alonso J.L., Yague E., Pivel J.P., Lopez-Cabrera M., Fernandez-Ruiz E., Sanchez-Madrid F. Induction of tumor necrosis factor alpha production by human hepatocytes in chronic viral hepatitis. J Exp Med 1994; 179:841-848.
24. Daniels H.M., Meager A., Eddleston A.L., Alexander G.J., Williams R. Spontaneous production of tumour necrosis factor alpha and interleukin-1 beta during interferon-alpha treatment of chronic HBV infection. Lancet 1990; 335:875-877.
25. Unutmaz D., Pileri P., Abrignani S. Antigen-independent activation of naive and memory resting T cells by a cytokine combination. J Exp Med 1994; 180:1159-1164.
26. Zinkernagel R.M., Haenseler E., Leist T., Cerny A., Hengartner H., Althage A. T cell-mediated hepatitis in mice infected with lymphocytic choriomeningitis virus. Liver cell destruction by H-2 class I-restricted virus-specific cytotoxic T cells as a physiological correlate of the 51Cr-release assay? J Exp Med 1986; 164:1075-1092.
27. Moriyama T., Guilhot S., Klopchin K., Moss B., Pinkert C.A., Palmiter R.D., Brinster R.L., Kanagawa O., Chisari F.V. Immunobiology and pathogenesis of

hepatocellular injury in hepatitis B virus transgenic mice. Science 1990; 248:361-364.

28. Guidotti L.G., Ando K., Hobbs M.V., Ishikawa T., Runkel L., Schreiber R.D., Chisari F.V. Cytotoxic T lymphocytes inhibit hepatitis B virus gene expression by a noncytolytic mechanism in transgenic mice. Proc Natl Acad Sci USA 1994;91:3764-3768.
29. Price D.A., Klenerman P., Booth B.L., Phillips R.E., Sewell A.K. Cytotoxic T lymphocytes, chemokines and antiviral immunity. Immunol Today 1999; 20:212-216.
30. Guidotti L.G., Chisari F.V. To kill or to cure: options in host defense against viral infection. Curr Opin Immunol 1996; 8:478-483.
31. Guidotti L.G., Borrow P., Hobbs M.V., Matzke B., Gresser I., Oldstone M.B., Chisari F.V. Viral cross talk: intracellular inactivation of the hepatitis B virus during an unrelated viral infection of the liver. Proc Natl Acad Sci USA 1996; 93:4589-4594.
32. Rossol S., Marinos G., Carucci P., Singer M.V., Williams R., Naoumov N.V. Interleukin-12 induction of Th1 cytokines is important for viral clearance in chronic hepatitis B. J Clin Invest 1997; 99:3025-3033.
33. Penna A., Del Prete G., Cavalli A., Bertoletti A., D'Elios M.M., Sorrentino R., D'Amato M., Boni C., Pilli M., Fiaccadori F., Ferrari C. Predominant T-helper 1 cytokine profile of hepatitis B virus nucleocapsid-specific T cells in acute self-limited hepatitis B. Hepatology 1997; 25:1022-1027.
34. Tsai S.L., Liaw Y.F., Chen M.H., Huang C.Y., Kuo G.C. Detection of type 2-like T-helper cells in hepatitis C virus infection: implications for hepatitis C virus chronicity. Hepatology 1997; 25:449-458.
35. Kusters S., Gantner F., Kunstle G., Tiegs G. Interferon gamma plays a critical role in T cell-dependent liver injury in mice initiated by concanavalin A. Gastroenterology 1996; 111:462-471.
36. Ohta A., Sekimoto M., Sato M., Koda T., Nishimura S., Iwakura Y., Sekikawa K., Nishimura T. Indispensable role for TNF-alpha and IFN-gamma at the effector phase of liver injury mediated by Th1 cells specific to hepatitis B virus surface antigen. J Immunol 2000; 165:956-961.
37. Toyonaga T., Hino O., Sugai S., Wakasugi S., Abe K., Shichiri M., Yamamura K. Chronic active hepatitis in transgenic mice expressing interferon-gamma in the liver. Proc Natl Acad Sci USA 1994; 91:614-618.
38. Feinstone S.M., Kapikian A.Z., Purceli R.H. Hepatitis A: detection by immune electron microscopy of a viruslike antigen associated with acute illness. Science 1973; 182:1026-1028.
39. Dotzauer A., Gebhardt U., Bieback K., Gottke U., Kracke A., Mages J., Lemon S.M., Vallbracht A. Hepatitis A virus-specific immunoglobulin A mediates infection of hepatocytes with hepatitis A virus via the asialoglycoprotein receptor. J Virol 2000; 74:10950-10957.
40. Mathiesen L.R., Moller A.M., Purcell R.H., London W.T., Feinstone S.M. Hepatitis A virus in the liver and intestine of marmosets after oral inoculation. Infect Immun 1980; 28:45-48.
41. Hashimoti E., Kojimahara N., Noguchi S., Taniai M., Ishiguro N., Hayashi N. Immunohistochemical characterization of hepatic lymphocytes in acute hepatitis A, B, and C. J Clin Gastroenterol 1996; 23:199-202.
42. Muller C., Godl I., Gottlicher J., Wolf H.M., Eibel M.M. Phenotypes of peripheral blood lymphocytes during acute hepatitis A. Acta Paediatr Scand 1991; 80:931-937.

43. Baba M., Hasegawa H., Nakayabu M., Fukai K., Suzuki S. Cytolytic activity of natural killer cells and lymphokine activated killer cells against hepatitis A virus infected fibroblasts. J Clin Lab Immunol 1993; 40:47-60.
44. Vallbracht A., Gabriel P., Maier K., Hartmann F., Steinhardt H.J., Muller C., Wolf A., Manncke K.H., Flehmig B. Cell-mediated cytotoxicity in hepatitis A virus infection. Hepatology 1986;6:1308-1314.
45. Glikson M., Galun E., Oren R., Tur-Kaspa R., Shouval D. Relapsing hepatitis A. Review of 14 cases and literature survey. Medicine (Baltimore) 1992; 71:14-23.
46. Liaw Y.F., Yang C.Y., Chu C.M., Huang M.J. Appearance and persistence of hepatitis A IgM antibody in acute clinical hepatitis A observed in an outbreak. Infection 1986; 14:156-158.
47. Kao H.W., Ashcavai M., Redeker A.G. The persistence of hepatitis A IgM antibody after acute clinical hepatitis A. Hepatology 1984; 4:933-936.
48. Gauss-Muller V., Lottspeich F., Deinhardt F. Characterization of hepatitis A virus structural proteins. Virology 1986; 155:732-736.
49. Gauss-Muller V., Deinhardt F. Immunoreactivity of human and rabbit antisera to hepatitis A virus. J Med Virol 1988; 24:219-228.
50. Wang C.H., Tschen S.Y., Heinricy U., Weber M., Flehmig B. Immune response to hepatitis A virus capsid proteins after infection. J Clin Microbiol 1996; 34:707-713.
51. Aggarwal R., Kini D., Sofat S., Naik S.R., Krawczynski K. Duration of viraemia and faecal viral excretion in acute hepatitis E. Lancet 2000; 356:1081-1082.
52. Longer C.F., Denny S.L., Caudill J.D., Miele T.A., Asher L.V., Myint K.S., Huang C.C., Engler W.F., LeDuc J.W., Binn L.N., et al. Experimental hepatitis E: pathogenesis in cynomolgus macaques (Macaca fascicularis). J Infect Dis 1993; 168:602-609.
53. Chau K.H., Dawson G.J., Bile K.M., Magnius L.O., Sjogren M.H., Mushahwar I.K. Detection of IgA class antibody to hepatitis E virus in serum samples from patients with hepatitis E virus infection. J Med Virol 1993; 40:334-338.
54. Riddell M.A., Li F., Anderson D.A. Identification of immunodominant and conformational epitopes in the capsid protein of hepatitis E virus by using monoclonal antibodies. J Virol 2000; 74:8011-8017.
55. Befeler A.S., Di Bisceglie A.M. Hepatitis B. Infect Dis Clin North Am 2000; 14:617-632.
56. Lok A. Acute viral hepatitis in chronic carriers of hepatitis B virus: different patterns in different places. Hepatology 1989; 10:252-253.
57. Beasley R.P., Hwang L.Y., Lin C.C., Chien C.S. Hepatocellular carcinoma and hepatitis B virus. A prospective study of 22 707 men in Taiwan. Lancet 1981; 2:1129-1133.
58. Chang M.H. Chronic hepatitis virus infection in children. J Gastroenterol Hepatol 1998; 13:541-548.
59. Chisari F.V. Rous-Whipple Award Lecture. Viruses, immunity, and cancer: lessons from hepatitis B. Am J Pathol 2000; 156:1117-1132.
60. Webster G.J., Reignat S., Maini M.K., Whalley S.A., Ogg G.S., King A., Brown D., Amlot P.L., Williams R., Vergani D., Dusheiko G.M., Bertoletti A. Incubation phase of acute hepatitis B in man: dynamic of cellular immune mechanisms. Hepatology 2000; 32:1117-1124.
61. Guidotti L.G., Rochford R., Chung J., Shapiro M., Purcell R., Chisari F.V. Viral clearance without destruction of infected cells during acute HBV infection. Science 1999; 284:825-829.
62. Ferrari C., Bertoletti A., Penna A., Cavalli A., Valli A., Missale G., Pilli M., Fowler P., Giuberti T., Chisari F.V., et al. Identification of immunodominant T cell epitopes of the hepatitis B virus nucleocapsid antigen. J Clin Invest 1991; 88: 214-222.

63. Tsai S.L., Chen P.J., Lai M.Y., Yang P.M., Sung J.L., Huang J.H., Hwang L.H., Chang T.H., Chen D.S. Acute exacerbations of chronic type B hepatitis are accompanied by increased T cell responses to hepatitis B core and e antigens. Implications for hepatitis B e antigen seroconversion. J Clin Invest 1992; 89:87-96.
64. Koziel M.J. What once was lost, now is found: restoration of hepatitis B-specific immunity after treatment of chronic hepatitis B. Hepatology 1999; 29:1331-1333.
65. Wild J., Grusby M.J., Schirmbeck R., Reimann J. Priming MHC-I-restricted cytotoxic T lymphocyte responses to exogenous hepatitis B surface antigen is CD4+ T cell dependent. J Immunol 1999; 163:1880-1887.
66. Diepolder H.M., Jung M.C., Keller E., Schraut W., Gerlach J.T., Gruner N., Zachoval R., Hoffmann R.M., Schirren C.A., Scholz S., Pape G.R. A vigorous virus-specific CD4+ T cell response may contribute to the association of HLA-DR13 with viral clearance in hepatitis B. Clin Exp Immunol 1998; 113:244-251.
67. Penna A., Chisari F.V., Bertoletti A., Missale G., Fowler P., Giuberti T., Fiaccadori F., Ferrari C. Cytotoxic T lymphocytes recognize an HLA-A2-restricted epitope within the hepatitis B virus nucleocapsid antigen. J Exp Med 1991; 174:1565-1570.
68. Rehermann B., Fowler P., Sidney J., Person J., Redeker A., Brown M., Moss B., Sette A., Chisari F.V. The cytotoxic T lymphocyte response to multiple hepatitis B virus polymerase epitopes during and after acute viral hepatitis. J Exp Med 1995; 181:1047-1058.
69. Rehermann B., Chang K.M., McHutchison J.G., Kokka R., Houghton M., Chisari F.V. Quantitative analysis of the peripheral blood cytotoxic T lymphocyte response in patients with chronic hepatitis C virus infection. J Clin Invest 1996; 98:1432-1440.
70. Rehermann B., Ferrari C., Pasquinelli C., Chisari F.V. The hepatitis B virus persists for decades after patients' recovery from acute viral hepatitis despite active maintenance of a cytotoxic T-lymphocyte response. Nat Med 1996;2:1104-1108.
71. Curry M.P., Koziel M. The dynamics of the immune response in acute hepatitis B: new lessons using new techniques. Hepatology 2000; 32:1177-1179.
72. Altman J.D,. Moss P.A., Goulder P.J., Barouch D.H., McHeyzer-Williams M.G., Bell J.I., McMichael A.J., Davis M.M. Phenotypic analysis of antigen-specific T lymphocytes. Science 1996; 274:94-96.
73. Maini M.K., Boni C., Ogg G.S., King A.S., Reignat S., Lee C.K., Larrubia J.R., Webster G.J., McMichael A.J., Ferrari C., Williams R., Vergani D., Bertoletti A. Direct ex vivo analysis of hepatitis B virus-specific CD8(+) T cells associated with the control of infection. Gastroenterology 1999; 117:1386-1396.
74. Hsu H.Y., Chang M.H., Hsieh K.H., Lee C.Y., Lin H.H., Hwang L.H., Chen P.J., Chen D.S. Cellular immune response to HBcAg in mother-to-infant transmission of hepatitis B virus. Hepatology 1992; 15:770-776.
75. Wang J.S., Zhu Q.R. Infection of the fetus with hepatitis B e antigen via the placenta. Lancet 2000; 355:989.
76. Raimondo G., Tanzi E., Brancatelli S., Campo S., Sardo M.A., Rodino G., Pernice M., Zanetti A.R. Is the course of perinatal hepatitis B virus infection influenced by genetic heterogeneity of the virus? J Med Virol 1993; 40:87-90.
77. Kurose K., Akbar S.M., Yamamoto K., Onji M. Production of antibody to hepatitis B surface antigen (anti-HBs) by murine hepatitis B virus carriers: neonatal tolerance versus antigen presentation by dendritic cells. Immunology 1997; 92:494-500.
78. Bertoletti A., Costanzo A., Chisari F.V., Levrero M., Artini M., Sette A., Penna A., Giuberti T., Fiaccadori F., Ferrari C. Cytotoxic T lymphocyte response to a wild type hepatitis B virus epitope in patients chronically infected by variant viruses carrying substitutions within the epitope. J Exp Med 1994; 180:933-943.

79. Barnaba V., Franco A., Alberti A., Balsano C., Benvenuto R., Balsano F. Recognition of hepatitis B virus envelope proteins by liver-infiltrating T lymphocytes in chronic HBV infection. J Immunol 1989; 143:2650-2655.

80. Barnaba V., Franco A., Alberti A., Benvenuto R., Balsano F. Selective killing of hepatitis B envelope antigen-specific B cells by class I-restricted, exogenous antigen-specific T lymphocytes. Nature 1990; 345:258-260.

81. Abrignani S. Bystander activation by cytokines of intrahepatic T cells in chronic viral hepatitis. Semin Liver Dis 1997; 17:319-322

82. Maini M.K., Boni C., Lee C.K., Larrubia J.R., Reignat S., Ogg G.S., King A.S., Herberg J., Gilson R., Alisa A., Williams R., Vergani D., Naoumov N.V., Ferrari C., Bertoletti A. The role of virus-specific CD8(+) cells in liver damage and viral control during persistent hepatitis B virus infection. J Exp Med 2000; 191:1269-1280.

83. Boni C., Bertoletti A., Penna A., Cavalli A., Pilli M., Urbani S., Scognamiglio P., Boehme R., Panebianco R., Fiaccadori F., Ferrari C. Lamivudine treatment can restore T cell responsiveness in chronic hepatitis B. J Clin Invest 1998; 102:968-975.

84. Boni C., Penna A., Ogg G.S., Bertoletti A., Pilli M., Cavallo C., Cavalli A., Urbani S., Boehme R., Panebianco R., Fiaccadori F., Ferrari C. Lamivudine treatment can overcome cytotoxic T-cell hyporesponsiveness in chronic hepatitis B: new perspectives for immune therapy. Hepatology 2001; 33:963-971.

85. Sjogren M., Hoofnagle J.H. Immunoglobulin M antibody to hepatitis B core antigen in patients with chronic type B hepatitis. Gastroenterology 1985; 89:252-258.

86. Lemon S.M., Thomas D.L. Vaccines to prevent viral hepatitis. N Engl J Med 1997; 336:196-204.

87. Chang M.H., Chen C.J., Lai M.S., Hsu H.M., Wu T.C., Kong M.S., Liang D.C., Shau W.Y, Chen D.S. Universal hepatitis B vaccination in Taiwan and the incidence of hepatocellular carcinoma in children. Taiwan Childhood Hepatoma Study Group. N Engl J Med 1997; 336:1855-1859.

88. Galle P.R., Hofmann W.J., Walczak H., Schaller H., Otto G., Stremmel W., Krammer P.H., Runkel L. Involvement of the CD95 (APO-1/Fas) receptor and ligand in liver damage. J Exp Med 1995; 182:1223-1230.

89. Taylor J.M. The structure and replication of hepatitis delta virus. Annu Rev Microbiol 1992; 46:253-276

90. Cole S.M., Gowans E.J., Macnaughton T.B., Hall P.D., Burrell C.J. Direct evidence for cytotoxicity associated with expression of hepatitis delta virus antigen. Hepatology 1991; 13:845-851.

91. Macnaughton T.B., Gowans E.J., Jilbert A.R., Burrell C.J. Hepatitis delta virus RNA, protein synthesis and associated cytotoxicity in a stably transfected cell line. Virology 1990; 177:692-698.

92. Verme G., Amoroso P., Lettieri G., Pierri P., David E., Sessa F., Rizzi R., Bonino F., Recchia S., Rizzetto M. A histological study of hepatitis delta virus liver disease. Hepatology 1986; 6:1303-1307.

93. Lefkowitch J.H., Goldstein H., Yatto R., Gerber M.A. Cytopathic liver injury in acute delta virus hepatitis. Gastroenterology 1987; 92:1262-1266.

94. David E., Rahier J., Pucci A., Camby P., Scevens M., Salizzoni M., Otte J.B., Galmarini D., Marinucci G., Ottobrelli A., et al. Recurrence of hepatitis D (delta) in liver transplants: histopathological aspects. Gastroenterology 1993; 104:1122-1128.

95. Guilhot S., Huang S.N., Xia Y.P., La Monica N., Lai M.M, Chisari F.V. Expression of the hepatitis delta virus large and small antigens in transgenic mice. J Virol 1994; 68:1052-1058.

96. Samuel D., Zignego A.L., Reynes M., Feray C., Arulnaden J.L., David M.F., Gigou M., Bismuth A., Mathieu D., Gentilini P., et al. Long-term clinical and virological outcome after liver transplantation for cirrhosis caused by chronic delta hepatitis. Hepatology 1995; 21:333-339.
97. Ottobrelli A., Marzano A., Smedile A., Recchia S., Salizzoni M., Cornu C., Lamy M.E., Otte J.B., De Hemptinne B., Geubel A., et al. Patterns of hepatitis delta virus reinfection and disease in liver transplantation. Gastroenterology 1991; 101:1649-1655.
98. Negro F., Rizzetto M. Pathobiology of hepatitis delta virus. J Hepatol 1993; 17:S149-153.
99. Nisini R., Paroli M., Accapezzato D., Bonino F., Rosina F., Santantonio T., Sallusto F., Amoroso A., Houghton M., Barnaba V. Human CD4+ T-cell response to hepatitis delta virus: identification of multiple epitopes and characterization of T-helper cytokine profiles. J Virol 1997; 71:2241-2251.
100. Karayiannis P., Saldanha J., Jackson A.M., Luther S., Goldin R., Monjardino J., Thomas H.C. Partial control of hepatitis delta virus superinfection by immunisation of woodchucks (Marmota monax) with hepatitis delta antigen expressed by a recombinant vaccinia or baculovirus. J Med Virol 1993; 41:210-214.
101. Cheney C.P., Chopra S., Graham C. Hepatitis C. Infect Dis Clin North Am 2000; 14:633-667.
102. Koup R.A., Safrit J.T., Cao Y., Andrews C.A., McLeod G., Borkowsky W., Farthing C., Ho D.D. Temporal association of cellular immune responses with the initial control of viremia in primary human immunodeficiency virus type 1 syndrome. J Virol 1994; 68:4650-4655.
103. Gerlach J.T., Diepolder H.M., Jung M.C., Gruener N.H., Schraut W.W., Zachoval R., Hoffmann R., Schirren C.A., Santantonio T., Pape G.R. Recurrence of hepatitis C virus after loss of virus-specific CD4(+) T-cell response in acute hepatitis C. Gastroenterology 1999; 117:933-941.
104. Borrow P., Lewicki H., Wei X., Horwitz M.S., Peffer N., Meyers H., Nelson J.A., Gairin J.E., Hahn B.H., Oldstone M.B., Shaw G.M. Antiviral pressure exerted by HIV-1-specific cytotoxic T lymphocytes (CTLs) during primary infection demonstrated by rapid selection of CTL escape virus. Nat Med 1997; 3:205-211.
105. Grakoui A., Hanson H.L., Rice C.M. Bad time for Bonzo? Experimental models of hepatitis C virus infection, replication, and pathogenesis. Hepatology 2001; 33:489-495.
106. Kawamura T., Furusaka A., Koziel M.J., Chung R.T., Wang T.C., Schmidt E.V., Liang T.J. Transgenic expression of hepatitis C virus structural proteins in the mouse. Hepatology 1997; 25:1014-1021.
107. Pasquinelli C., Shoenberger J.M., Chung J., Chang K.M., Guidotti L.G., Selby M., Berger K., Lesniewski R., Houghton M., Chisari F.V. Hepatitis C virus core and E2 protein expression in transgenic mice. Hepatology 1997; 25:719-727.
108. Moradpour D., Kary P., Rice C.M., Blum H.E. Continuous human cell lines inducibly expressing hepatitis C virus structural and nonstructural proteins. Hepatology 1998; 28:192-201.
109. Moriya K., Yotsuyanagi H., Shintani Y., Fujie H., Ishibashi K., Matsuura Y., Miyamura T., Koike K. Hepatitis C virus core protein induces hepatic steatosis in transgenic mice. J Gen Virol 1997; 78:1527-1531.
110. Lemon S.M., Lerat H., Weinman S.A., Honda M. A transgenic mouse model of steatosis and hepatocellular carcinoma associated with chronic hepatitis C virus infection in humans. Trans Am Clin Climatol Assoc 2000; 111:146-156
111. Koziel M.J. The role of immune responses in the pathogenesis of hepatitis C virus infection. J Viral Hepat 1997; 4:31-41.

112. Diepolder H.M., Zachoval R., Hoffmann R.M., Wierenga E.A., Santantonio T., Jung M.C., Eichenlaub D., Pape G.R. Possible mechanism involving T-lymphocyte response to non-structural protein 3 in viral clearance in acute hepatitis C virus infection. Lancet 1995; 346:1006-1007.

113. Lechmann M., Ihlenfeldt H.G., Braunschweiger I., Giers G., Jung G., Matz B., Kaiser R., Sauerbruch T., Spengler U. T- and B-cell responses to different hepatitis C virus antigens in patients with chronic hepatitis C infection and in healthy anti-hepatitis C virus--positive blood donors without viremia. Hepatology 1996; 24:790-795.

114. Missale G., Bertoni R., Lamonaca V., Valli A., Massari M., Mori C., Rumi M.G., Houghton M., Fiaccadori F., Ferrari C. Different clinical behaviors of acute hepatitis C virus infection are associated with different vigor of the anti-viral cell-mediated immune response. J Clin Invest 1996; 98:706-714.

115. Hoffmann R.M., Diepolder H.M., Zachoval R., Zwiebel F.M., Jung M.C., Scholz S., Nitschko H., Riethmuller G., Pape G.R. Mapping of immunodominant CD4+ T lymphocyte epitopes of hepatitis C virus antigens and their relevance during the course of chronic infection. Hepatology 1995; 21:632-638.

116. Cramp M.E., Rossol S., Chokshi S., Carucci P., Williams R., Naoumov N.V. Hepatitis C virus-specific T-cell reactivity during interferon and ribavirin treatment in chronic hepatitis C. Gastroenterology 2000; 118:346-355.

117. Lasarte J.J., Garcia-Granero M., Lopez A., Casares N., Garcia N., Civeira M.P., Borras-Cuesta F., Prieto J. Cellular immunity to hepatitis C virus core protein and the response to interferon in patients with chronic hepatitis C. Hepatology 1998; 28:815-822.

118. Cramp M.E., Carucci P., Rossol S., Chokshi S., Maertens G., Williams R., Naoumov N.V. Hepatitis C virus (HCV) specific immune responses in anti-HCV positive patients without hepatitis C viraemia. Gut 1999; 44:424-429.

119. Cooper S., Erickson A.L., Adams E.J., Kansopon J., Weiner A.J., Chien D.Y., Houghton M., Parham P., Walker C.M. Analysis of a successful immune response against hepatitis C virus. Immunity 1999; 10:439-449.

120. Chang K.M., Gruener N.H., Southwood S., Sidney J., Pape G.R., Chisari F.V., Sette A. Identification of HLA-A3 and -B7-restricted CTL response to hepatitis C virus in patients with acute and chronic hepatitis C. J Immunol 1999; 162:1156-1164.

121. Scognamiglio P., Accapezzato D., Casciaro M.A., Cacciani A., Artini M., Bruno G., Chircu M.L., Sidney J., Southwood S., Abrignani S., Sette A., Barnaba V. Presence of effector CD8+ T cells in hepatitis C virus-exposed healthy seronegative donors. J Immunol 1999; 162:6681-6689.

122. Surcel H.M., Troye-Blomberg M., Paulie S., Andersson G., Moreno C., Pasvol G., Ivanyi J. Th1/Th2 profiles in tuberculosis, based on the proliferation and cytokine response of blood lymphocytes to mycobacterial antigens. Immunology 1994; 81:171-176.

123. Gruner N.H., Gerlach T.J., Jung M.C., Diepolder H.M., Schirren C.A., Schraut W.W., Hoffman R., Zachoval R., Santantonio T., Cucchiarini M., Cerny A., Pape G.R. Association of hepatitis C virus-specific CD8+ T cells with viral clearance in acute hepatitis C. J Infect Dis 2000; 181:1528-1536.

124. Lechner F., Wong D.K., Dunbar P.R., Chapman R., Chung R.T., Dohrenwend P., Robbins G., Phillips R., Klenerman P., Walker B.D. Analysis of successful immune responses in persons infected with hepatitis C virus. J Exp Med 2000; 191:1499-1512.

125. Koziel M.J., Wong D.K., Dudley D., Houghton M., Walker B.D. Hepatitis C virus-specific cytolytic T lymphocyte and T helper cell responses in seronegative persons. J Infect Dis 1997; 176:859-986.

126. Diepolder H.M., Gerlach J.T., Zachoval R., Hoffmann R.M., Jung M.C., Wierenga E.A., Scholz S., Santantonio T., Houghton M., Southwood S., Sette A., Pape G.R. Immunodominant CD4+ T-cell epitope within nonstructural protein 3 in acute hepatitis C virus infection. J Virol 1997; 71:6011-6019.

127. Koziel M.J., Dudley D., Wong J.T., Dienstag J., Houghton M, Ralston R, Walker BD. Intrahepatic cytotoxic T lymphocytes specific for hepatitis C virus in persons with chronic hepatitis. J Immunol 1992; 149:3339-3344.

128. Wong D.K., Dudley D.D., Afdhal N.H., Dienstag J., Rice C.M., Wang L., Houghton M., Walker B.D., Koziel M.J. Liver-derived CTL in hepatitis C virus infection: breadth and specificity of responses in a cohort of persons with chronic infection. J Immunol 1998; 160:1479-1488.

129. Rehermann B., Lau D., Hoofnagle J.H., Chisari F.V. Cytotoxic T lymphocyte responsiveness after resolution of chronic hepatitis B virus infection. J Clin Invest 1996; 97:1655-1665.

130. Rehermann B., Chang K.M., McHutchinson J., Kokka R., Houghton M., Rice C.M., Chisari F.V. Differential cytotoxic T-lymphocyte responsiveness to the hepatitis B and C viruses in chronically infected patients. J Virol 1996; 70:7092-7102.

131. Nelson D.R., Marousis C.G., Davis G.L., Rice C.M., Wong J., Houghton M., Lau J.Y. The role of hepatitis C virus-specific cytotoxic T lymphocytes in chronic hepatitis C. J Immunol 1997; 158:1473-1481.

132. Hiroishi K., Kita H., Kojima M., Okamoto H., Moriyama T., Kaneko T., Ishikawa T., Ohnishi S., Aikawa T., Tanaka N., Yazaki Y., Mitamura K., Imawari M. Cytotoxic T lymphocyte response and viral load in hepatitis C virus infection. Hepatology 1997; 25:705-712.

133. Ando K., Hiroishi K., Kaneko T., Moriyama T., Muto Y., Kayagaki N., Yagita H., Okumura K., Imawari M. Perforin, Fas/Fas ligand, and TNF-alpha pathways as specific and bystander killing mechanisms of hepatitis C virus-specific human CTL. J Immunol 1997; 158:5283-5291.

134. Chang K.M., Thimme R., Melpolder J.J., Oldach D., Pemberton J., Moorhead-Loudis J., McHutchison J.G., Alter H.J., Chisari F.V. Differential CD4(+) and CD8(+) T-cell responsiveness in hepatitis C virus infection. Hepatology 2001; 33:267-276.

135. Rosen H.R., Hinrichs D.J., Gretch D.R., Koziel M.J., Chou S., Houghton M., Rabkin J., Corless C.L., Bouwer H.G. Association of multispecific CD4(+) response to hepatitis C and severity of recurrence after liver transplantation. Gastroenterology 1999; 117:926-932.

136. Kamal S., Rasenack J., Bianchi L., Taweel A., Khalifa K., Peters T., Koziel M. A Prospective study of the Outcome of Acute Hepatitis C with and without Schistomiasis: Correlation with Hepatitis C specific CD4+ T-cell and cytokine responses. Gastroenterology; In Press

137. Koike K., Moriya K., Ishibashi K., Matsuura Y., Suzuki T., Saito I., Iino S., Kurokawa K., Miyamura T. Expression of hepatitis C virus envelope proteins in transgenic mice. J Gen Virol 1995; 76:3031-3038.

138. Imawari M., Kita H., Moriyama T., Kaneko T., Hiroishi K. Cytotoxic T lymphocyte responses in hepatitis C virus infection. Princess Takamatsu Symp 1995; 25:221-225

139. Jin Y., Fuller L., Carreno M., Zucker K., Roth D., Esquenazi V., Karatzas T., Swanson S.J., 3rd, Tzakis A.G., Miller J. The immune reactivity role of HCV-induced liver infiltrating lymphocytes in hepatocellular damage. J Clin Immunol 1997; 17:140-153.

140. Liaw Y.F., Lee C.S., Tsai S.L., Liaw B.W., Chen T.C., Sheen I.S., Chu C.M. T-cell--mediated autologous hepatocytotoxicity in patients with chronic hepatitis C virus infection. Hepatology 1995; 22:1368-1373.

141. Napoli J., Bishop G.A., McGuinness P.H., Painter D.M., McCaughan G.W. Progressive liver injury in chronic hepatitis C infection correlates with increased intrahepatic expression of Th1-associated cytokines. Hepatology 1996; 24:759-765.
142. Graham C.S., Herren T., Baden L., Mrus J., Yu E., Koziel M.J. Influence of human immunodeficiency virus (HIV) infection on the course of hepatitis C virus (HCV) infection: a meta-analysis. Clinical Infectious Diseases; In Press
143. Rosen H.R., Martin P. Viral hepatitis in the liver transplant recipient. Infect Dis Clin North Am 2000; 14:761-784.
144. Rubin R.A., Falestiny M., Malet P.F. Chronic hepatitis C. Advances in diagnostic testing and therapy. Arch Intern Med 1994; 154:387-392.
145. Quinti I., Hassan N.F., El Salman D., Shalaby H., El Zimatty D., Monier M.K., Arthur R.R. Hepatitis C virus-specific B cell activation: IgG and IgM detection in acute and chronic hepatitis C. J Hepatol 1995; 23:640-647.
146. Chien D.Y., Choo Q.L., Tabrizi A., Kuo C., McFarland J., Berger K., Lee C., Shuster J.R., Nguyen T., Moyer D.L., et al. Diagnosis of hepatitis C virus (HCV) infection using an immunodominant chimeric polyprotein to capture circulating antibodies: reevaluation of the role of HCV in liver disease. Proc Natl Acad Sci USA 1992; 89:10011-10015.
147. Beld M., Penning M., van Putten M., Lukashov V., van den Hoek A., McMorrow M., Goudsmit J. Quantitative antibody responses to structural (Core) and nonstructural (NS3, NS4, and NS5) hepatitis C virus proteins among seroconverting injecting drug users: impact of epitope variation and relationship to detection of HCV RNA in blood. Hepatology 1999; 29:1288-1298.
148. Kobayashi M., Tanaka E., Matsumoto A., Ichijo T., Kiyosawa K. Antibody response to E2/NS1 hepatitis C virus protein in patients with acute hepatitis C. J Gastroenterol Hepatol 1997; 12:73-76.
149. Lesniewski R., Okasinski G., Carrick R., Van Sant C., Desai S., Johnson R., Scheffel J., Moore B., Mushahwar I. Antibody to hepatitis C virus second envelope (HCV-E2) glycoprotein: a new marker of HCV infection closely associated with viremia. J Med Virol 1995; 45:415-422.
150. Bassett S.E., Thomas D.L., Brasky K.M., Lanford R.E. Viral persistence, antibody to E1 and E2, and hypervariable region 1 sequence stability in hepatitis C virus-inoculated chimpanzees. J Virol 1999; 73:1118-1126.
151. Prince A.M., Brotman B., Lee D.H., Ren L., Moore B.S., Scheffel J.W. Significance of the anti-E2 response in self-limited and chronic hepatitis C virus infections in chimpanzees and in humans. J Infect Dis 1999; 180:987-991.
152. Chen M., Sallberg M., Sonnerborg A., Weiland O., Mattsson L., Jin L., Birkett A., Peterson D., Milich D.R. Limited humoral immunity in hepatitis C virus infection. Gastroenterology 1999; 116:135-143.
153. Baumert T.F., Wellnitz S., Aono S., Satoi J., Herion D., Gerlach J.T., Pape G.R., Lau J.Y., Hoofnagle J.H., Blum H.E., Liang T.J. Antibodies against hepatitis C virus-like particles and viral clearance in acute and chronic hepatitis C. Hepatology 2000; 32:610-617.
154. Farci P., Alter H.J., Govindarajan S., Wong D.C., Engle R., Lesniewski R.R., Mushahwar I.K., Desai S.M., Miller R.H., Ogata N., et al. Lack of protective immunity against reinfection with hepatitis C virus. Science 1992; 258:135-140.
155. Choo Q.L., Kuo G., Ralston R., Weiner A., Chien D., Van Nest G., Han J., Berger K., Thudium K., Kuo C., et al. Vaccination of chimpanzees against infection by the hepatitis C virus. Proc Natl Acad Sci U S A 1994; 91:1294-1298.
156. Farci P., Alter H.J., Wong D.C., Miller R.H., Govindarajan S., Engle R., Shapiro M., Purcell R.H. Prevention of hepatitis C virus infection in chimpanzees after antibody-mediated in vitro neutralization. Proc Natl Acad Sci USA 1994; 91:7792-7796.

157. Kato N., Ootsuyama Y., Sekiya H., Ohkoshi S., Nakazawa T., Hijikata M., Shimotohno K. Genetic drift in hypervariable region 1 of the viral genome in persistent hepatitis C virus infection. J Virol 1994; 68:4776-4784.

158. Agnello V. Hepatitis C virus infection and type II cryoglobulinemia: an immunological perspective. Hepatology 1997; 26:1375-1379.

159. Pozzato G., Moretti M., Crovatto M., Modolo M.L., Gennari D., Santini G. Lymphocyte subsets in HCV-positive chronic liver disease. Immunol Today 1994; 15:137-138.

160. Curry M.P., Golden-Mason L., Nolan N., Parfrey N.A., Hegarty J.E., O'Farrelly C. Expansion of peripheral blood CD5+ B cells is associated with mild disease in chronic hepatitis C virus infection. J Hepatol 2000; 32:121-125.

161. Sansonno D., De Vita S., Iacobelli A.R., Cornacchiulo V., Boiocchi M., Dammacco F. Clonal analysis of intrahepatic B cells from HCV-infected patients with and without mixed cryoglobulinemia. J Immunol 1998; 160:3594-3601.

162. Magalini A.R., Facchetti F., Salvi L., Fontana L., Puoti M., Scarpa A. Clonality of B-cells in portal lymphoid infiltrates of HCV-infected livers. J Pathol 1998; 185:86-90.

163. Chan C.H., Hadlock K.G., Foung S.K., Levy S. V(H)1-69 gene is preferentially used by hepatitis C virus-associated B cell lymphomas and by normal B cells responding to the E2 viral antigen. Blood 2001; 97:1023-1026.

164. Minutello M.A., Pileri P., Unutmaz D., Censini S., Kuo G., Houghton M., Brunetto M.R., Bonino F., Abrignani S. Compartmentalization of T lymphocytes to the site of disease: intrahepatic CD4+ T cells specific for the protein NS4 of hepatitis C virus in patients with chronic hepatitis C. J Exp Med 1993; 178:17-25.

165. Schirren C.A., Jung M.C., Gerlach J.T, Worzfeld T., Baretton G., Mamin M., Gruener N.H., Houghton M., Pape G.R. Liver-derived hepatitis C virus (HCV)-specific CD4(+) T cells recognize multiple HCV epitopes and produce interferon gamma. Hepatology 2000; 32:597-603.

166. Kao J.H., Chen P.J., Lai M.Y., Wang T.H., Chen D.S. Positive and negative strand of hepatitis C virus RNA sequences in peripheral blood mononuclear cells in patients with chronic hepatitis C: no correlation with viral genotypes 1b, 2a, and 2b. J Med Virol 1997; 52:270-274.

167. Rehermann B. Interaction between the hepatitis C virus and the immune system. Semin Liver Dis 2000; 20:127-141.

168. Kanto T., Hayashi N., Takehara T., Tatsumi T., Kuzushita N., Ito A., Sasaki Y., Kasahara A., Hori M. Impaired allostimulatory capacity of peripheral blood dendritic cells recovered from hepatitis C virus-infected individuals. J Immunol 1999; 162:5584-5591.

169. Farci P., Shimoda A., Coiana A., Diaz G., Peddis G., Melpolder J.C., Strazzera A., Chien D.Y., Munoz S.J., Balestrieri A., Purcell R.H., Alter H.J. The outcome of acute hepatitis C predicted by the evolution of the viral quasispecies. Science 2000; 288:339-344.

170. Ray S.C., Wang Y.M., Laeyendecker O., Ticehurst J.R., Villano S.A., Thomas D.L. Acute hepatitis C virus structural gene sequences as predictors of persistent viremia: hypervariable region 1 as a decoy. J Virol 1999; 73:2938-2946.

171. Weiner A., Erickson A.L., Kansopon J., Crawford K., Muchmore E., Hughes A.L., Houghton M., Walker C.M. Persistent hepatitis C virus infection in a chimpanzee is associated with emergence of a cytotoxic T lymphocyte escape variant. Proc Natl Acad Sci U S A 1995; 92:2755-2759.

172. Odeberg J., Yun Z., Sonnerborg A., Bjoro K., Uhlen M., Lundeberg J. Variation of hepatitis C virus hypervariable region 1 in immunocompromised patients. J Infect Dis 1997; 175:938-943.

173. Major M.E., Mihalik K., Fernandez J., Seidman J., Kleiner D., Kolykhalov A.A., Rice C.M., Feinstone S.M. Long-term follow-up of chimpanzees inoculated with the first infectious clone for hepatitis C virus. J Virol 1999; 73:3317-3325.

174. Kaneko T., Moriyama T., Udaka K., Hiroishi K., Kita H., Okamoto H., Yagita H., Okumura K., Imawari M. Impaired induction of cytotoxic T lymphocytes by antagonism of a weak agonist borne by a variant hepatitis C virus epitope. Eur J Immunol 1997; 27:1782-1787.

175. Chang K.M., Rehermann B., McHutchison J.G., Pasquinelli C., Southwood S., Sette A., Chisari F.V. Immunological significance of cytotoxic T lymphocyte epitope variants in patients chronically infected by the hepatitis C virus. J Clin Invest 1997; 100:2376-2385.

176. Christie J.M., Chapel H., Chapman R.W., Rosenberg W.M. Immune selection and genetic sequence variation in core and envelope regions of hepatitis C virus. Hepatology 1999; 30:1037-1044.

177. Eckels D.D., Tabatabail N., Bian T.H., Wang H., Muheisen S.S., Rice C.M., Yoshizawa K., Gill J. In vitro human Th-cell responses to a recombinant hepatitis C virus antigen: failure in IL-2 production despite proliferation. Hum Immunol 1999; 60:187-99.

178. Wang H., Eckels D.D. Mutations in immunodominant T cell epitopes derived from the nonstructural 3 protein of hepatitis C virus have the potential for generating escape variants that may have important consequences for T cell recognition. J Immunol 1999; 162:4177-4183.

179. Xu X.N., Screaton G.R., Gotch F.M., Dong T., Tan R., Almond N., Walker B., Stebbings R., Kent K., Nagata S., Stott J.E., McMichael A.J. Evasion of cytotoxic T lymphocyte (CTL) responses by nef-dependent induction of Fas ligand (CD95L) expression on simian immunodeficiency virus-infected cells. J Exp Med 1997; 186:7-16.

180. Cerny A., Chisari F.V. Pathogenesis of chronic hepatitis C: immunological features of hepatic injury and viral persistence. Hepatology 1999; 30:595-601.

181. Lechner F., Gruener N.H, Urbani S., Uggeri J., Santantonio T., Kammer A.R., Cerny A., Phillips R., Ferrari C., Pape G.R., Klenerman P. CD8+ T lymphocyte responses are induced during acute hepatitis C virus infection but are not sustained. Eur J Immunol 2000; 30:2479-2487.

Chapter 9

PATHOGENESIS OF HEPATOCELLULAR CARCINOMA BY HEPATITIS VIRUSES

Marcus W. Wiedmann and Jack R. Wands
Liver Research Center, Rhode Island Hospital, Brown University, Providence, RI 02903

1. INTRODUCTION

Hepatocellular carcinoma (HCC), diagnosed with increased frequency, contributes to 4.1% of all known human carcinomas, and accounts for 350,000 new cases per year worldwide (1-4). The incidence of histology-proven HCC rose from 1.4 per 100,000 population (1976-1980) to 2.4 per 100,000 (1991-1995) associated with higher mortality (41%), increased hospitalization (46%), and rise in hepatitis viral infection (5). Hepatitis B (HBV) and hepatitis C viral (HCV) infections account for 75 - 85% of HCC, with the remainder related to ethanol, oral contraceptives, and aflatoxin B1 exposure (2). Approximately 500-600 million people worldwide are HBV- or HCV- positive (6), with the United States accounting for 3.9 million (HCV), and 1.25 million (HBV) (4). Severe forms of HBV/HCV infection, encompassing hepatocyte inflammation, and increased cell turnover, place individuals at higher risk for HCC. Chronic hepatitis leads to liver necrosis/fibrosis, inflammation, and increased cytokine synthesis, resulting in enhanced cell proliferation followed by DNA mutation(s), cellular transformation, and the development of HCC (7). HBV- and HCV-coinfection progresses to liver cirrhosis at a higher rate (33%) than in HCV infection alone (19%), and both are important in the pathogenesis of HCC (8–11). Not less important is the HCV interaction with ethanol; approximately half of those diagnosed with alcoholic cirrhosis or HCC are anti-HCV positive (40-50%) when compared to those with minimal liver damage (20%). The same correlation between the severity of the liver disease and the HCV-infection rate was not found in HBV, i.e. both alcoholic cirrhotics and alcoholics with minimal liver damage were anti-HBs positive at the same level (20%), but HCC combined with ethanol accounted for 50% anti-HBs positivity further underlining the increased risk of ethanol-induced HBV or HCV co-infection (3, 12-17).

The pathogenic role of HBV- and HCV-infection in HCC will be reviewed as described below with particular emphasis on viral-cellular protein interactions.

2. HEPATITIS B VIRUS INFECTION

HBV, contributing to 60% of all HCCs in high-risk areas and seen in 10-15% of the population (Southeast Asia, China, sub-Saharan Africa), may be oncogenically involved at multiple levels i.e., inflammation, increased cell turnover, and cirrhosis (2). Sustained cell proliferation in chronic infection, combined with impaired pathways (P450), leads to dysplasia. Viral DNA-integration into the host genome occurs frequently following prolonged infection (20-30 years), promotes HCC in the presence of HBsAg positivity (90%) causing genomic instability and results in insertional mutagenesis (cis-activation) (18). Tumor cells frequently lack viral DNA-replication, HBcAg-expression, but may be weakly HBsAg positive (20%). Of interest is the observation that HBx expression may be retained. Both HBx and truncated preS2/S proteins have strong *in vitro* transactivator activities (7).

2.1 HBV X (HBx)

The viral gene protein, HBx, when induced *in vivo* causes tumorigenic transformation in some transgenic murine livers, promotes hepatocyte sensitivity to other hepatocarcinogens (i.e., diethylnitrosamine), and accelerates c-myc-induced HCC formation (19-26). An important mechanism of hepatic oncogenesis is the role of HBx as a transactivator of cellular genes controlling cell growth (27). In addition, HBx may activate cytosolic signal transduction pathways including the Src family kinases (28, 29), c-jun terminal kinase (28, 30), Jak-STAT (31), and the PKC pathway(s) which are further stimulated by HBx bound to a cellular protein (XAP-3) (32).

HBx may also interact with growth factor-induced signal transduction pathways i.e., the human insulin receptor substrate signal (hIRS-1) cascade is phosphorylated on tyrosine residues following cellular stimulation by ligands i.e., insulin, insulin-like growth factor (IGF-1), interleukins (Il-4, 9, 13), interferon α, β, γ, and ω, growth hormone (GH), leukemia inhibitory factor (LIF), and tumor necrosis factor (TNF) (33-38) (Figure 1). Tyrosyl phosphorylated hIRS-1 is a key docking protein transmitting mitogenic/ metabolic signals through interaction with downstream Src 2, 3 homology (SH2, SH3) domain containing molecules (39). Important hIRS-1 binding adaptor molecules are: growth factor receptor-bound protein 2 (Grb2) (40), Syp-phosphatase (SH-

PTP2) (41), and the p85 subunit of phosphatidyl-inositol-3 kinase (PI3K) (42-43).

The hIRS-1 N-terminal sequence involves three important pleckstrin homology (PH) regions which interact with other signaling molecules such as: the G-protein(s) β and γ-subunit(s), phosphatidylinositol-4,5-bisphosphate and Janus tyrosin kinase (Tyk-2), and therefore hIRS-1 provides a linkage with additional signal transduction pathways (37, 44-48). The growth signal is transmitted when Grb2 binds to a hIRS-1 phosphotyrosine residue via its SH2 domain. Next, the SH3 domain of Grb2 binds to a proline-rich region of son-of-sevenless (Sos) protein characterized as a Ras-specific GDP/GTP exchange factor (49, 50). The Grb2/Sos complex interacts at the cell surface with Ras-GDP, catalyses a Ras GDP/GTP exchange, activating the down-stream Raf/MAPKK/MAPK cascade further enhanced by PI3K (51-53). MAPK translocates into the nucleus activating transcription factors i.e., ELK-1, ribosomal S6 kinase pp90rsk (RSK), cyclic adenosine monophosphate response element binding protein (CREB) via CREB kinase (RSK2), and serum response factor (SRF) via pp90rsk (54, 55) (Figure 1). These transcription factors stimulate a group of 50 to 100 immediate-early genes (IEGs) and induce the reentry of resting cells into the cell cycle leading to DNA replication, and cell division (56-58). IEGs, as exemplified by c-fos and c-jun which form the AP-1 complex are important in the transcription of cellular genes involved in growth regulation.

The biologic importance of the hIRS-1 signal pathway in HCC is derived from the observation that this protein is overexpressed in 15 of 22 (70%) tumors, examined by Westernblot analysis (59-61). Moreover, overexpression of at least one signal transduction protein (Syp or Grb2) has been found in 95% of tumors compared to adjacent normal tissue; most important, enzyme activities of P13K and MAPK were substantially increased in HCC (80%) indicating *in vivo* activation.

A possible role of HBx in activation of the hIRS-1 signal transduction cascade lies in its proven role as direct stimulator of the serine/threonine kinase Raf-1 and the Ras-GTP-complex formation, both are essential components of this growth factor cascade (62, 63) (Figure 2). Furthermore, HBx induces cell proliferation in quiescent fibroblasts and deregulates cell-cycle check-points by increasing cyclin-dependent kinase levels (CDK2, CDC2) leading to genetically unstable, mutated cells (64, 65) (Figure 2). Further, HBx inhibits DNA repair factors like ERCC3, XAP-1, and UV damaged DNA binding protein (UVDDB) and sensitizes liver cells to UV irradiation by binding to damaged DNA, thus increasing the likelihood of further mutation(s) (66-69) (Figure 2).

Although HBx has no direct DNA binding activity, it maintains protein-protein interaction with RPB5 of RNA polymerases (70), TATA-binding protein (TBP, TFIID) (71), octamer binding protein-1

(Oct-1) (72), and ATF-2/CREB (73). Cis-activating sequences, influenced by HBx transactivation properties, have been identified for AP-1, AP-2, NFκB, NF-AT and CRE sites (Figure 2). Interestingly, two novel inhibitory proteins, RMP and XAP-2, appear to inhibit HBx transactivation (74-76).

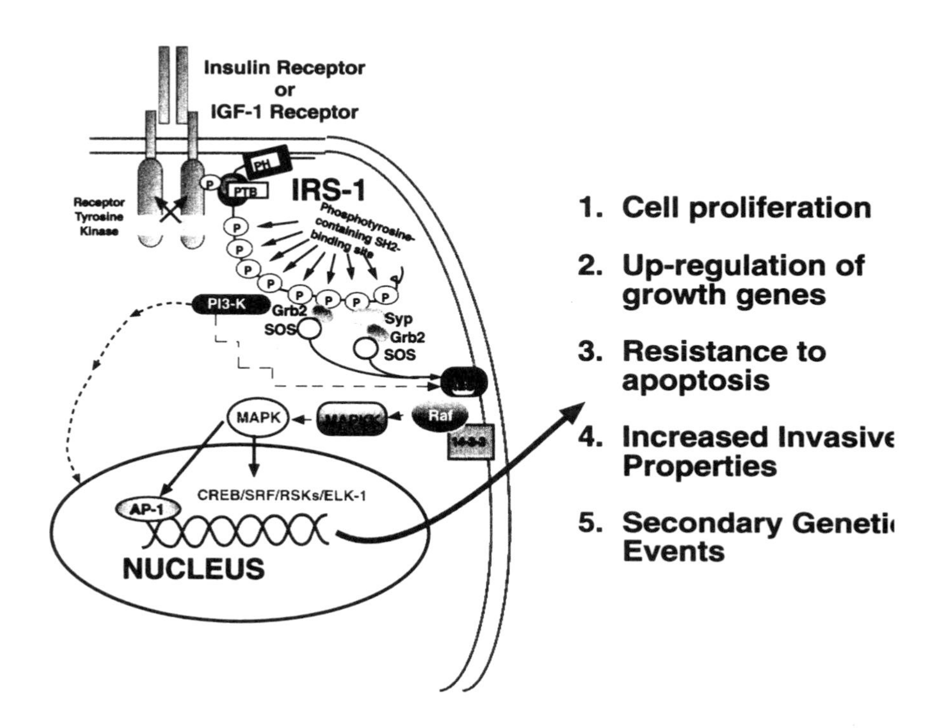

Figure 1. Diagram depicting the major features and proposed role(s) of the IRS-1 signal transduction pathway during hepatocyte growth and transformation.

Finally, HBx appears to indirectly participate in the activation of nuclear factor kappa B (NFκB) by either phosphorylation and thus degrading IκB inhibitor or direct binding to IκB (Figure 2). HBx has been found to interact with certain proteases and the proteasome, thus increasing transcription factors as well as nuclear regulatory proteins half-life. The cell growth suppressing translation initiation factors, hu-Sui I and $p55^{sen}$ are also inhibited by HBx (77-81) (Figure 2).

Other pleotropic HBx effects include modulation of programmed cell death pathways, and reduction of cell colonies promoting G1/S cell cycle arrest (82, 83) favoring HBV-DNA replication (84), Ha-ras, v-src, v-myc, v-fos, and E1a oncogene inhibition (85), and the induction/sensitization to p53 independent apoptosis (86-88). The transactivation of TNFα and TGFβ (cytokine encoded genes) by HBx might also contribute to programmed cell death pathways (89, 90) (Figure 2). HBx binding inactivates p53-dependent apoptosis by blocking entry into the nucleus (91-97). Finally, HBx induces Fas-ligand expression, inhibits caspase 3 activity, and associates with mitochondria that then aggregate at the nucleus and show increased cytochrom C release, a decreased membrane potential, and membrane blebbing. The first phenomenon may promote a viral mechanism from the activities of the cellular immune response, mediated by Fas expressing cytotoxic T-cells (98-100). One possible explanation for the diverse and, seemingly contradictory, effects of HBx is as follows: high HBx levels during acute infection are associated with apoptosis and G1/S blockage, whereas low levels in chronic infection may permit cell proliferation.

1. Transactivates cellular genes controlling cell growth
2. Directly binds to and stimulates multiple transcription factors
3. Can abolish or prevoke apoptosis (p53 dependent or not)
4. Deregulates cell cycle checkpoints
5. cis activating sequences for AP-1, AP-2, NFκB, NF-AT and CRE sites
6. Inhibits DNA repair and sensitizes liver cells to UV radiation
7. Nuclear translocation and activation of NFκB
8. Interacts with certain proteases and the proteasome
9. Inhibits hu-Sui 1 and p55^{sen} proteins

Figure 2. The different targets of HBx, illustrating the complexity of the biological actions of HBx.

2.2 PreS2/S

The preS/S gene consists of a single open reading frame divided into preS1, preS2 and S coded regions, each having an in-frame ATG codon. A large (LHBs: PreS1 + PreS2 + S), mid-sized (MHBs: PreS2 + S), and a small (SHBs: S) envelope glycoprotein are synthesized by alternate translational AUG codon initiation. Approximately 25% of all HCCs contain integrated truncated preS/S sequences (81% preS/S or HBx) (101). In contrast to the constitutive activator function of HBx and LHBs, the MHBs^t (C-terminally truncated MHBs) activator is only generated after the deletion of 3` terminal sequences of the preS/S gene, as caused by viral-cellular-DNA recombination during the integration process. LHBs- and MHBs^t –dependent transcriptional activation appears based on the PreS2 domain, and triggered through the same signal transduction pathway (102, 103). Since truncated PreS/S proteins are exclusively cytoplasmic, anchored in endoplasmatic reticulum by the S sequence, the transactivating effect must be indirect (104). The PreS2 domain is PKC-dependent phosphorylated, binds to PKC α/β, and triggers the c-Raf/MAP2-kinase signal transduction cascade resulting in transcription factor activation (AP-1, NFκB). There also exists a co-operative effect with c-Ha ras in cell transformation (105).

2.3 HBV Spliced Protein (HBSP)

Spliced HBV-transcripts have been detected in human liver tissues and HBV-transfected hepatoma cell lines (106). Singly spliced HBV-RNA leads to *in vivo* synthesis of a new 10.4 kDa protein (HBSP) which induces massive apoptosis *in vitro*, but leaves viral DNA replication/transcription intact (107). Thus, HBSP may regulate liver-cell viability and play a role for viral particle dissemination *in vivo.*

2.4 Insertional Mutagenesis (Cis-Activation)

The insertion of viral DNA into the cellular genome, as well as downstream activities from viral promoters of cellular genes associated with growth and transformation (i.e. cis-activation), are other potential conse-quences of HBV-DNA integration, but such events are unusual. Rarely, insertion of viral DNA into the cellular genome produces a transformation event by this mechanism. Indeed, the literature cites only five examples: HBV-RAR B (retinoic acid receptor beta gene) (108), the HBV-RAR B-construct [transforms erythrocyte progenitor cells in vitro]

(109), HBV-cyclin A2 (110), HBV-carboxypeptidase N-like gene (111), the HBV-SERCA1 gene (sarcoplasmatic calcium ATPase-dependent pump) (112), and the HBV-EGF receptor gene (113).

2.5 Hepatitis B Virus Mutants

Mutant HBV could, upon replication, display enhanced virulence, antiviral therapy resistance (i.e., interferon-α or nucleoside analogues), facilitated cell attachment/penetration, or epitope alteration important in the host immune response and, thus, play a role in hepatocyte transformation (114). Mutations in the precore/core gene region(s) (nt 1816 - 2458), pre/surface genes (nt 2856 - 843), X gene (nt 1376 - 1837), polymerase gene (nt 2309 - 1622), and the basal core promoter (nt 1744 - 1804) have been described.

Precore messenger RNA (mRNA) encodes for HBeAg, and a well-defined mutation produces a stop codon at nucleotide (nt) 1896 (codon 28), resulting in cessation of HBeAg expression. The nt 1896 mutation is found in 47-60% of chronic active hepatitis isolates in Asia, Africa, Southern Europe, and the Middle East, but in only 12-27% of U.S. or Northern European isolates (115); individuals with this mutation may be predisposed to chronic infection, and a negative interferon response (116). A C-1858 variant, in the absence of a stop codon mutation at nt 1896, contributes to increased inflammation and fibrosis as compared to "wild type" viruses (115). The combination of precore gene and core gene mutations may enhance HBV virulence (117). Basal core promoter mutations (BCP) at nt 1762 and nt 1764 are present in 88-100% of those with chronic hepatitis, cirrhosis, or HCC (118). Mutated BCP can not bind a liver-enriched transcription factor, and decreases HBeAg synthesis (70%); it, however, does not influence HBV pregenomic RNA transcription or core protein translation (119, 120) since there is an alternate initiation site for core RNA downstream from the precore initiation site. However, viral replication may increase due to the altered relative proportion of precore/core RNA. Finally, the combined effects of X gene mutations at codons 130 and 131 (which are affiliated with the BCP mutations due to the overlapping reading frames) lead to the generation of a new hepatocyte nuclear factor 1 (HNF-1) transcription factor binding site that may affect viral replication (121).

Core gene deletions may reduce viremia due to their effect on the overlapping polymerase and pregenomic RNAs. Such deletions may also inhibit the cytotoxic T-lymphocyte response to HBcAg due to removal of critical CTL epitopes, and promote chronic HBV infection. In addition, core gene mutations may affect the interferon response and seroconversion rate (122, 123). Finally, PreS2 gene mutations have been associated with enhanced HBV virulence inducing fulminant hepatitis, and subsequent liver failure (124).

3. HEPATITIS C VIRUS INFECTION

Persistent hepatitis C infection is one of the most important causes of chronic liver disease including HCC. Individuals with more active and severe disease seem to be at higher risk for developing HCC particularly in the setting of fibrosis/cirrhosis, whereas healthy HCV carriers have little risk of HCC. It has been suggested that the severity of inflammation is an important factor determining the prognosis and development of HCC. Accelerated hepatocyte DNA synthesis and cell turnover are major factors in the development of HCC. Unlike HBV, there is no evidence that HCV integrates into the host-genome since it is an RNA virus, but HCV proteins may modulate cell proliferation and CTL-mediated inflammation (6). Typically, HCV exposure leads to chronic infection in 80 – 90% of cases (2). Persistent infection is associated with cirrhosis (20%), and 1.9 – 6.7% develop HCC after two to three decades of infection (125). It has been shown that patients with HCV infection have a 69.1 increased risk ratio to develop HCC (126). It has also been demonstrated that HCC can develop in patients without cirrhosis (127), and that the genotype 1b of HCV may be a risk factor for the development of HCC (128, 129). More importantly, recent studies suggest that the interferon treatment of patients with chronic hepatitis C with cirrhosis may reduce the subsequent risk of HCC (130, 131).

3.1 HCV Core Protein

Oncogenic functions for the HCV core protein, which can be located in the cytoplasm and also the nucleus (132), have been shown in transgenic mice and certain cell lines. Some lines of HCV core transgenic mice with high levels of protein expression develop HCC. However, the levels of HCV core protein are much higher than what one would observe during natural infection; thus, the biological relevance of this model is not clear, and requires further study. The histologic changes in transgenic mice range from normal liver histology to steatosis hepatitis and HCC; additionally, rat fibroblasts were transformed following co-infection of HCV core and H-ras cDNAs (133-140) (Figure 3). HCV core induced Ras/Raf kinase cascade activation via interaction with the 14-3-3 protein occurs synergistically with tumor promoter TPA leading to Ets transcription activation, cell proliferation, differentiation, and oncogenic transformation (141-144) (Figure 3). Further, other HCV core binding proteins have been identified as: apolipoprotein AII (145), cytoplasmic tails of lymphotoxin-β receptor (146), tumor necrosis factor (TNF) receptor (147), heterogeneous nuclear riboprotein K (148), and cellular RNA helicases (149) (Figure 3). The oncogenic/mitogenic effect(s) of such interaction on HCV core remain to be determined. Fas- and TNF-induced apoptosis can be enhanced or diminished, for instance by NFκB activation (147, 150-153) (Figure 3). The regulation of p53 by HCV core remains unclear; whereas one group reported HCV core activation of p53 and p21/waf1, others reported either no, or repressed activity (154-157) (Figure 3). Finally, HCV core reportedly modulates the activity of certain viral and cellular genes i.e., HBV, HIV, β-actin, rb susceptibility, β-interferon, SV40 early region, and the c-fos oncogene (158-160) (Figure 3). Their role in hepatic oncogenesis remains to be determined.

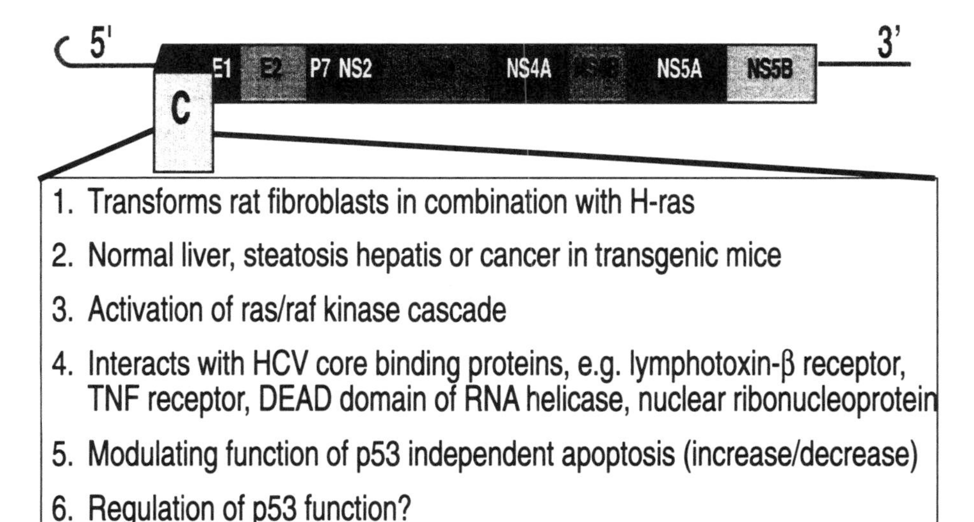

Figure 3. Cartoon illustration of the proposed biologic effects of HCV core protein

3.2 Structural Protein E2

Most HCV strains are IFN-resistant prompting evaluation of viral/cellular protein interactions in the signal transduction cascade. Clinical studies with IFN-2α reported an 8 - 10% sustained response rate to genotype 1a and 1b; when combined with ribavirin, the rate increased to 30 - 40% (161-163). Viral proteins may modulate the biological properties of IFN-induced effector proteins (164, 165). Double-stranded RNA activated protein kinase (PKR), induced by IFN, phosphorylates the eucaryotic translation initiation factor (eIF-2α), thus inhibiting protein synthesis and cell growth. Two HCV proteins inhibit PKR kinase activity i.e., envelope protein E2 (containing a sequence identical with PKR and eIF-2α phosphorylation sites) and NS5A (as discussed below) (166) (Fig. 4). Recent *in vitro* investigations indicate that HCV protein expression strongly inhibited IFN-α-induced signal transduction, downstream of STAT tyrosine phosphorylation, through the Jak-STAT pathway. The HCV proteins involved in IFN-α therapeutic resistance remain to be defined (167-170).

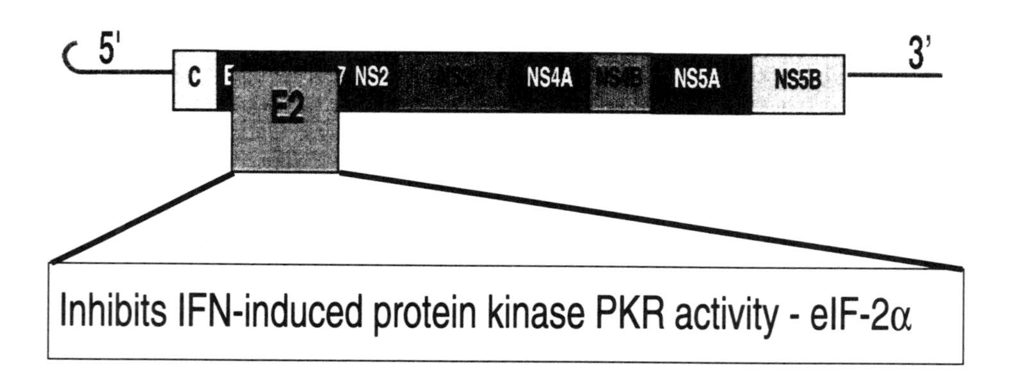

Figure 4. Interaction of envelope protein E2 with cellular proteins that may influence interferon resistance

3.3 Non-Structural Proteins NS3, NS4B and NS5A

NS3 transforms transient transfected NIH 3T3 cells, and these cells form tumors in nude mice. There is some evidence that NS3 complexes with p53, and this may inactivate the tumor suppressor gene leading to enhanced cell proliferation (171, 172) (Figure 5). Interestingly, NS4B, in combination with Ha-ras, transforms NIH 3T3 cells (173)

(Figure 5). NS5A specifically binds to, and modulates, Grb2 function, a critical adaptor protein in the growth factor signal transduction pathway described above (174). NS5A reportedly serves as a transcriptional transactivator, and inhibits PKR activity leading to enhanced protein synthesis, cell growth and inhibition of programmed cell death (166, 175-179). Current research questions whether mutations (esp. H2218R) in the NS5A interferon sensitivity determining region (ISDR; aa 2209-48) promote a positive response to interferon in genotype 1b individuals (180-186). Additionally, NS5A interacts with SNARE-like proteins further postulating their potential role in intracellular membrane trafficking (187) (Figure 5). These studies emphasize the potential interaction(s) of the non-structural viral proteins with cellular proteins that may eventually be shown to have importance in the transformation process.

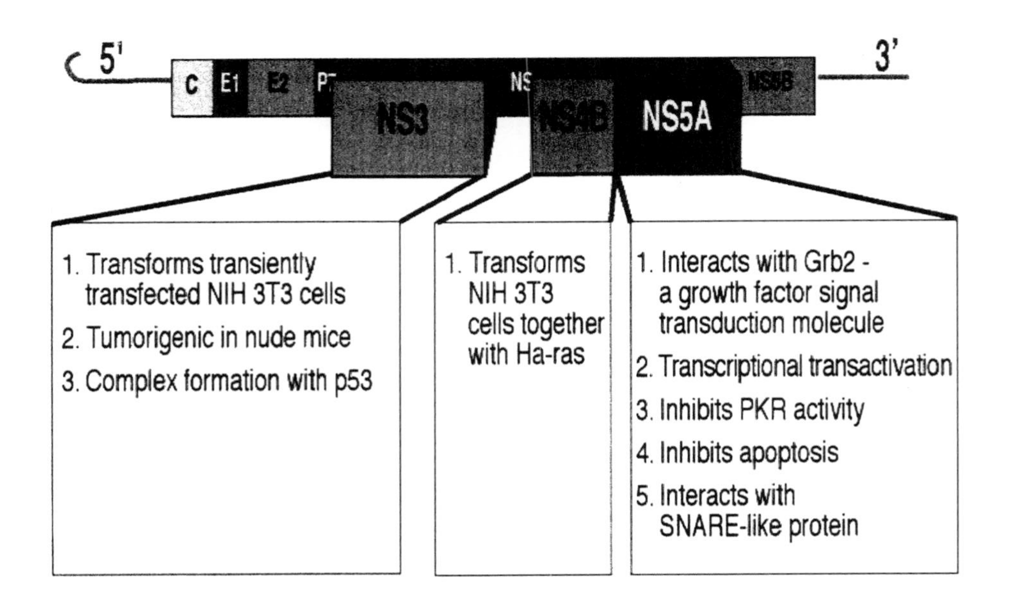

***Figure** 5.* Summary of HCV NS3, NS4B, NS5/cellular protein interactions and proposed cellular effects.

4. SUMMARY

Hepatocellular carcinoma (HCC), a malignant tumor that currently ranks eighth world-wide, has been linked to persistent HBV and

HCV infection. Chronic hepatitis leads to liver cell necrosis/fibrosis, inflammation, and increased cytokine synthesis resulting in enhanced liver cell proliferation and, ultimately, hepatic transformation. HBV may be oncogenically involved at multiple levels in the disease process i.e., inflammation, increased cell-turnover, viral DNA-integration into the host genome, HBx expression (transactivating proteins), and truncated preS/S proteins. Precore and basal core promoter (BCP) mutations may predispose to chronic infection and fibrosis.

HCV infection induces chronic hepatitis, cirrhosis, and HCC. Although HCV-RNA does not integrate into the host genome, HCV core protein and non-structural proteins (NS3, NS4B, NS5A) may contribute to hepatocyte transformation via interaction with cellular proteins involved in growth factor regulated signal transduction pathways.

REFERENCES

1. Schafer D.F., Sorrell M.F. Hepatocellular carcinoma [see comments]. Lancet 1999; 353(9160):1253-1257.
2. Bradley D.W. Hepatitis viruses: their role in human cancer. Proc Assoc Am Physicians 1999; 111(6):588-593.
3. Poynard T., Aubert A., Lazizi Y., Bedossa P., Hamelin B., Terris B. et al. Independent risk factors for hepatocellular carcinoma in French drinkers. Hepatology 1991; 13(5):896-901.
4. El Serag H.B., Mason A.C. Rising incidence of hepatocellular carcinoma in the United States [see comments]. N Engl J Med 1999; 340(10):745-750.
5. Ince N., Wands J.R. The increasing incidence of hepatocellular carcinoma [editorial; comment] [see comments]. N Engl J Med 1999; 340(10):798-799.
6. Shimotohno K.. Hepatitis C virus and its pathogenesis [In Process Citation]. Semin Cancer Biol 2000; 10(3):233-240.
7. Brechot C., Gozuacik D., Murakami Y., Paterlini-Brechot P. Molecular bases for the development of hepatitis B virus (HBV)-related hepatocellular carcinoma (HCC) [In Process Citation]. Semin Cancer Biol 2000; 10(3):211-231.
8. Cacciola I., Pollicino T., Squadrito G., Cerenzia G., Orlando M.E., Raimondo G. Occult hepatitis B virus infection in patients with chronic hepatitis C liver disease. N Engl J Med 1999; 341(1):22-26.
9. Pontisso P., Gerotto M., Benvegnu L., Chemello L., Alberti A. Coinfection by hepatitis B virus and hepatitis C virus. Antivir Ther 1998; 3(Suppl 3):137-142.
10. Ohkawa K., Hayashi N., Yuki N., Masuzawa M., Kato M., Yamamoto K. et al. Long-term follow-up of hepatitis B virus and hepatitis C virus replicative levels in chronic hepatitis patients coinfected with both viruses. J Med Virol 1995; 46(3):258-264.
11. Donato F., Boffetta P., Puoti M. A meta-analysis of epidemiological studies on the combined effect of hepatitis B and C virus infections in causing hepatocellular carcinoma. Int J Cancer 1998; 75(3):347-354.
12. Nalpas B., Thiers V., Pol S., Driss F., Thepot V., Berthelot P. et al. Hepatitis C viremia and anti-HCV antibodies in alcoholics. J Hepatol 1992; 14(2-3):381-384.

13. Nalpas B., Driss F., Pol S., Hamelin B., Housset C., Brechot C. et al. Association between HCV and HBV infection in hepatocellular carcinoma and alcoholic liver disease. J Hepatol 1991; 12(1):70-74.
14. Nalpas B., Berthelot P., Thiers V., Duhamel G., Courouce A.M., Tiollais P. et al. Hepatitis B virus multiplication in the absence of usual serological markers. A study of 146 chronic alcoholics. J Hepatol 1985; 1(2):89-97.
15. Mendenhall C.L., Seeff L., Diehl A.M., Ghosn S.J., French S.W., Gartside P.S. et al. Antibodies to hepatitis B virus and hepatitis C virus in alcoholic hepatitis and cirrhosis: their prevalence and clinical relevance. The VA Cooperative Study Group (No. 119) [see comments]. Hepatology 1991; 14(4 Pt 1):581-589.
16. Pares A., Barrera J.M., Caballeria J., Ercilla G., Bruguera M., Caballeria L. et al. Hepatitis C virus antibodies in chronic alcoholic patients: association with severity of liver injury [see comments]. Hepatology 1990; 12(6):1295-1299.
17. Saunders J.B., Wodak A.D., Morgan-Capner P., White Y.S., Portmann B., Davis M. et al. Importance of markers of hepatitis B virus in alcoholic liver disease. Br Med J (Clin Res Ed) 1983; 286(6381):1851-1854.
18. Brechot C. Hepatitis B virus (HBV) and hepatocellular carcinoma. HBV DNA status and its implications. J Hepatol 1987; 4(2):269-279.
19. Hohne M., Schaefer S., Seifer M., Feitelson M.A., Paul D., Gerlich W.H. Malignant transformation of immortalized transgenic hepatocytes after transfection with hepatitis B virus DNA. EMBO J 1990; 9(4):1137-1145.
20. Shirakata Y., Kawada M., Fujiki Y., Sano H., Oda M., Yaginuma K. et al. The X gene of hepatitis B virus induced growth stimulation and tumorigenic transformation of mouse NIH3T3 cells. Jpn J Cancer Res 1989; 80(7):617-621.
21. Gottlob K., Pagano S., Levrero M., Graessmann A. Hepatitis B virus X protein transcription activation domains are neither required nor sufficient for cell transformation. Cancer Res 1998; 58(16):3566-3570.
22. Kim C.M., Koike K., Saito I., Miyamura T., Jay G. HBx gene of hepatitis B virus induces liver cancer in transgenic mice. Nature 1991; 351(6324):317-320.
23. Yu D.Y., Moon H.B., Son J.K., Jeong S., Yu S.L., Yoon H. et al. Incidence of hepatocellular carcinoma in transgenic mice expressing the hepatitis B virus X-protein. J Hepatol 1999; 31(1):123-132.
24. Lee T.H., Finegold M.J., Shen R.F., DeMayo J.L., Woo S.L., Butel J.S. Hepatitis B virus transactivator X protein is not tumorigenic in transgenic mice. J Virol 1990; 64(12):5939-5947.
25. Slagle B.L., Lee T.H., Medina D., Finegold M.J., Butel J.S. Increased sensitivity to the hepatocarcinogen diethylnitrosamine in transgenic mice carrying the hepatitis B virus X gene. Mol Carcinog 1996; 15(4):261-269.
26. Terradillos O., Billet O., Renard C.A., Levy R., Molina T., Briand P. et al. The hepatitis B virus X gene potentiates c-myc-induced liver oncogenesis in transgenic mice. Oncogene 1997; 14(4):395-404.
27. Murakami S. Hepatitis B virus X protein: structure, function and biology. Intervirology 1999; 42(2-3):81-99.
28. Benn J., Su F., Doria M., Schneider R.J. Hepatitis B virus HBx protein induces transcription factor AP-1 by activation of extracellular signal-regulated and c-Jun N-terminal mitogen-activated protein kinases. J Virol 1996; 70(8):4978-4985.
29. Klein N.P., Schneider R.J. Activation of Src family kinases by hepatitis B virus HBx protein and coupled signaling to Ras. Mol Cell Biol 1997; 17(11):6427-6436.
30. Natoli G., Avantaggiati M.L., Chirillo P., Puri P.L., Ianni A., Balsano C. et al. Ras- and Raf-dependent activation of c-jun transcriptional activity by the hepatitis B virus transactivator pX. Oncogene 1994; 9(10):2837-2843.

31. Lee Y.H., Yun Y. HBx protein of hepatitis B virus activates Jak1-STAT signaling. J Biol Chem 1998; 273(39):25510-25515.
32. Cong Y.S., Yao Y.L., Yang W.M., Kuzhandaivelu N., Seto E. The hepatitis B virus X-associated protein, XAP3, is a protein kinase C- binding protein. J Biol Chem 1997; 272(26):16482-16489.
33. Argetsinger L.S., Hsu G.W., Myers M.G., Jr., Billestrup N., White M.F., Carter-Su C. Growth hormone, interferon-gamma, and leukemia inhibitory factor promoted tyrosyl phosphorylation of insulin receptor substrate-1 . J Biol Chem 1995; 270(24):14685-14692.
34. Backer J.M., Wjasow C., Zhang Y. In vitro binding and phosphorylation of insulin receptor substrate 1 by the insulin receptor. Role of interactions mediated by the phosphotyrosine-binding domain and the pleckstrin-homology domain. Eur J Biochem 1997; 245(1):91-96.
35. Guo D., Donner D.B. Tumor necrosis factor promotes phosphorylation and binding of insulin receptor substrate 1 to phosphatidylinositol 3-kinase in 3T3-L1 adipocytes. J Biol Chem 1996; 271(2):615-618.
36. Myers M.G., Jr., Sun X.J., White M.F. The IRS-1 signaling system. Trends Biochem Sci 1994; 19(7):289-293.
37. Platanias L.C., Uddin S., Yetter A., Sun X.J., White M.F. The type I interferon receptor mediates tyrosine phosphorylation of insulin receptor substrate 2. J Biol Chem 1996; 271(1):278-282.
38. Welham M.J., Learmonth L., Bone H., Schrader J.W. Interleukin-13 signal transduction in lymphohemopoietic cells. Similarities and differences in signal transduction with interleukin-4 and insulin. J Biol Chem 1995; 270(20):12286-12296.
39. Sun X.J., Crimmins D.L., Myers M.G., Jr., Miralpeix M., White M.F. Pleiotropic insulin signals are engaged by multisite phosphorylation of IRS-1. Mol Cell Biol 1993; 13(12):7418-7428.
40. Baltensperger K., Kozma L.M., Cherniack A.D., Klarlund J.K., Chawla A., Banerjee U. et al. Binding of the Ras activator son of sevenless to insulin receptor substrate-1 signaling complexes. Science 1993; 260(5116):1950-1952.
41. Kuhne M.R., Pawson T., Lienhard G.E., Feng G.S. The insulin receptor substrate 1 associates with the SH2-containing phosphotyrosine phosphatase Syp. J Biol Chem 1993; 268(16):11479-11481.
42. Backer J.M., Myers M.G., Jr., Shoelson S.E., Chin D.J., Sun X.J., Miralpeix M. et al. Phosphatidylinositol 3'-kinase is activated by association with IRS-1 during insulin stimulation. EMBO J 1992; 11(9):3469-3479.
43. Myers M.G., Jr., Backer J.M., Sun X.J., Shoelson S., Hu P., Schlessinger J. et al. IRS-1 activates phosphatidylinositol 3'-kinase by associating with src homology 2 domains of p85. Proc Natl Acad Sci U S A 1992; 89(21):10350-10354.
44. Musacchio A., Gibson T., Rice P., Thompson J., Saraste M. The PH domain: a common piece in the structural patchwork of signalling proteins. Trends Biochem Sci 1993; 18(9):343-348.
45. Sun X.J., Wang L.M., Zhang Y., Yenush L., Myers M.G., Jr., Glasheen E. et al. Role of IRS-2 in insulin and cytokine signalling. Nature 1995; 377(6545):173-177.
46. Gustafson T.A., He W., Craparo A., Schaub C.D., O'Neill T.J. Phosphotyrosine-dependent interaction of SHC and insulin receptor substrate 1 with the NPEY motif of the insulin receptor via a novel non- SH2 domain. Mol Cell Biol 1995; 15(5):2500-2508.
47. Touhara K., Inglese J., Pitcher J.A., Shaw G., Lefkowitz R.J. Binding of G protein beta gamma-subunits to pleckstrin homology domains. J Biol Chem 1994; 269(14):10217-10220.

48. Harlan J.E., Hajduk P.J., Yoon H.S., Fesik S.W. Pleckstrin homology domains bind to phosphatidylinositol-4,5- bisphosphate. Nature 1994; 371(6493):168-170.
49. Skolnik E.Y., Lee C.H., Batzer A., Vicentini L.M., Zhou M., Daly R. et al. The SH2/SH3 domain-containing protein GRB2 interacts with tyrosine-phosphorylated IRS1 and Shc: implications for insulin control of ras signalling. EMBO J 1993; 12(5):1929-1936.
50. Mansour S.J., Matten W.T., Hermann A.S., Candia J.M., Rong S., Fukasawa K. et al. Transformation of mammalian cells by constitutively active MAP kinase kinase. Science 1994; 265(5174):966-970.
51. Cobb M.H., Goldsmith E.J. How MAP kinases are regulated. J Biol Chem 1995; 270(25):14843-14846.
52. Li W., Nishimura R., Kashishian A., Batzer A.G., Kim W.J., Cooper J.A. et al. A new function for a phosphotyrosine phosphatase: linking GRB2-Sos to a receptor tyrosine kinase. Mol Cell Biol 1994; 14(1):509-517.
53. Hu Q., Klippel A., Muslin A.J., Fantl W.J., Williams L.T. Ras-dependent induction of cellular responses by constitutively active phosphatidylinositol-3 kinase. Science 1995; 268(5207):100-102.
54. Rivera V.M., Miranti C.K., Misra R.P., Ginty D.D., Chen R.H., Blenis J. et al. A growth factor-induced kinase phosphorylates the serum response factor at a site that regulates its DNA-binding activity. Mol Cell Biol 1993; 13(10):6260-6273.
55. Xing J., Ginty D.D., Greenberg M.E. Coupling of the RAS-MAPK pathway to gene activation by RSK2, a growth factor-regulated CREB kinase. Science 1996; 273(5277):959-963.
56. Chen R.H., Sarnecki C., Blenis J. Nuclear localization and regulation of erk- and rsk-encoded protein kinases. Mol Cell Biol 1992; 12(3):915-927.
57. Ginty D.D., Bonni A., Greenberg M.E. Nerve growth factor activates a Ras-dependent protein kinase that stimulates c-fos transcription via phosphorylation of CREB. Cell 1994; 77(5):713-725.
58. Zhao Y., Bjorbaek C., Weremowicz S., Morton C.C., Moller D.E. RSK3 encodes a novel pp90rsk isoform with a unique N-terminal sequence: growth factor-stimulated kinase function and nuclear translocation. Mol Cell Biol 1995; 15(8):4353-4363.
59. Ito T., Sasaki Y., Wands J.R. Overexpression of human insulin receptor substrate 1 induces cellular transformation with activation of mitogen-activated protein kinases. Mol Cell Biol 1996; 16(3):943-951.
60. Tanaka S., Mohr L., Schmidt E.V., Sugimachi K., Wands J.R. Biological effects of human insulin receptor substrate-1 overexpression in hepatocytes. Hepatology 1997; 26(3):598-604.
61. Ito Y., Sasaki Y., Horimoto M., Wada S., Tanaka Y., Kasahara A. et al. Activation of mitogen-activated protein kinases/extracellular signal- regulated kinases in human hepatocellular carcinoma. Hepatology 1998; 27(4):951-958.
62. Cross J.C., Wen P., Rutter W.J. Transactivation by hepatitis B virus X protein is promiscuous and dependent on mitogen-activated cellular serine/threonine kinases. Proc Natl Acad Sci U S A 1993; 90(17):8078-8082.
63. Benn J., Schneider R.J. Hepatitis B virus HBx protein activates Ras-GTP complex formation and establishes a Ras, Raf, MAP kinase signaling cascade . Proc Natl Acad Sci U S A 1994; 91(22):10350-10354.
64. Koike K., Moriya K., Yotsuyanagi H., Iino S., Kurokawa K. Induction of cell cycle progression by hepatitis B virus HBx gene expression in quiescent mouse fibroblasts. J Clin Invest 1994; 94(1):44-49.
65. Benn J., Schneider R.J. Hepatitis B virus HBx protein deregulates cell cycle checkpoint controls. Proc Natl Acad Sci U S A 1995; 92(24):11215-11219.

66. Groisman I.J., Koshy R., Henkler F., Groopman J.D., Alaoui-Jamali M.A. Downregulation of DNA excision repair by the hepatitis B virus-x protein occurs in p53-proficient and p53-deficient cells. Carcinogenesis 1999; 20(3):479-483.
67. Sitterlin D., Lee T.H., Prigent S., Tiollais P., Butel J.S., Transy C. Interaction of the UV-damaged DNA-binding protein with hepatitis B virus X protein is conserved among mammalian hepadnaviruses and restricted to transactivation-proficient X-insertion mutants. J Virol 1997; 71(8):6194-6199.
68. Lee T.H., Elledge S.J., Butel J.S. Hepatitis B virus X protein interacts with a probable cellular DNA repair protein. J Virol 1995; 69(2):1107-1114.
69. Capovilla A., Carmona S., Arbuthnot P. Hepatitis B virus X-protein binds damaged DNA and sensitizes liver cells to ultraviolet irradiation. Biochem Biophys Res Commun 1997; 232(1):255-260.
70. Cheong J.H., Yi M., Lin Y., Murakami S. Human RPB5, a subunit shared by eukaryotic nuclear RNA polymerases, binds human hepatitis B virus X protein and may play a role in X transactivation. EMBO J 1995; 14(1):143-150.
71. Qadri I., Maguire H.F., Siddiqui A. Hepatitis B virus transactivator protein X interacts with the TATA- binding protein. Proc Natl Acad Sci U S A 1995; 92(4):1003-1007.
72. Antunovic J., Lemieux N., Cromlish J.A. The 17 kDa HBx protein encoded by hepatitis B virus interacts with the activation domains of Oct-1, and functions as a coactivator in the activation and repression of a human U6 promoter. Cell Mol Biol Res 1993; 39(5):463-482.
73. Maguire H.F., Hoeffler J.P., Siddiqui A. HBV X protein alters the DNA binding specificity of CREB and ATF-2 by protein-protein interactions. Science 1991; 252(5007):842-844.
74. Dorjsuren D., Lin Y., Wei W., Yamashita T., Nomura T., Hayashi N. et al. RMP, a novel RNA polymerase II subunit 5-interacting protein, counteracts transactivation by hepatitis B virus X protein. Mol Cell Biol 1998; 18(12):7546-7555.
75. Kashuba E., Kashuba V., Pokrovskaja K., Klein G., Szekely L. Epstein-Barr virus encoded nuclear protein EBNA-3 binds XAP-2, a protein associated with Hepatitis B virus X antigen. Oncogene 2000; 19(14):1801-1806.
76. Kuzhandaivelu N., Cong Y.S., Inouye C., Yang W.M., Seto E. XAP2, a novel hepatitis B virus X-associated protein that inhibits X transactivation. Nucleic Acids Res 1996; 24(23):4741-4750.
77. Sirma H., Weil R., Rosmorduc O., Urban S., Israel A., Kremsdorf D. et al. Cytosol is the prime compartment of hepatitis B virus X protein where it colocalizes with the proteasome. Oncogene 1998; 16(16):2051-2063.
78. Fischer M., Runkel L., Schaller H. HBx protein of hepatitis B virus interacts with the C-terminal portion of a novel human proteasome alpha-subunit. Virus Genes 1995; 10(1):99-102.
79. Hu Z., Zhang Z., Doo E., Coux O., Goldberg A.L., Liang T.J. Hepatitis B virus X protein is both a substrate and a potential inhibitor of the proteasome complex. J Virol 1999; 73(9):7231-7240.
80. Lian Z., Pan J., Liu J., Zhang S., Zhu M., Arbuthnot P. et al. The translation initiation factor, hu-Sui1 may be a target of hepatitis B X antigen in hepatocarcinogenesis. Oncogene 1999; 18(9):1677-1687.
81. Sun B.S., Zhu X., Clayton M.M., Pan J., Feitelson M.A. Identification of a protein isolated from senescent human cells that binds to hepatitis B virus X antigen. Hepatology 1998; 27(1):228-239.
82. Oguey D., Dumenco L.L., Pierce R.H., Fausto N. Analysis of the tumorigenicity of the X gene of hepatitis B virus in a nontransformed hepatocyte cell line and

the effects of cotransfection with a murine p53 mutant equivalent to human codon 249. Hepatology 1996; 24(5):1024-1033.

83. Sirma H., Giannini C., Poussin K., Paterlini P., Kremsdorf D., Brechot C. Hepatitis B virus X mutants, present in hepatocellular carcinoma tissue abrogate both the antiproliferative and transactivation effects of HBx. Oncogene 1999; 18(34):4848-4859.

84. Ozer A., Khaoustov V.I., Mearns M., Lewis D.E., Genta R.M., Darlington G.J. et al. Effect of hepatocyte proliferation and cellular DNA synthesis on hepatitis B virus replication. Gastroenterology 1996; 110(5):1519-1528.

85. Kim H., Lee H., Yun Y. X-gene product of hepatitis B virus induces apoptosis in liver cells. J Biol Chem 1998; 273(1):381-385.

86. Su F., Schneider R.J. Hepatitis B virus HBx protein sensitizes cells to apoptotic killing by tumor necrosis factor alpha. Proc Natl Acad Sci U S A 1997; 94(16):8744-8749.

87. Shintani Y., Yotsuyanagi H., Moriya K., Fujie H., Tsutsumi T., Kanegae Y. et al. Induction of apoptosis after switch-on of the hepatitis B virus X gene mediated by the Cre/loxP recombination system. J Gen Virol 1999; 80 (Pt 12):3257-3265.

88. Terradillos O., Pollicino T., Lecoeur H., Tripodi M., Gougeon M.L., Tiollais P. et al. p53-independent apoptotic effects of the hepatitis B virus HBx protein in vivo and in vitro. Oncogene 1998; 17(16):2115-2123.

89. Yoo Y.D., Ueda H., Park K., Flanders K.C., Lee Y.I., Jay G. et al. Regulation of transforming growth factor-beta 1 expression by the hepatitis B virus (HBV) X transactivator. Role in HBV pathogenesis. J Clin Invest 1996; 97(2):388-395.

90. Lara-Pezzi E., Majano P.L., Gomez-Gonzalo M., Garcia-Monzon C., Moreno-Otero R., Levrero M. et al. The hepatitis B virus X protein up-regulates tumor necrosis factor alpha gene expression in hepatocytes. Hepatology 1998; 28(4):1013-1021.

91. Chirillo P., Pagano S., Natoli G., Puri P.L., Burgio V.L., Balsano C. et al. The hepatitis B virus X gene induces p53-mediated programmed cell death. Proc Natl Acad Sci U S A 1997; 94(15):8162-8167.

92. Wang X.W., Forrester K., Yeh H., Feitelson M.A., Gu J.R., Harris C.C. Hepatitis B virus X protein inhibits p53 sequence-specific DNA binding, transcriptional activity, and association with transcription factor ERCC3. Proc Natl Acad Sci U S A 1994; 91(6):2230-2234.

93. Ueda H., Ullrich S.J., Gangemi J.D., Kappel C.A., Ngo L., Feitelson M.A. et al. Functional inactivation but not structural mutation of p53 causes liver cancer. Nat Genet 1995; 9(1):41-47.

94. Truant R., Antunovic J., Greenblatt J., Prives C., Cromlish J.A. Direct interaction of the hepatitis B virus HBx protein with p53 leads to inhibition by HBx of p53 response element-directed transactivation. J Virol 1995; 69(3):1851-1859.

95. Greenblatt M.S., Feitelson M.A., Zhu M., Bennett W.P., Welsh J.A., Jones R. et al. Integrity of p53 in hepatitis B x antigen-positive and -negative hepatocellular carcinomas. Cancer Res 1997; 57(3):426-432.

96. Feitelson M.A., Zhu M., Duan L.X., London W.T. Hepatitis B x antigen and p53 are associated in vitro and in liver tissues from patients with primary hepatocellular carcinoma. Oncogene 1993; 8(5):1109-1117.

97. Elmore L.W., Hancock A.R., Chang S.F., Wang X.W., Chang S., Callahan C.P. et al. Hepatitis B virus X protein and p53 tumor suppressor interactions in the modulation of apoptosis. Proc Natl Acad Sci U S A 1997; 94(26):14707-14712.

98. Shin E.C., Shin J.S., Park J.H., Kim H., Kim S.J. Expression of fas ligand in human hepatoma cell lines: role of hepatitis-B virus X (HBX) in induction of Fas ligand. Int J Cancer 1999; 82 (4):587-591.

99. Gottlob K., Fulco M., Levrero M., Graessmann A. The hepatitis B virus HBx protein inhibits caspase 3 activity. J Biol Chem 1998; 273(50):33347-33353.

100. Takada S., Shirakata Y., Kaneniwa N., Koike K. Association of hepatitis B virus X protein with mitochondria causes mitochondrial aggregation at the nuclear periphery, leading to cell death. Oncogene 1999; 18(50):6965-6973.

101. Schluter V., Meyer M., Hofschneider P.H., Koshy R., Caselmann W.H. Integrated hepatitis B virus X and 3' truncated preS/S sequences derived from human hepatomas encode functionally active transactivators. Oncogene 1994; 9(11):3335-3344.

102. Hildt E., Hofschneider P.H. The PreS2 activators of the hepatitis B virus: activators of tumour promoter pathways. Recent Results Cancer Res 1998; 154:315-329.

103. Lauer U., Weiss L., Lipp M., Hofschneider P.H., Kekule A.S. The hepatitis B virus preS2/St transactivator utilizes AP-1 and other transcription factors for transactivation. Hepatology 1994; 19(1):23-31.

104. Hildt E., Urban S., Lauer U., Hofschneider P.H., Kekule A.S. ER-localization and functional expression of the HBV transactivator MHBst. Oncogene 1993; 8(12):3359-3367.

105. Luber B., Arnold N., Sturzl M., Hohne M., Schirmacher P., Lauer U. et al. Hepatoma-derived integrated HBV DNA causes multi-stage transformation in vitro. Oncogene 1996; 12(8):1597-1608.

106. Terre S., Petit M.A., Brechot C. Defective hepatitis B virus particles are generated by packaging and reverse transcription of spliced viral RNAs in vivo. J Virol 1991; 65(10):5539-5543.

107. Soussan P., Garreau F., Zylberberg H., Ferray C., Brechot C., Kremsdorf D. In vivo expression of a new hepatitis B virus protein encoded by a spliced RNA. J Clin Invest 2000; 105(1):55-60.

108. Dejean A., Bougueleret L., Grzeschik K.H., Tiollais P. Hepatitis B virus DNA integration in a sequence homologous to v-erb-A and steroid receptor genes in a hepatocellular carcinoma. Nature 1986; 322(6074):70-72.

109. Garcia M., de The H., Tiollais P., Samarut J., Dejean A. A hepatitis B virus pre-S-retinoic acid receptor beta chimera transforms erythrocytic progenitor cells in vitro. Proc Natl Acad Sci U S A 1993; 90(1):89-93.

110. Wang J., Chenivesse X., Henglein B., Brechot C. Hepatitis B virus integration in a cyclin A gene in a hepatocellular carcinoma. Nature 1990; 343(6258):555-557.

111. Pineau P., Marchio A., Terris B., Mattei M.G., Tu Z.X., Tiollais P. et al. A t(3;8) chromosomal translocation associated with hepatitis B virus intergration involves the carboxypeptidase N locus. J Virol 1996; 70(10):7280-7284.

112. Chami M., Gozuacik D., Saigo K., Capiod T., Falson P., Lecoeur H. et al. Hepatitis B virus-related insertional mutagenesis implicates SERCA1 gene in the control of apoptosis. Oncogene 2000; 19(25):2877-2886.

113. Zhang X.K., Egan J.O., Huang D., Sun Z.L., Chien V.K., Chiu J.F. Hepatitis B virus DNA integration and expression of an erb B-like gene in human hepatocellular carcinoma. Biochem Biophys Res Commun 1992; 188(1):344-351.

114. Hunt C.M., McGill J.M., Allen M.I., Condreay L.D. Clinical relevance of hepatitis B viral mutations . Hepatology 2000; 31(5):1037-1044.

115. Lindh M., Horal P., Dhillon A.P., Furuta Y., Norkrans G. Hepatitis B virus carriers without precore mutations in hepatitis B e antigen-negative stage show more severe liver damage. Hepatology 1996; 24(3):494-501.

116. Zhang X., Zoulim F., Habersetzer F., Xiong S., Trepo C. Analysis of hepatitis B virus genotypes and pre-core region variability during interferon treatment of HBe antigen negative chronic hepatitis B. J Med Virol 1996; 48(1):8-16.

117. Hur G.M., Lee Y.I., Suh D.J., Lee J.H., Lee Y.I. Gradual accumulation of mutations in precore core region of HBV in patients with chronic active hepatitis: implications of clustering changes in a small region of the HBV core region. J Med Virol 1996; 48(1):38-46.
118. Takahashi K., Aoyama K., Ohno N., Iwata K., Akahane Y., Baba K. et al. The precore/core promoter mutant (T1762A1764) of hepatitis B virus: clinical significance and an easy method for detection. J Gen Virol 1995; 76 (Pt 12):3159-3164.
119. Li J., Buckwold V.E., Hon M.W., Ou J.H. Mechanism of suppression of hepatitis B virus precore RNA transcription by a frequent double mutation. J Virol 1999; 73(2):1239-1244.
120. Buckwold V.E., Xu Z., Chen M., Yen T.S., Ou J.H. Effects of a naturally occurring mutation in the hepatitis B virus basal core promoter on precore gene expression and viral replication. J Virol 1996; 70(9):5845-5851.
121. Hsia C.C., Yuwen H., Tabor E. Hot-spot mutations in hepatitis B virus X gene in hepatocellular carcinoma [letter]. Lancet 1996; 348(9027):625-626.
122. Naoumov N.V., Thomas M.G., Mason A.L., Chokshi S., Bodicky C.J., Farzaneh F. et al. Genomic variations in the hepatitis B core gene: a possible factor influencing response to interferon alfa treatment. Gastroenterology 1995; 108(2):505-514.
123. Marinos G., Torre F., Gunther S., Thomas M.G., Will H., Williams R. et al. Hepatitis B virus variants with core gene deletions in the evolution of chronic hepatitis B infection. Gastroenterology 1996; 111(1):183-192.
124. Pollicino T., Zanetti A.R., Cacciola I., Petit M.A., Smedile A., Campo S. et al. Pre-S2 defective hepatitis B virus infection in patients with fulminant hepatitis. Hepatology 1997; 26(2):495-499.
125. Di Bisceglie A.M. Hepatitis C and hepatocellular carcinoma. Hepatology 1997; 26(3 Suppl 1):34S-38S.
126. Caselmann W.H., Alt M. Hepatitis C virus infection as a major risk factor for hepatocellular carcinoma. J Hepatol 1996; 24(2 Suppl):61-66.
127. el Refaie A., Savage K., Bhattacharya S., Khakoo S., Harrison T.J., el Batanony M. et al. HCV-associated hepatocellular carcinoma without cirrhosis. J Hepatol 1996; 24(3):277-285.
128. Tanaka K., Ikematsu H., Hirohata T., Kashiwagi S. Hepatitis C virus infection and risk of hepatocellular carcinoma among Japanese: possible role of type 1b (II) infection. J Natl Cancer Inst 1996; 88(11):742-746.
129. Zein N.N., Poterucha J.J., Gross J.B., Jr., Wiesner R.H., Therneau T.M., Gossard A.A. et al. Increased risk of hepatocellular carcinoma in patients infected with hepatitis C genotype 1b. Am J Gastroenterol 1996; 91(12):2560-2562.
130. Shiratori Y., Imazeki F., Moriyama M., Yano M., Arakawa Y., Yokosuka O. et al. Histologic improvement of fibrosis in patients with hepatitis C who have sustained response to interferon therapy [see comments]. Ann Intern Med 2000; 132(7):517-524.
131. Yoshida H., Shiratori Y., Moriyama M., Arakawa Y., Ide T., Sata M. et al. Interferon therapy reduces the risk for hepatocellular carcinoma: national surveillance program of cirrhotic and noncirrhotic patients with chronic hepatitis C in Japan. IHIT Study Group. Inhibition of Hepatocarcinogenesis by Interferon Therapy. Ann Intern Med 1999; 131(3):174-181.
132. Yasui K., Wakita T., Tsukiyama-Kohara K., Funahashi S.I., Ichikawa M., Kajita T. et al. The native form and maturation process of hepatitis C virus core protein. J Virol 1998; 72(7):6048-6055.
133. Ray R.B., Lagging L.M., Meyer K., Ray R. Hepatitis C virus core protein cooperates with ras and transforms primary rat embryo fibroblasts to tumorigenic phenotype. J Virol 1996; 70(7):4438-4443.

134. Honda A., Arai Y., Hirota N., Sato T., Ikegaki J., Koizumi T. et al. Hepatitis C virus structural proteins induce liver cell injury in transgenic mice. J Med Virol 1999; 59(3):281-289.

135. Moriya K., Fujie H., Yotsuyanagi H., Shintani Y., Tsutsumi T., Matsuura Y. et al. Subcellular localization of hepatitis C virus structural proteins in the liver of transgenic mice. Jpn J Med Sci Biol 1997; 50(4-5):169-177.

136. Moriya K., Yotsuyanagi H., Shintani Y., Fujie H., Ishibashi K., Matsuura Y. et al. Hepatitis C virus core protein induces hepatic steatosis in transgenic mice. J Gen Virol 1997; 78 (Pt 7):1527-1531.

137. Moriya K., Fujie H., Shintani Y., Yotsuyanagi H., Tsutsumi T., Ishibashi K. et al. The core protein of hepatitis C virus induces hepatocellular carcinoma in transgenic mice. Nat Med 1998; 4(9):1065-1067.

138. Matsuda J., Suzuki M., Nozaki C., Shinya N., Tashiro K., Mizuno K. et al. Transgenic mouse expressing a full-length hepatitis C virus cDNA. Jpn J Cancer Res 1998; 89 (2):150-158.

139. Kawamura T., Furusaka A., Koziel M.J., Chung R.T., Wang T.C., Schmidt E.V. et al. Transgenic expression of hepatitis C virus structural proteins in the mouse. Hepatology 1997; 25(4):1014-1021.

140. Pasquinelli C., Shoenberger J.M., Chung J., Chang K.M., Guidotti L.G., Selby M. et al. Hepatitis C virus core and E2 protein expression in transgenic mice. Hepatology 1997; 25(3):719-727.

141. Tsuchihara K., Hijikata M., Fukuda K., Kuroki T., Yamamoto N., Shimotohno K. Hepatitis C virus core protein regulates cell growth and signal transduction pathway transmitting growth stimuli. Virology 1999; 258(1):100-107.

142. Hayashi J., Aoki H., Kajino K., Moriyama M., Arakawa Y., Hino O. Hepatitis C virus core protein activates the MAPK/ERK cascade synergistically with tumor promoter TPA, but not with epidermal growth factor or transforming growth factor alpha [In Process Citation]. Hepatology 2000; 32(5):958-961.

143. Aoki H., Hayashi J., Moriyama M., Arakawa Y., Hino O. Hepatitis C virus core protein interacts with 14-3-3 protein and activates the kinase Raf-1. J Virol 2000; 74(4):1736-1741.

144. Wasylyk B., Hagman J., Gutierrez-Hartmann A. Ets transcription factors: nuclear effectors of the Ras-MAP-kinase signaling pathway. Trends Biochem Sci 1998; 23(6):213-216.

145. Sabile A., Perlemuter G., Bono F., Kohara K., Demaugre F., Kohara M. et al. Hepatitis C virus core protein binds to apolipoprotein AII and its secretion is modulated by fibrates. Hepatology 1999; 30(4):1064-1076.

146. Matsumoto M., Hsieh T.Y., Zhu N., VanArsdale T., Hwang S.B., Jeng K.S. et al. Hepatitis C virus core protein interacts with the cytoplasmic tail of lymphotoxin-beta receptor. J Virol 1997; 71(2):1301-1309.

147. Zhu N., Khoshnan A., Schneider R., Matsumoto M., Dennert G., Ware C. et al. Hepatitis C virus core protein binds to the cytoplasmic domain of tumor necrosis factor (TNF) receptor 1 and enhances TNF-induced apoptosis. J Virol 1998; 72(5):3691-3697.

148. Hsieh T.Y., Matsumoto M., Chou H.C., Schneider R., Hwang S.B., Lee A.S. et al. Hepatitis C virus core protein interacts with heterogeneous nuclear ribonucleoprotein K. J Biol Chem 1998; 273(28):17651-17659.

149. Mamiya N., Worman H.J. Hepatitis C virus core protein binds to a DEAD box RNA helicase. J Biol Chem 1999; 274(22):15751-15756.

150. Ruggieri A., Harada T., Matsuura Y., Miyamura T. Sensitization to Fas-mediated apoptosis by hepatitis C virus core protein. Virology 1997; 229(1):68-76.

151. Ray R.B., Meyer K., Ray R. Suppression of apoptotic cell death by hepatitis C virus core protein. Virology 1996; 226(2):176-182.

152. Ray R.B., Meyer K., Steele R., Shrivastava A., Aggarwal B.B., Ray R. Inhibition of tumor necrosis factor (TNF-alpha)-mediated apoptosis by hepatitis C virus core protein. J Biol Chem 1998; 273(4):2256-2259.

153. Marusawa H., Hijikata M., Chiba T., Shimotohno K. Hepatitis C virus core protein inhibits Fas- and tumor necrosis factor alpha-mediated apoptosis via NF-kappaB activation. J Virol 1999; 73(6):4713-4720.

154. Lu W., Lo S.Y., Chen M., Wu K., Fung Y.K., Ou J.H. Activation of p53 tumor suppressor by hepatitis C virus core protein. Virology 1999; 264(1):134-141.

155. Dumoulin F.L., vsn dem B.A., Sohne J., Sauerbruch T., Spengler U. Hepatitis C virus core protein does not inhibit apoptosis in human hepatoma cells. Eur J Clin Invest 1999; 29(11):940-946.

156. Ray R.B., Steele R., Meyer K., Ray R. Hepatitis C virus core protein represses p21WAF1/Cip1/Sid1 promoter activity. Gene 1998; 208(2):331-336.

157. Ray R.B., Steele R., Meyer K., Ray R. Transcriptional repression of p53 promoter by hepatitis C virus core protein. J Biol Chem 1997; 272(17):10983-10986.

158. Srinivas R.V., Ray R.B., Meyer K., Ray R. Hepatitis C virus core protein inhibits human immunodeficiency virus type 1 replication. Virus Res 1996; 45(2):87-92.

159. Shih C.M., Lo S.J., Miyamura T., Chen S.Y., Lee Y.H. Suppression of hepatitis B virus expression and replication by hepatitis C virus core protein in HuH-7 cells. J Virol 1993; 67(10):5823-5832.

160. Kim D.W., Suzuki R., Harada T., Saito I., Miyamura T. Trans-suppression of gene expression by hepatitis C viral core protein. Jpn J Med Sci Biol 1994; 47(4):211-220.

161. Hoofnagle J.H., Di Bisceglie A.M. The treatment of chronic viral hepatitis. N Engl J Med 1997; 336(5):347-356.

162. Moradpour D., Blum H.E. Current and evolving therapies for hepatitis C. Eur J Gastroenterol Hepatol 1999; 11(11):1199-1202.

163. Poynard T., Leroy V., Cohard M., Thevenot T., Mathurin P., Opolon P. et al. Meta-analysis of interferon randomized trials in the treatment of viral hepatitis C: effects of dose and duration. Hepatology 1996; 24(4):778-789.

164. Sen G.C., Lengyel P. The interferon system. A bird's eye view of its biochemistry. J Biol Chem 1992; 267(8):5017-5020.

165. Meurs E., Chong K., Galabru J., Thomas N.S., Kerr I.M., Williams B.R. et al. Molecular cloning and characterization of the human double-stranded RNA-activated protein kinase induced by interferon. Cell 1990; 62(2):379-390.

166. Taylor D.R., Shi S.T., Romano P.R., Barber G.N., Lai M.M. Inhibition of the interferon-inducible protein kinase PKR by HCV E2 protein [see comments]. Science 1999; 285(5424):107-110.

167. Darnell J.E., Jr., Kerr I.M., Stark G.R. Jak-STAT pathways and transcriptional activation in response to IFNs and other extracellular signaling proteins. Science 1994; 264(5164):1415-1421.

168. Ihle J.N. STATs: signal transducers and activators of transcription. Cell 1996; 84(3):331-334.

169. Heim M.H., Moradpour D., Blum H.E. Expression of hepatitis C virus proteins inhibits signal transduction through the Jak-STAT pathway. J Virol 1999; 73(10):8469-8475.

170. Heim M.H. The Jak-STAT pathway: specific signal transduction from the cell membrane to the nucleus. Eur J Clin Invest 1996; 26(1):1-12.

171. Sakamuro D., Furukawa T., Takegami T. Hepatitis C virus nonstructural protein NS3 transforms NIH 3T3 cells. J Virol 1995; 69(6):3893-3896.

172. Ishido S., Hotta H. Complex formation of the nonstructural protein 3 of hepatitis C virus with the p53 tumor suppressor. FEBS Lett 1998; 438(3):258-262.

173. Park J.S., Yang J.M., Min M.K. Hepatitis C virus nonstructural protein NS4B transforms NIH3T3 cells in cooperation with the Ha-ras oncogene. Biochem Biophys Res Commun 2000; 267(2):581-587.

174. Tan S.L., Nakao H., He Y., Vijaysri S., Neddermann P., Jacobs B.L. et al. NS5A, a nonstructural protein of hepatitis C virus, binds growth factor receptor-bound protein 2 adaptor protein in a Src homology 3 domain/ligand-dependent manner and perturbs mitogenic signaling. Proc Natl Acad Sci U S A 1999; 96(10):5533-5538.

175. Kato N., Lan K.H., Ono-Nita S.K., Shiratori Y., Omata M. Hepatitis C virus nonstructural region 5A protein is a potent transcriptional activator. J Virol 1997; 71(11):8856-8859.

176. Chung K.M., Song O.K., Jang S.K. Hepatitis C virus nonstructural protein 5A contains potential transcriptional activator domains. Mol Cells 1997; 7(5):661-667.

177. Gale M.J., Jr., Korth M.J., Katze M.G. Repression of the PKR protein kinase by the hepatitis C virus NS5A protein: a potential mechanism of interferon resistance. Clin Diagn Virol 1998; 10(2-3):157-162.

178. Gale M.J., Jr., Korth M.J., Tang N.M., Tan S.L., Hopkins D.A., Dever T.E. et al. Evidence that hepatitis C virus resistance to interferon is mediated through repression of the PKR protein kinase by the nonstructural 5A protein. Virology 1997; 230(2):217-227.

179. Gale M., Jr., Kwieciszewski B., Dossett M., Nakao H., Katze M.G. Antiapoptotic and oncogenic potentials of hepatitis C virus are linked to interferon resistance by viral repression of the PKR protein kinase. J Virol 1999; 73(8):6506-6516.

180. Enomoto N., Sakuma I., Asahina Y., Kurosaki M., Murakami T., Yamamoto C. et al. Mutations in the nonstructural protein 5A gene and response to interferon in patients with chronic hepatitis C virus 1b infection. N Engl J Med 1996; 334(2):77-81.

181. Enomoto N., Sakuma I., Asahina Y., Kurosaki M., Murakami T., Yamamoto C. et al. Comparison of full-length sequences of interferon-sensitive and resistant hepatitis C virus 1b. Sensitivity to interferon is conferred by amino acid substitutions in the NS5A region. J Clin Invest 1995; 96(1):224-230.

182. Chung R.T., Monto A., Dienstag J.L., Kaplan L.M. Mutations in the NS5A region do not predict interferon-responsiveness in american patients infected with genotype 1b hepatitis C virus. J Med Virol 1999; 58(4):353-358.

183. Frangeul L., Cresta P., Perrin M., Lunel F., Opolon P., Agut H. et al. Mutations in NS5A region of hepatitis C virus genome correlate with presence of NS5A antibodies and response to interferon therapy for most common European hepatitis C virus genotypes. Hepatology 1998; 28(6):1674-1679.

184. Ibarrola N., Moreno-Monteagudo J.A., Saiz M., Garcia-Monzon C., Sobrino F., Garcia-Buey L. et al. Response to retreatment with interferon-alpha plus ribavirin in chronic hepatitis C patients is independent of the NS5A gene nucleotide sequence. Am J Gastroenterol 1999; 94(9):2487-2495.

185. Murakami T., Enomoto N., Kurosaki M., Izumi N., Marumo F., Sato C. Mutations in nonstructural protein 5A gene and response to interferon in hepatitis C virus genotype 2 infection. Hepatology 1999; 30(4):1045-1053.

186. Squadrito G., Orlando M.E., Cacciola I., Rumi M.G., Artini M., Picciotto A. et al. Long-term response to interferon alpha is unrelated to "interferon sensitivity determining region" variability in patients with chronic hepatitis C virus-1b infection [see comments]. J Hepatol 1999; 30(6):1023-1027.

187. Tu H., Gao L., Shi S.T., Taylor D.R., Yang T., Mircheff A.K. et al. Hepatitis C virus RNA polymerase and NS5A complex with a SNARE-like protein. Virology 1999; 263 (1):30-41.

Chapter 10

ANTIVIRAL AGENTS FOR HEPATITIS VIRUSES

Robert K. Hamatake, Zhi Hong, Johnson Y. N. Lau and Weidong Zhong

ICN Pharmaceuticals, 3300 Hyland Avenue, Costa Mesa, CA 92626
Corresponding author: W. Zhong, Tel: (714) 545-0100, ext. 2201; Fax: (714) 668-3141;
E-Mail: wzhong@icnpharm.com

1. THERAPY FOR HAV

HAV is a positive-sense single-stranded RNA virus belonging to the genus hepatovirus of the *Picornaviridae* family (1, 2). HAV infection occurs worldwide through the fecal-oral route and is an important cause of acute viral hepatitis, particularly in many developing countries. Although most patients recover completely from this disease, elderly patients have a substantial high mortality risk. Fulminant hepatitis is a rare complication of acute HAV infection (3). Superinfection by HAV in patients with pre-existing liver diseases is often associated with more severe liver disease and increased morbidity and mortality (1).

Inactivated whole-virus vaccines, Havrix (SmithKline Beecham) and Vaqta (Merck & Co.), are available and highly efficacious in prevention of HAV infection (4-6). These highly immunogenic vaccines contain formalin-killed whole-virus particles and empty capsids produced in infected cell culture and elicit protective response in >90% of individuals. The vaccines have good safety profile, with reactogenicity similar to that of HBV. Multivalent vaccines directed against HAV antigens as well as other viral or nonviral antigens are currently under development. Immune globulin can be given before exposure for short-term protection and for individuals who have recently been exposed to HAV. In addition, ribavirin was found to be effective in treating HAV infection in a small study (7).

2. THERAPIES FOR HBV

Hepatitis B Virus (HBV) is the causative agent for hepatitis B, which infects an estimated 140,000-320,000 people annually in the United States. Despite a high rate of viral clearance by immunocompetent adult individuals, a large proportion of the world's population is chronically infected with HBV. This is due to the fact that vertical transmission of HBV in neonates leads to a high rate of chronic infection (>90%). Currently, more than 350 million people are chronically infected with HBV. The pathology of HBV infection is attributed to immune-mediated liver damage rather than viral cytopathology. Repeated cycles of HBV replication and immune lysis of infected hepatocytes is associated with an increased risk of fibrosis, cirrhosis, and hepatocellular carcinoma. The medical and scientific communities have devoted an enormous effort to reduce the 250,000 deaths each year resulting from hepatitis B. This section will describe the current approved therapies for treatment of hepatitis B and the antiviral strategies being pursued by research scientists, clinicians, and the pharmaceutical industry to develop novel chemotherapies for hepatitis B.

2.1 Current Therapies

Strategies for treating hepatitis B have focused on clearance of active HBV infection through suppression of viral replication. The efficacies of these treatments have been determined by monitoring the levels of HBV DNA in the serum, serum ALT levels, loss of viral antigens (HBsAg and HBeAg), seroconversion (HBsAb and HBeAb), and ultimately by improvements in liver histology. Interferon-α (IFN-α) and the nucleoside analog lamivudine have proven their effectiveness by these clinical markers and are currently approved in the Western world for the treatment of HBV infection. The immunomodulator thymosin alpha 1 has not been approved worldwide due to the lack of well demonstrated activity but is approved for chronic hepatitis B in a few countries including China.

Treatment with IFN-α consists of a 4-month course of 5 million units daily or 9-10 million units 3 times a week. Side-effects are common. This regimen results in a 30-40% response rate with reduction in serum HBV DNA, normalization of ALT levels, and loss of HBeAg (8, 9). Most patients who respond develop a sustained response and many responders in the Western world lose HBsAg within a decade after treatment. An individual's response to IFN-α treatment may depend on his pretherapy clinical markers. A favorable outcome is more likely in the sub-population of patients that present with low serum HBV DNA and elevated serum ALT levels (10). IFN-α is not indicated for patients infected with precore mutants or who have decompensated cirrhosis.

The nucleoside analog lamivudine acts as a chain terminator after conversion to the triphosphate form and specifically targets the viral polymerase (11). Treatment with a 100 mg oral daily dose is well tolerated and inhibition of the HBV polymerase results in a rapid drop in serum viral DNA levels. After one year of treatment, almost all patients have reduced levels of HBV DNA, 41-72% have sustained normalized ALT levels, and 52-56% show improvement in liver biopsy (12, 13). There is no consensus on length of treatment although increased seroconversion from HBeAg to HBeAb has been reported to occur in patients with increasing length of treatment (13, 14). Lamivudine is also effective in cases where IFN-α treatment is not indicated. In patients with chronic hepatitis B who were negative for HBeAg (precore variants), a patient population that normally responds poorly to IFN-α treatment, loss of HBV DNA and normalization of ALT levels was seen in 63% of patients after 24 weeks of treatment, a response rate identical to that observed in patients positive for HBeAg (15). Lamivudine treatment also shows some efficacy in patients with decompensated cirrhosis (16).

Although lamivudine has shown itself to be both safe and effective in inhibiting HBV replication, some issues related to its mechanism of action have limited its widespread use. The first issue is the durability of the response to lamivudine treatment. For patients who have not seroconverted during treatment, rebound of HBV DNA to pretreatment levels occurs soon after cessation of lamivudine treatment (17, 18). For patients who have seroconverted, the sustained response rate can be high (13). The lack of durability for non-seroconverters is attributed to the continued presence of covalently closed circular DNA (cccDNA) within infected hepatocytes. cccDNA resides in the nuclei of infected cells and is the key template for viral transcripts and genomic material. The copy number of cccDNA is maintained by recycling of nucleocapsids containing viral DNA from the cytoplasm to the nucleus (19). Lamivudine can block this cycle by inhibiting the synthesis of viral DNA within the nucleocapsid but lamivudine cannot directly affect the resident copies of cccDNA. While the half-life of cccDNA in duck hepatocytes has been measured in days (20), very little loss of cccDNA was seen in woodchuck hepatocytes (21, 22). This *in vitro* observation has also been confirmed *in vivo* where it was concluded that the half-life of cccDNA during antiviral treatment of infected woodchucks was 33-50 days, similar to the half-life of infected hepatocytes (23). Therefore, in the absence of immune-mediated clearance of infected hepatocytes, long term chemotherapy may be necessary to eliminate cccDNA by turnover of infected hepatocytes and thereby effect a durable seroconversion. The cost for long-term treatment may be prohibitive in the developing countries.

Another serious issue with lamivudine therapy is the emergence of lamivudine resistant HBV mutants. Lamivudine resistance for HBV is very similar to lamivudine resistance for HIV in that it arises from a mutation of the methionine residue in the YMDD motif of the viral polymerase (24, 25). The number of patients with lamivudine resistant HBV increases significantly with longer treatment and may reach 66% of patients after 4 years (26). The emergence of lamivudine resistant HBV is accompanied by the reappearance of serum HBV DNA and a rise in serum ALT levels. In one study, hepatic flares occurred more frequently after emergence of resistant HBV (40%) compared to patients with wild type HBV (4%). However, the patients who experienced exacerbations had higher seroconversion rates than those who did not have hepatic flares (27). The full clinical implications of the emergence of lamivudine resistant HBV are not yet understood but it is clear that this is an issue that may impact the long-term efficacy of lamivudine treatment and has created a strong desire for additional antiviral agents.

2.2 Molecular Targets for HBV Therapy

Polymerase. HBV encodes a multifunctional viral polymerase that is involved throughout the replication reaction. Binding of HBV polymerase to a specific RNA stem-loop structure on the pre-genomic RNA, called epsilon or ε, is responsible for packaging of this complex into core particles (28, 29). The ε sequence also defines the site of minus-strand DNA initiation (30, 31) which occurs through a unique self-priming mechanism employing the amino-terminal portion of HBV polymerase (32, 33). HBV polymerase has both RNA- and DNA-dependent polymerase activities (34) as well as an RNase H activity that degrades pre-genomic RNA during reverse transcription (35). Although HBV polymerase activity can be measured in several *in vitro* systems (36-38), the type of detailed structural and biochemical information available for HIV reverse transcriptase is not available for HBV polymerase due to the inability to express large quantities of functional, non-capsid associated polymerase in heterologous systems. Nevertheless, targeting the HBV polymerase has been achieved, as it has for the HIV reverse transcriptase, by the development of nucleoside analogs. The availability of stable cell lines capable of producing HBV particles (39-41) has been critical for screening nucleoside analogs for antiviral activity (42). Nucleoside analogs represent the largest class of HBV inhibitors. The status of nucleoside analogs recently or currently being developed is shown in Table 1. The structures of these compounds are shown in Figure 1. Both names are given if the nucleoside analog is known by more than one name.

In addition to the nucleoside analogs listed in Table 1, Triangle's DAPD, which is currently being developed for HIV, also has anti-HBV activity (43). Many of the nucleoside analogs listed in Table 1 are potent and selective inhibitors of HBV polymerase. However, the main issue for lamivudine therapy, whether a sustained response can be achieved before the emergence of drug resistant mutants, must still be addressed with the newer treatments. In this regard, combinations of different nucleoside analogs may be helpful or even necessary. Adefovir Dipivoxil was shown to be active in patients with lamivudine resistant HBV (44) and therefore has potential for use in combination therapy with lamivudine. The clinical efficacy can be explained by the *in vitro* activity of PMEA, to which adefovir dipivoxil is metabolized, against lamivudine resistant HBV mutants (52, 53). However, FTC and L-FMAU are cross-resistant to lamivudine resistant HBV mutants *in vitro* (41, 53) so their utility in pre-treated patients or in combination with lamivudine may be limited. Whether the HBV polymerase will have the plasticity of the HIV reverse transcriptase in tolerating additional mutations to acquire resistance to multiple drugs remains to be determined.

Table 1. Nucleoside analogs with anti-HBV activity

Nucleoside Analog	Company	Status
Lamivudine/3TC	GlaxoSmithKline	Marketed
Famvir	SmithKline Beecham	Discontinued for HBV
Lobucavir	Bristol-Myers Squibb	Discontinued
Adefovir Dipivoxil	Gilead	Phase III
Entecavir/BMS-200475	Bristol-Myers Squibb	Phase III
Coviricil/FTC	Triangle	Phase II
Clevudine/L-FMAU	Triangle	Phase I/II
L-dT	Novirio	Phase I/II

Core. The core gene encodes the 21.5 kDa viral capsid protein also known as hepatitis B core antigen (HBcAg). Overlapping with the core open reading frame is the precore open reading frame which encodes a secreted form of core protein known as hepatitis B e antigen (HBeAg). HBeAg is not required for viral replication (54, 55) but it may affect the immunological response to infected hepatocytes (56). Core protein assembles into nucleocapsids, or core particles, containing the polymerase and pre-genomic RNA which are either transported to the nucleus to amplify the cccDNA pool or enveloped at an internal cellular membrane and secreted as virus particles (19).

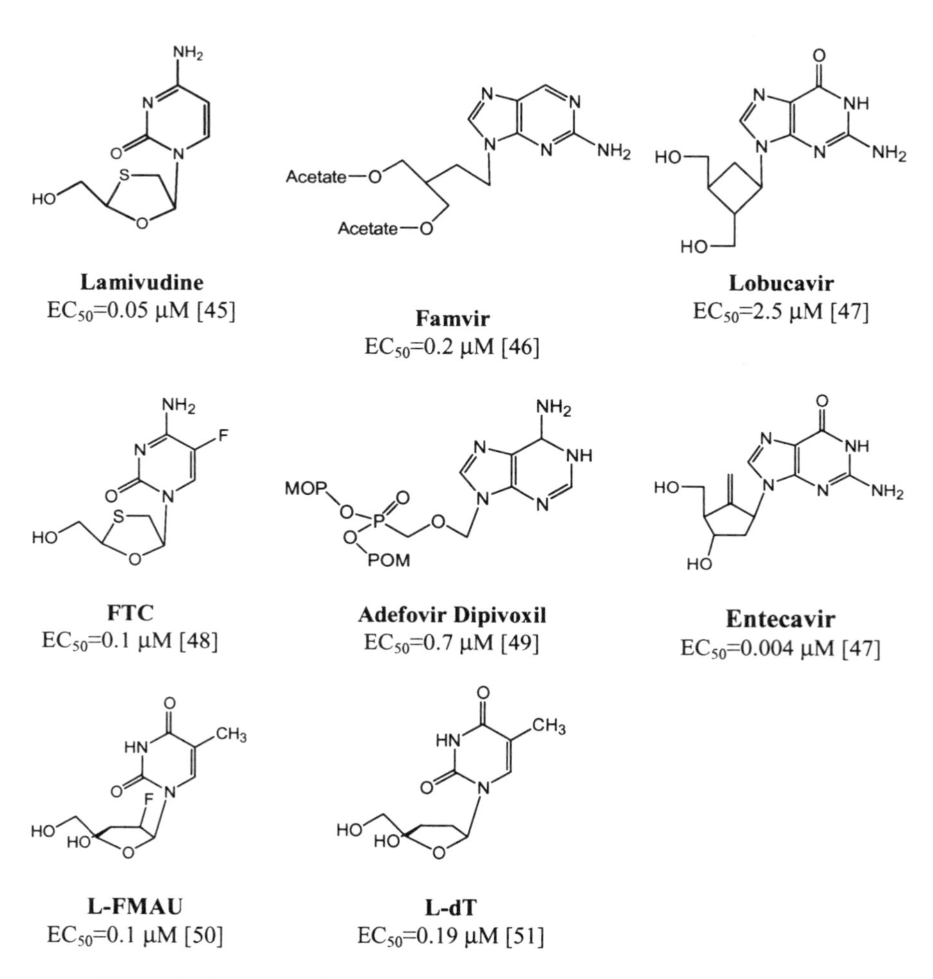

Figure 1. Structures of nucleoside analogs and their *in vitro* HBV antiviral activity

The role of core protein in viral assembly is complex and involves core-core interactions to form the icosahedral core particle (for review, see (57), functions affecting DNA synthesis (58-60), and interactions that direct envelopment and secretion (61, 62). Given the central role that the core protein plays in virus maturation, agents that interfere with the intricate interactions involving core protein may possess antiviral activity. The interference could be direct (e.g. binding to core) or indirect, such as affecting the phosphorylation state at any of the 3 known sites for core protein phosphorylation (63). Although no small molecules are presently known to inhibit capsid assembly or maturation, the existence of dominant negative core mutants that inhibit viral replication (64) and of peptides capable of blocking the association between core particle and surface antigens and inhibiting virus production (65, 66) indicates that disruption of assembly does affect viral replication. Furthermore, AT-61, a small molecule inhibitor of HBV replication from Avid Therapeutics (67), reduced the amount of core particles containing pre-genomic RNA suggesting that assembly was being affected, although the exact mechanism remains unknown. Thus, targeting core protein assembly or maturation may have potential as a therapeutic modality.

Envelope Proteins. Infectious HBV particles contain three forms of envelope protein termed L, M, and S. These are produced from a single ORF through alternative transcription and translation initiation sites. All three forms of envelope protein are inserted into the endoplasmic reticulum membrane and become glycosylated as they are transported through the secretory pathway (68). They are the only virally encoded component in the envelope surrounding the viral nucleocapsid and are therefore involved in binding to the cellular receptor for viral uptake. The identity of the receptor for HBV remains elusive but carboxypeptidase D has been identified as the receptor for the duck hepatitis B virus (DHBV) (69-73). Additional, as yet unidentified, factors are needed after the initial binding event to allow infection to occur and these may be functionally analogous to the co-receptors identified for HIV (74, 75). The identification of the human HBV receptor and the establishment of an *in vitro* infection system will allow HBV particle entry to be targeted. Until these difficult hurdles are overcome, the prospects are poor for therapies that target HBV particle entry.

The envelope proteins are post-translationally modified, and inhibitors of α-glucosidase I, a cellular enzyme involved in oligosaccharide processing, have been found to inhibit the secretion of M protein but not S or L proteins (76). This inhibition results in reduced secretion of viral particles both *in vitro* (76, 77) and *in vivo* (78). Egress of the envelope protein, in addition to ingress via receptor binding, may therefore be a target for antiviral therapy. In this case, inhibition of the viral process is affected through inhibition of a cellular target. The resistance profile of the α-glucosidase inhibitors will clearly differ from the polymerase inhibitors and will enable these two classes of inhibitors to be used in combination. However, toxicity arising from inhibiting cellular processes is an issue that needs to be addressed.

X Protein. The hepatitis B X protein (HBx) is not required for viral replication *in vitro* (79, 80) but in the woodchuck hepatitis virus system it is required to establish an infection *in vivo* (81, 82). Despite a plethora of activities attributed to X, its function during infection remains unknown. The lack of an *in vitro* viral replication system dependent upon HBx and the inability to express and purify X protein have hampered the dissection of its function. These difficulties also seriously diminish the feasibility of developing antivirals targeting HBx protein or pathways affected by X.

2.3 Immunotherapy

The host immune system is central to controlling HBV infection and to the pathogenesis of the disease (for reviews, see [83, 84]. Individuals who clear HBV infection have a vigorous polyclonal class I-restricted cytotoxic T lymphocyte (CTL) and class II-restricted CD4+ T-helper response to HBV proteins (85-88). In contrast, the CTL responses in chronically infected patients are difficult to detect and narrowly focused (89). Extensive work with the transgenic mouse model of HBV has shown that CTLs can abolish HBV expression and replication by a noncytopathic mechanism (90). Clearance of HBV DNA in the liver of infected chimpanzees during an acute infection also occurs by a noncytopathic mechanism (91). A major role for the innate immune response in viral clearance, possibly by natural killer cells, was hypothesized from the delayed influx of T cells into the liver. A recent study of a single-source outbreak of HBV infection showed that control of HBV DNA levels in the serum also occurred before the onset of clinical disease (92). The amount of natural killer cells was highest at the earliest points post-infection and decreased as the HBV DNA levels fell, consistent with their proposed role in antiviral defense (93). Although these measurements were made in the serum compartment and not in the liver, they support the notion that noncytopathic clearance can occur before cytolytic processes cause liver damage. Contrasting with these results is a study performed on woodchucks transiently infected with woodchuck hepatitis virus where viral clearance in the liver was associated with a CTL response and hepatocyte turnover due to apoptosis and regeneration (94). Since every hepatocyte is infected in these animals (95), clearance of virus during transient infection requires that nuclear cccDNA be lost from the regenerating hepatocytes and that they be protected from reinfection. Therefore, some fraction of the infected hepatocytes must be cleared of cccDNA in a noncytopathic manner. An understanding of the identity and role of the components of the immune system involved in these processes may reveal the mechanism for viral persistence and chronic infection and aid in the rational development of immunomodulatory agents for treatment.

Vaccine Strategies. Vaccines that can stimulate a broad-based immune response or generate a CTL response may clear chronic infections. Vaccines based on HBV-specific peptides or DNA sequences have been proposed for the therapeutic treatment of chronic HBV. Since clearance during acute infection is associated with a CTL response, administration of a vaccine containing a CTL epitope may effect resolution of infection. Cytel Corp. is developing such a vaccine and has reported that their vaccine containing a CTL epitope to HBV core protein elicited a CTL response and helper T lymphocyte activity in a Phase I trial (96). Unfortunately, the vaccine showed no efficacy when tested against individuals with chronic hepatitis B, perhaps because the magnitude of the CTL response was lower in infected vs non-infected individuals (97). Another vaccine approach has been taken with the current prophylactic HBV vaccines and using them in a therapeutic setting. Although the study is not yet published, Pol et al report that vaccination of patients with chronic hepatitis B with GenHevac B (preS2/S vaccine from Pasteur-Merieux) or Recombivax (S vaccine from Merck) resulted in a significant loss of viral DNA in the vaccinated group (98). These vaccines normally elicit a strong B cell response so their effect on T-cell response in responders and non-responders is of interest.

DNA based vaccines have the potential to induce both humoral and cell-mediated immune responses against epitopes expressed by cloned DNA (for review, see [99]). Studies with DHBV have shown that immunizations with plasmid DNA containing the preS/S region were protective in uninfected ducks and were therapeutic in infected ducks (100). An advance in DNA vaccine technology is the delivery of DNA coated gold beads directly into the cells of the skin with a gene gun-like device. PowderJect Vaccines has reported that their device, using DNA encoding HBV S antigen, produced protective antibody and a CTL response in uninfected individuals (101). Whether these promising results can be extended to chronically infected individuals remains to be determined.

Immunomodulators. Immunomodulators are agents that effect the host immune response in a non-specific manner. These agents may, therefore, be active against more than one virus but may also induce more systemic side effects. An example is IFN-α which is used to treat both hepatitis B and hepatitis C but is not well tolerated. A combination of IFN-β and IFN-γ has also been used to treat chronic HBV infection (102). In this study the response rate was similar to that of IFN-α but was achieved with 4 weeks of treatment vs 4 months of treatment for IFN-α. Recombinant human IFN-γ had little or no effect when used as a monotherapy (103). Further studies are needed to determine if the advantages of combination IFN-β and IFN-γ treatment warrant further development.

In addition to the interferons, interleukin-12 (IL-12) has been evaluated for anti-HBV activity. IL-12 is known to induce endogenous IFN-γ and administration of IL-12 in the transgenic mouse model resulted in the inhibition of HBV replication at doses as low as 100 ng (104). Moderate, dose-dependent elevations in serum ALT levels were observed but foci of inflammation were seen at only the highest dose, 1 μg of IL-12 given for 5 days. A phase I/II study in chronically infected patients reported a dose-dependent decrease in serum HBV DNA levels upon subcutaneous administration of IL-12 (105). Treatment with 0.5 μg/kg resulted in HBV DNA clearance in 25% and loss of HBeAg in 16% of treated patients. These results show promise but more work is needed to optimize the dose and evaluate the safety of this therapy.

2.4 Combination Therapy

Combination therapy for HIV infection has become the standard method for suppressing the emergence of drug resistant mutants. Experience with the many different HIV therapies has taught clinicians that concurrent treatment with a multi-drug regimen is preferable to sequential treatment. These 'cocktails' have been further refined to include two or three drugs acting against a single target, two or three nucleoside analogs for example, rather than one nucleoside analog and one protease inhibitor to prevent the sequential selection of virus that are resistant in multiple targets. GlaxoSmithKline's lamivudine is the only approved drug available for treatment of chronic HBV infection that acts as a direct antiviral. The knowledge gained from the HIV experience should be applied to combinations of future HBV antivirals with lamivudine and would dictate their concurrent administration for combination therapy.

Interestingly, a trial that combined the two currently approved therapies for HBV used only sequential treatment with lamivudine and IFN-α (106). The combination treatment consisted of 8 weeks of lamivudine treatment followed by 16 weeks of IFN-α treatment and was compared to monotherapy with 16 weeks of IFN-α or 52 weeks of lamivudine. The per protocol analysis showed a significantly higher rate of HBeAg seroconversion at week 52 for the combination of lamivudine and IFN-α (36%) than for lamivudine monotherapy (19%) or IFN-α monotherapy

(22%). Sequential treatment appears to be justified in this case since two different treatment modalities were used. The rationales for lamivudine therapy preceding the IFN-α therapy were to reduce the initial viral load, since high pretreatment levels HBV DNA were associated with interferon treatment failure (10), and to release the suppressive effects of high viral load on the immune system, since inhibition of HBV DNA levels by lamivudine may partially restore T-cell responsiveness in chronically infected patients (107). The potential benefits of combining lamivudine and IFN-α treatment should be further investigated with different regimens.

Ribavirin has also been used in combination with IFN-α for treatment of chronic hepatitis B. Ribavirin is a nucleoside analog with broad-spectrum antiviral activity (108) that is currently approved for treatment of chronic hepatitis C in combination with IFN-α (see the section on hepatitis C treatment). In addition to its antiviral activity against certain viruses ribavirin may act as an immunomodulator by affecting the balance between Th1 and Th2 responses (109-111). Ribavirin's immunomodulatory activity may be responsible for its activity in the woodchuck hepatitis system (112) since it has very little activity against HBV *in vitro* (113). Two studies with ribavirin as monotherapy for chronic hepatitis B infection showed only a modest decrease in HBV DNA levels (114, 115) while a more recent study reported that 33% of patients became HBV DNA negative by PCR analysis and 50% had HBeAg seroconversion (116). A pilot study of ribavirin used in combination with IFN-α reported a 50% virological response and 21% biochemical response rate after 12 months of treatment (117). The patient population in this study had failed previous interferon therapy and had also acquired precore mutations so the efficacy of interferon treatment was expected to be low. Unfortunately, ribavirin or IFN-α monotherapy were not included in this study so the benefits arising from the combination treatment cannot be adequately evaluated. Nevertheless, the combination of two immunomodulators for treatment of chronic hepatitis B shows some promise and deserves further investigation. In this regard, the L-enantiomer of ribavirin (ICN 17261) that retains the immunomodulatory activity of ribavirin but has no toxicity (118) may also prove useful for HBV therapy (see the section on hepatitis C treatment).

2.5 Conclusions

Current therapies for the treatment of chronic hepatitis B are still limited in their general utility: IFN-α because of its low response rate and lamivudine because of its low sustained response and the emergence of resistant mutants. The expansion of treatment options by the development of future therapies will allow combinations of antivirals, an antiviral and an immunomodulator, or a combination of immunomodulators to suppress viral replication and to control viral infection by the host immune system.

3. THERAPIES FOR HCV

HCV infection is an important public health problem worldwide and is recognized as the major cause of non-A, non-B hepatitis (119, 120). Although HCV infection resolves in some cases, the virus establishes chronic infection in up to 80% of the infected individuals and persists for decades. Recent studies of the natural history of HCV infection indicate that the majority of people with chronic HCV infection have relatively mild disease with slow progression. However, an estimated 20% of these infected individuals will go on to develop cirrhosis and 1 to 5% will develop liver failure and hepatocellular carcinoma (121-123). It is estimated that HCV infection affects around 4 million people in the United States, 8 million in Europe, and 170 million worldwide. Chronic hepatitis C is the leading cause of chronic liver disease and the leading indication for liver transplantation in the United States. The Centers for Disease Control and Prevention estimates that hepatitis C currently is responsible for approximately 8,000 to 10,000 deaths in the United States annually. This number is projected to increase significantly over the next decade. Currently, there is no vaccine for HCV infection due to the high degree of heterogeneity of this virus. The low fidelity of HCV-encoded replication enzyme is believed to be responsible for the high mutation rate and the development of quasispecies.

The objectives of treatment of chronic hepatitis C are to achieve a complete and sustained clearance of HCV RNA in serum and normalization of serum alanine aminotransferase (ALT) levels. The current treatment options for chronic hepatitis C include IFN-α monotherapy and IFN-α and ribavirin combination therapy, with sustained virological response rates of about 10% for IFN-α monotherapy and slightly over 40% for the combination therapy (124-127). Clearly, more effective and direct antiviral interventions are necessary for further prevention of the life-threatening complications caused by HCV infection. This section summarizes the progress that has been made in the treatment of chronic hepatitis C and in the search for novel anti-HCV agents.

3.1 Current therapies

Interferon-α monotherapy. Interferons (IFNs) are a large family of multifunctional secreted proteins involved in antiviral defense, cell growth regulation and immune activation. IFNs have many biological effects and can inhibit replication of a variety of DNA and RNA viruses. Type I IFNs are produced in direct response to virus infection and consists of the IFN-α multigene family, which are synthesized predominantly by leukocytes, and IFN-β which is synthesized by most cell types but particularly by fibroblasts (128). IFN-α is the only interferon thus far approved for the treatment of hepatitis C virus infection. However, the effectiveness of IFN-α monotherapy is unsatisfactory. Only about 40% of patients respond to IFN-α treatment and, after cessation of therapy, about 70% of the initial responders relapse (129, 130). This leads to a sustained response rate for IFN-α monotherapy around 10-15%, with negative PCR for viral RNA in serum and normal serum ALT values sustained for 6 months after cessation of treatment. Factors associated with the low response rate include infection with HCV genotype 1, high viral load (> 1 million copies/ml) and presence of advanced fibrosis-cirrhosis on liver biopsy (131).

Pegylated-interferon. Pegylated-interferons (PEG-IFNs) are chemically modified IFN with a covalently attached polyethylene glycol (PEG) molecule. This modification renders IFN with a decreased systemic clearance rate and approximately 10-fold increase in serum half-life compared with IFN. Therefore, PEG-IFN offers a significant improvement in dosing regimen and allows the standard thrice-weekly injection with IFN-α to be reduced to once a week injection. In a recent report at the 51st Annual Meeting of the American Association for the Study of Liver Diseases (AASLD) (Dallas, TX, 11/2000), PEG-IFN-α-2b (PEG-INTRON™, Schering-Plough Corp.), a formulation consisting of a conjugate of straight-chain PEG (molecular weight ~12,000) and IFN-α-2b in a 1:1 ratio, with once-weekly injection at 1.5 μg/kg and daily ribavirin capsules (REBETOL, ICN Pharmaceuticals, Inc.) at 800 mg/daily, achieved a 54% overall sustained response rate in previously untreated adult patients with chronic hepatitis C, compared with 47% for REBETRON (IFN-α-2b/REBETOL) in the same study. Sustained virological response across HCV genotypes ranged from 42% to 82% for patients who received PEG-INTRON™ plus ribavirin combination therapy, with genotype 1 having the lowest response. In addition, another form of PEG-IFN, PEG-IFN-α-2a (PEGASYS™, Roche Corp.), also showed much improved sustained virological response rate (~36%) administrated once-weekly over the current standard thrice weekly IFN-α treatment (3%) in a phase II study. Combination study of PEGASYS™ and ribavirin is currently ongoing (Roche Media Release, June 1999).

Ribavirin. Ribavirin (1-β-D-ribofuranosyl-1,2,4-triazole-3-carboxyamide) is a purine nucleoside analogue with broad-spectrum antiviral activity against various RNA and DNA viruses (108, 132). In combination with IFN-α, ribavirin improves the sustained virological response rate in HCV patients to over 40% compared with 10-15 % for IFN-α monotherapy (124-127), although treatment with ribavirin alone revealed no significant effect on viral load or quasispecies in a number of small studies (133-135). The antiviral mechanism of ribavirin in patients with chronic hepatitis C remains unclear. Several possible mechanisms of action have been proposed, including (i) depletion of intracellular GTP pools via inhibition of inosine monophosphate dehydrogenase (IMPDH), a key enzyme involved in *de novo* biosynthesis of GTP; (ii) modulation of immune system by enhancing antiviral Th1 response while suppressing Th2 response; (iii) inhibition of viral polymerase activity; and more recently, (iv) acting as an RNA virus mutagen by mis-incorporating into nascent viral RNA.

The therapeutic use of ribavirin is limited by its adverse effects. Prolonged administration of ribavirin is frequently associated with hemolytic anemia, whose severity correlates with dose level and can be reversed upon dose reduction or cessation of treatment. Compounds with improved toxicity profiles and similar or enhanced antiviral/immunomodulatory properties may thus offer advantages over ribavirin in the treatment of HCV infection. Two compounds of such nature, ICN 17261 (Levovirin, ICN Pharmaceuticals) and VX-497 (Vertex Pharmaceuticals), are currently under development based on their improved biological properties over ribavirin.

ICN 17261. It has been suggested that stimulating host immune response towards a Th1 rather than Th2 response may aid viral clearance. Ribavirin has been shown to act as an immunomodulator by promoting T-cell-mediated antiviral type I cytokine expression such as interleukin-2 (IL-2), IFN-γ and tumor necrosis factor alpha (TNF-α) and concomitant suppression of type II cytokine expression (IL-4, IL-5 and IL-10) (109, 111). The major side effect of ribavirin is anemia, which results from the accumulation of ribavirin triphosphates in erythrocytes. In an effort to improve the toxicity profile of ribavirin, while retaining its immunomodulatory function, a new chemical entity, ICN 17261 (1-β-L-ribofuranosyl-1, 2, 4-triazole-3-carboxyamide), was identified. ICN 17261 is the L-enantiomer of ribavirin currently under development by ICN Pharmaceuticals (Figure 2). ICN 17261 has very similar immunomodulatory activities as ribavirin in a variety of assays that evaluate the enhancement of type I cytokine response by activated T cells. These assays include both human and murine systems and antigen-dependent activation *in vitro*. In addition, ICN 17261 was similar to ribavirin in two *in vivo* assays of type I cytokine activation (118, 136).

More importantly, ICN 17261 has a much improved toxicity profile over ribavirin in animal studies. It is less toxic and is inactive against a number of viruses whose replication is normally inhibited by ribavirin *in vitro*, presumably through ribavirin triphosphate. ICN 17261 did not have apparent toxicity in rats following a 4-week multi-dose toxicity study. In contrast, ribavirin dosing led to hemolytic anemia (118). The lack of antiviral activity of ICN 17261 is likely due to its inability to be phosphorylated. Therefore ICN17261, unlike ribavirin, is unable to interfere with viral polymerase function or inhibit IMPDH activity.

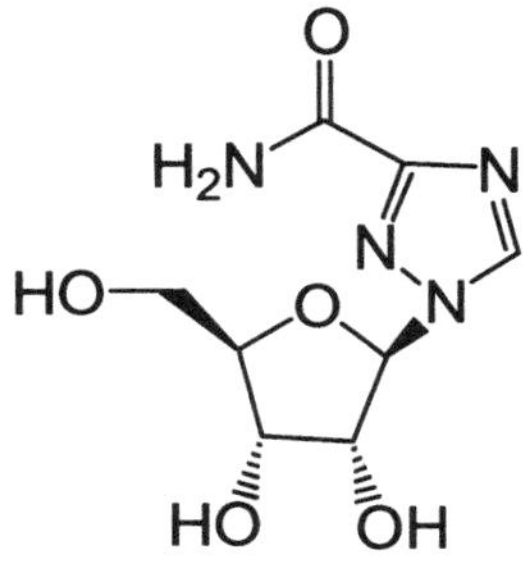

Ribavirin 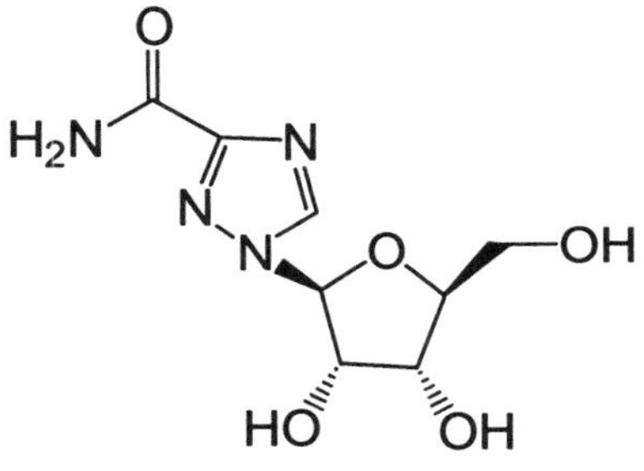   ICN 17261

Figure 2. Chemical structures of ribavirin and ICN 17261

Clinical data have shown that a robust, multi-specific T-cell response is observed in the majority of patients who clear their HCV infection, either spontaneously or in response to IFN-α treatment. Such an immune response is much less frequent in patients who remain chronically infected following antiviral treatment (137). The favorable clinical outcome of IFN-α and ribavirin combination therapy is therefore consistent with ribavirin enhancement of the type I cytokine response. This provides further support for the potential of ICN 17261, or similar compounds, as a new therapeutic agent against HCV.

VX-497. IMPDH catalyzes the rate-limiting step in the *de novo* biosynthesis of guanine nucleotides, the NAD^+-dependent conversion of IMP to XMP. XMP is aminated in the next biosynthesis step to form GMP. The monophosphate form of ribavirin is an inhibitor of IMPDH (Ki = ~250 nM) and this inhibition and subsequent reduction in intracellular GTP pools is believed to be at least partially responsible for its efficacy against HCV infection. As reported previously, high concentrations of GTP are able to stimulate HCV RNA-dependent RNA polymerase activity *in vitro* (138).

VX-497 (molecular weight 452.5) is a selective, highly potent, reversible, and uncompetitive inhibitor of both forms of human IMPDH (*Ki* = 10 and 7 nM for isoform I and II, respectively) currently under development by Vertex Pharmaceuticals (113) (Figure 3). VX-497 is approximately 35-fold more potent than ribavirin in relative affinity for IMPDH. In a comparative study, VX-497 is 17- to 186-fold more potent than ribavirin against infections of HBV, human cytomegalovirus (HCMV), respiratory syncytial virus (RSV), herpes simplex virus 1 (HSV-1), parainfluenza-3 virus, encephalomyocarditis virus (EMCV) and Venezuelan equine encephalomyelitis virus (VEEV) in cell cultures (113). The improved antiviral activity of VX-497 can be attributed to the more effective inhibition of IMPDH. This study suggests that intracellular GTP pool depletion is an important component of the viral reduction mechanism and is consistent with a mechanism of action involving inhibition of IMPDH. The antiviral activity of VX-497 was found to be additive to that of IFN-α in the treatment of virally infected cells.

Given the improved antiviral activity of VX-497, it is possible that a VX-497 and IFN-α combination may prove to be an alternative therapy for treating chronic HCV infection. The outcome of such a combination will depend primarily on whether IMPDH inhibition is the major mechanism by which ribavirin inhibits HCV replication in treated patients. The efficacy of such a combination awaits further confirmation by on-going clinical studies.

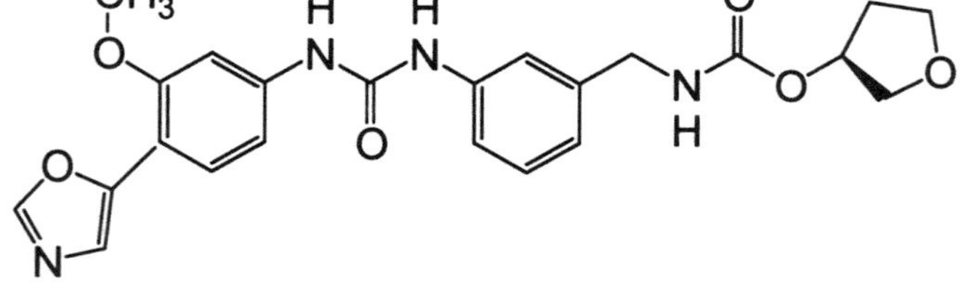

Figure 3. Chemical structure of VX-497.

Other compounds. Amantadine is used primarily for the treatment of influenza A virus infection. A study of amantadine monotherapy in chronic

HCV patients who had failed IFN therapy suggested that amantadine may have some inhibitory effect on HCV replication (139). However, further clinical studies showed that addition of amantadine to IFN and ribavirin, though well tolerated, had little, if any, impact on HCV RNA eradication in nonresponders or response/relapsers to previous IFN/ribavirin combination therapy (140-143). Consistent with this, amantadine was found to have no inhibitory activity against HCV IRES, NS3 protease, NS3 NTPase/helicase and NS5B polymerase *in vitro* (144).

Histamine dihydrochloride (Maxamine) is an immunomodulator and is currently being investigated as a part of a combination therapy for treating chronic hepatitis C patients who failed IFN therapy. In addition, recombinant human interleukin-12 (rIL-12) also showed anti-HCV activity in several preliminary studies (145, 146).

3.2 Molecular targets for developing novel anti-HCV therapies

3.2.1 Background

HCV is a positive-sense RNA virus belonging to the *Flaviviridae* family (147). The genome of HCV consists of approximately 9600 bases with a single open reading frame (ORF) encoding a polyprotein of around 3010 amino acids. The ORF is flanked by 5' and 3' untranslated region (UTR) of several hundred nucleotides, which are important for RNA translation and replication. The polyprotein is cleaved both co- and post-translationally by the cellular and virally-encoded proteases into at least 10 separate mature viral structural and nonstructural proteins, from the N to the C terminus: C-E1-E2-p7-NS2-NS3-NS4A-NS4B-NS5A-NS5B (119, 120, 148). The C protein (~22 kDa) is the major component of viral nucleocapsid and is believed to be able to bind with viral RNA during assembly. Glycoproteins E1 and E2 are the viral envelope proteins. P7 is a small protein of unknown function. NS2 is a protease which, in conjunction with the N-terminal region of NS3, cleaves the NS2/NS3 junction. NS3 is a bi-functional protein whose N-terminal one-third contains a serine protease, which is used to cleave the remaining junctions in the nonstructural region, and the C-terminal region possesses a NTPase/RNA helicase activity. NS4A is a co-factor of NS3 protease. The functions of NS4B and NS5A are unknown, but they presumably exist as part of the viral replicase complex. NS5A has been implicated in the sensitivity to IFN-α treatment. NS5B is the virally-encoded RNA-dependent RNA polymerase and its activity has been characterized by a number of groups. Despite the lack of direct experimental evidence in some cases, all the nonstructural proteins are believed to be essential for viral replication and therefore are suitable drug discovery targets in the search for improved anti-HCV therapeutic agents.

In the past decade, studies examining the replication of HCV have been hindered by the lack of a robust, reproducible cell culture system permissive to HCV replication *in vitro* and the lack of a manageable small animal model other than chimpanzee. Despite the success in obtaining infectious HCV cDNA clones that replicate in chimpanzee (149-153), the efforts to establish a tissue culture replication system have been problematic. This situation has changed recently with the development of a unique HCV subgenomic replicon system (genotype 1b) in human hepatoma cell line Huh7. In the subgenomic replicon, the HCV structural genes (C, E1 and E2) was replaced with a selection marker, neomycin phosphotransferase gene (Neo), and the translation of nonstructural genes were driven by the IRES from encephalomyocarditis virus (EMCV). When the subgenomic replicon RNA was transfected into Huh7 cells, colonies resistant to G418 selection were generated, though at very low frequency. Some of the neo-resistant cell lines carry the self-replicating HCV RNAs as high as a few thousand copies per cell (154, 155). More recently, adapted mutations that significantly improved the transduction efficiency of the replicon have been identified (156, 157). The exact mechanism of the cell culture adaptation, however, is not clear. Nonetheless, establishment of a robust, cell-based system finally makes the genetic and functional analysis of HCV replication feasible. More importantly, this system will allow novel anti-HCV compounds to be tested in a near "authentic" HCV cell culture.

So far, the only accepted animal model for HCV is chimpanzee. Due to its limited availability and high cost, *in vivo* studies of HCV replication and pathogenesis have been extremely limited. There is therefore an urgent need to identify small animal models, which can support HCV infection. A surrogate model system was recently established using GB virus B (158-160). GBV-B, a newly identified hepatitis virus that is closely related to HCV, can infect tamarin (*Saguinus* species) and causes acute hepatitis. An infectious cDNA clone of GBV-B has become available recently (161). The GBV-B/tamarin model may provide an effective *in vivo* system reminiscent of the use of the woodchuck hepatitis model, a strategy that has been successfully developed for anti-HBV drug development. In addition, bovine viral diarrhea virus (BVDV) has been extensively used as a surrogate system for evaluation of anti-HCV compounds based on its closeness to HCV (162-164).

3.2.2 Viral targets for development of novel anti-HCV agents

NS3-4A serine protease. The cleavage at junctions of NS3/NS4A, NS4A/NS4B, NS4B/NS5A and NS5A/NS5B of the polyprotein is catalyzed by NS3-4A serine protease. NS4A serves as the co-factor for the protease and is intercalated into the protease domain [165-171]. Validation of this target has been provided by the observation that HCV RNA bearing mutations ablating the activity of NS3-4A serine protease failed to replicate in chimpanzee (172). It is well documented for HIV that inhibitors of viral proteases can be effective antiviral drugs. The X-ray crystal structures, in the form of the protease domain alone or the entire NS3 protein, have been determined (166, 173-175). This structural information plays an important role in structure-based design of NS3 protease inhibitors.

Three possible targets have been suggested for the NS3 protease: the active site that is composed of His57/Asp81/Ser139; the NS4A co-factor binding site; and the metal ion binding site (176). In the case of NS4A binding, studies have revealed a very high affinity between NS4A and NS3 (177). This, coupled with the rather formidable effective concentration of NS4A in the proximity of NS3 required for this *cis* cleavage, predicts that it would be difficult to identify a small molecule that can effectively inhibit this interaction. It is even more challenging to identify inhibitors that can prevent metal ion binding to NS3 protease, as such inhibitors would require a high specificity to avoid toxicity issues.

So far, extensive drug discovery efforts have been focused on identifying inhibitors specific for the active site of NS3 serine protease. However, due to the unusually shallow and relatively non-polar substrate-binding pocket, highly specific and potent inhibitors suitable for clinical testing are yet to be reported. Two classes of inhibitors, substrate-based (peptidic) and nonsubstrate-based, have been pursued. The substrate-based peptidic inhibitors are usually modeled after the NS4A/NS4B, NS4B/NS5A and NS5A/NS5B cleavage sites. Some of them can achieve *in vitro* potency (IC_{50}) in the range of sub-nanomolar (nM), with the most potent at 1.5 nM (Ac-Asp-D-γ-carboxy-Glu-Leu-Ile-β-cyclohexyl-L-Ala-Cys-OH), and with high degrees of selectivity (179-181). A recent study showed that replacing the carboxylic acid moiety of the peptide inhibitors with the α-ketoacid moiety further improved their potency, with the overall *K*i* values between 10 pM and 67 nM (178) (Figure 4). The peptide α-ketoacids are slow binding inhibitors of HCV NS3 protease. Their inhibition mechanism involves a rapid formation of a noncovalent collision complex between the enzyme and the inhibitor that is followed by a slow formation of an extremely stable, reversible and covalent complex at the enzymatic active site. The His57 residue was found to be the key determinant for formation of the noncovalent collision complex and the Lys136 residue for formation of the covalent complex (178).

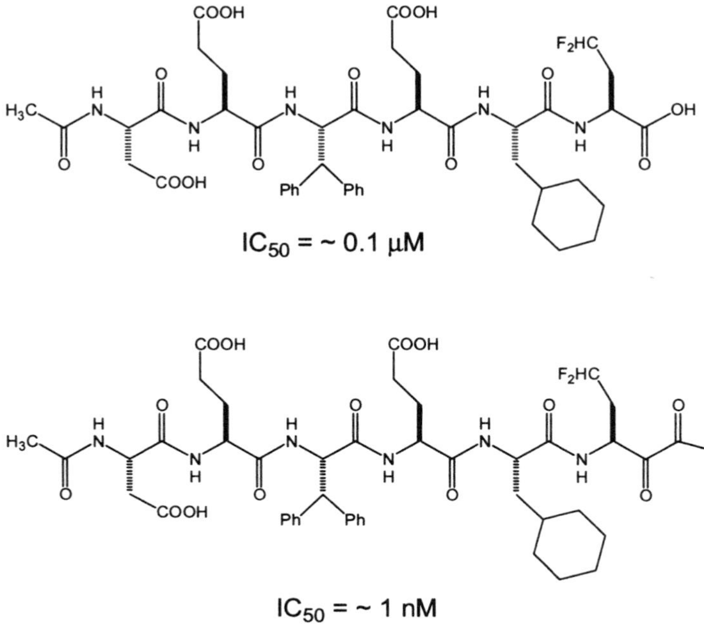

IC$_{50}$ = ~ 0.1 μM

IC$_{50}$ = ~ 1 nM

Figure 4. Homologous hexapeptide inhibitors of HCV NS3 protease and their potency: α-ketoacid (bottom) and carboxylic acid (top) (178).

In the case of nonsubstrate-based inhibitors, screening of random compound libraries has identified several classes of inhibitors as shown in Figure 5, including derivatives of 2,4,6-trihydroxy-3-nitro-benzamides (compound **1 & 2**), thiazolidines (compound **3, 4, 5, 6**) (182, 183), benzanilides (compound **7 & 8**) (184, 185), as well as products isolated from natural products (**9 & 10**) (186, 187). These small molecule inhibitors exhibit single- to double-digit micromolar (μM) potency against NS3 serine protease and were found to possess either non-competitive or a mixture of non-competitive and un-competitive enzymatic kinetics. However, these compounds were mostly non-selective and were able to inhibit the activity of host serine proteases such as chymotrypsin, trypsin and elastase, which will likely create toxicity issues. The difficulty in identifying potent small molecule inhibitors of HCV NS3 serine protease further reflects the unusual shallow and non-polar nature of its active site.

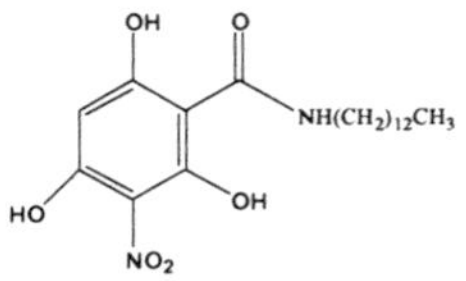

OH O NH(CH2)12CH3 HO OH NO2

$IC_{50} = 5.8\ \mu M$ (**1**)

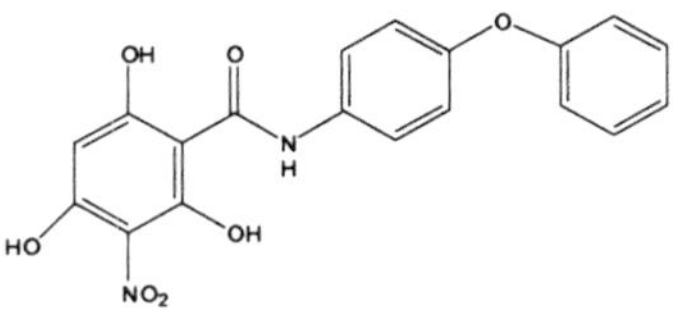

OH O N H HO OH NO2 O

$IC_{50} = 22.2\ \mu M$ (**2**)

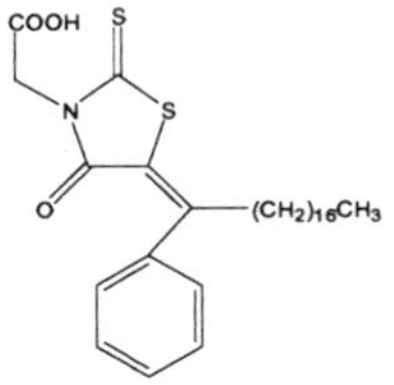

$IC_{50} = 8.35\ \mu M$ (**3**)

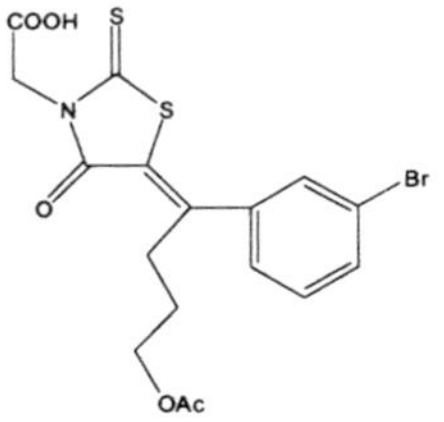

$IC_{50} = 14.5\ \mu M$ (**4**)

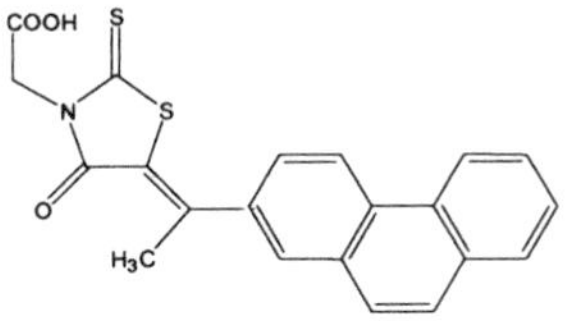

$IC_{50} = 17\ \mu M$ (**5**)

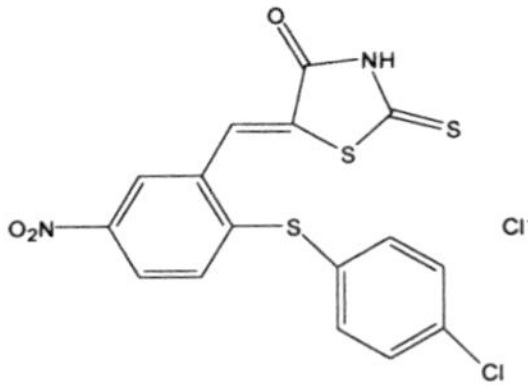

$IC_{50} = 3.2\ \mu M$ (**6**)

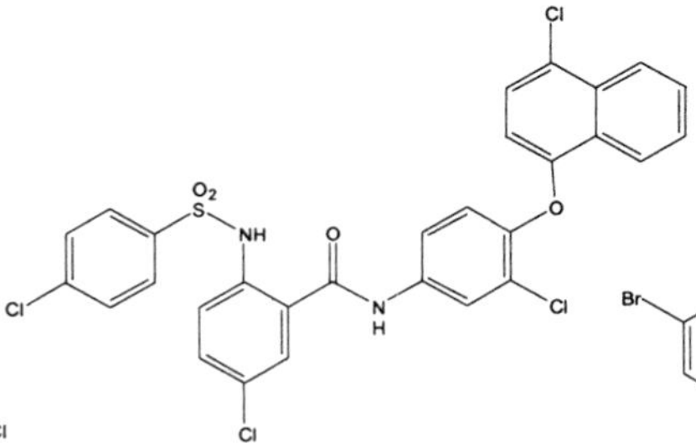

$IC_{50} = 6.5\ \mu M$ (**7**)

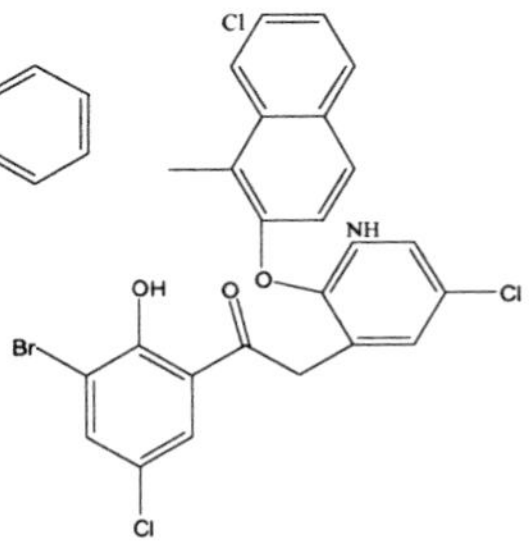

$IC_{50} = 6.2\ \mu M$ (**8**)

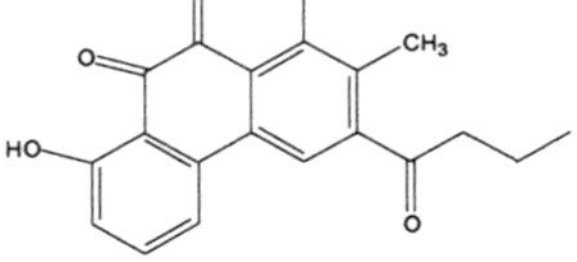

$IC_{50} = 7.7\ \mu M$ (**9**)

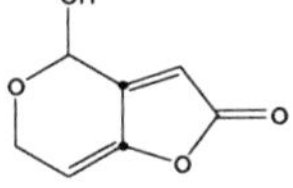

$IC_{50} = 17.4\ \mu M$ (**10**)

Figure 5. Non-substrate-based inhibitors of HCV NS3 serine protease

NS3 RNA helicase. The C-terminal two-thirds of NS3 contains the NTPase/helicase activities and offers another rational target for drug discovery. All helicases carry an intrinsic RNA-binding and an NTPase activity which hydrolyses nucleoside triphosphates to provide the energy source for unwinding (188-194). Although no helicase inhibitors currently exist as marketed drugs, the essentiality of this target in viral replication has been established by the failure of HCV RNA containing defective NTPase/helicase function to replicate in chimpanzee (172), as well as similar observations with a related flavivirus (BVDV) in cell culture (195, 196).

The X-ray crystal structure of NS3 helicase domain (the C-terminal 465 residues of NS3) shows that the NTPase/helicase consists of three distinct domains to form an Y-shaped triangular molecule. Domains I and II contain βαβ structures, which are alpha helices packed around a central beta sheet. Domain III contains primarily alpha helices. Domain I possesses the NTP-binding site. Domain II appears to have the flexibility to rotate as a rigid body and can pivot relative to the other two domains. Domain II also contains an arginine-rich region required for RNA unwinding. The RNA-binding site lies in a channel that separates Domain III from Domains I/II, based on a co-crystal of the helicase complexed with a $(dU)_8$ oligonucleotide. Based on this information, compounds that bind to the RNA-binding site, to the NTP-binding site, or perhaps to an allosteric site critical for the dynamic movements during the unwinding process, could potentially inhibit the activity of this enzyme (197, 198).

The NS3 helicase domain can be expressed in large quantities in an active form and various assays suitable for high throughput screening have been established. However, a number of key questions about this enzyme remain unanswered, including its mode of action on the viral RNA template, modulation of its activity by other proteins within the replicase complex and the detailed structure-function relationship. Despite extensive screening efforts, only a handful of small molecule inhibitors have been reported. Among these molecules, the series of compounds listed in Figure 6 (n=2,4,6,7,8) possesses the highest potency (IC_{50} = 0.7 μM) (199, 200). However, no mode of action has been characterized for this class of compounds. The lack of detailed mechanistic characterization of this enzyme and the inability so far to obtain the complex structure (enzyme-compound) will likely hinder further optimization of these compounds as well as additional compounds to be identified by future screening.

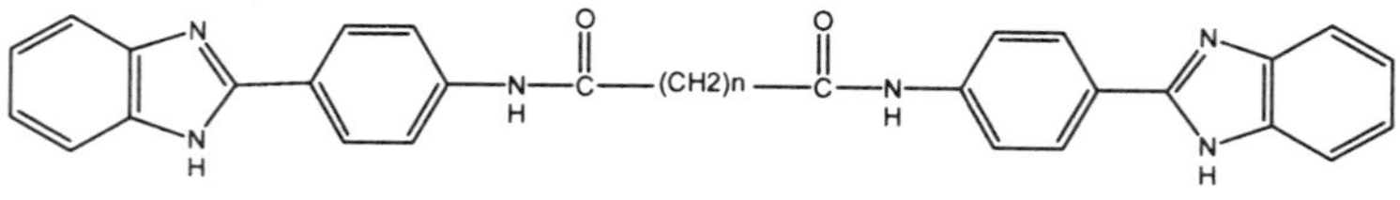

N N H N H O C (CH2)n O C N H N N H

Figure 6. The most potent NS3 helicase inhibitor series reported (IC_{50}=0.7 μM)

RNA-dependent RNA polymerase (RdRp)._ Another key enzyme of HCV essential for viral replication is the RNA-dependent RNA polymerase encoded by NS5B. A large number of currently marketed drugs against viral infections are polymerase inhibitors, which clearly validates this target for antiviral intervention. HCV NS5B polymerase has been expressed in large quantities and the crystal structure of this enzyme has been determined recently (201-203). The enzymatic activity of NS5B has been characterized in some detail and assays suitable for inhibitor screening have been established (204-211). Although the recombinant NS5B alone was used in these assays, it is expected that additional viral or host proteins within the replicase complex will enhance or modulate the activity, processivity, specificity, and perhaps fidelity of this enzyme *in vivo*. The X-ray crystal structure of NS5B revealed that this enzyme is similar to other classes of RNA or DNA polymerases in possessing characteristic fingers, palm and thumb subdomains. It has a fully encircled active site with a relatively rigid interdomain structure, resembling the nucleic acid-bound conformation of several other polymerases. The encircled overall structure of this enzyme is the result of extensive interactions between the fingers and thumb subdomains that are likely to be unique to viral RdRps. In addition, an HCV-specific β-loop structure ("flap"), located in the thumb subdomain, protrudes towards the active site. Recently, this β-loop structure was proposed to impose a steric barrier that prevents binding of double-stranded RNA molecules to ensure that terminal initiation, rather than internal initiation, of RNA synthesis is catalyzed by this enzyme. This may represent a novel mechanism used by viral RdRps, which initiate RNA synthesis via a *de novo* mechanism (primer-independent), to correctly initiate and copy the entire viral genome (212).

So far, a number of weak small molecule inhibitors have been reported, including nucleotide analogues, and the natural products cerulenin ($IC_{50} \sim 500\ \mu M$) and gliotoxin ($IC_{50} \sim 230\ \mu M$) (204, 208). Another series of compounds, represented by VP50406 developed by ViroPharma Inc., is currently under clinical studies and is believed to inhibit NS5B polymerase.

NS2/3 protease. This protease consists of NS2 and the N-terminal portion of NS3 and is responsible for the cleavage of the NS2/NS3 junction. It has been proposed to be either a metalloprotease, based on its metal ion requirement for activity (170, 213, 214), or a cysteine protease based on the limited mutational analyses (170, 215). NS2 protein is found to be dispensable for viral RNA replication in the HCV subgenomic replicon cell culture (154, 156). Additionally, the difficulty encountered in the generation of a soluble, recombinant NS2 for large-scale inhibitor screenings further limits the efforts in developing inhibitors against this viral target.

Cell entry. Although the exact mechanism employed by HCV to enter cells is not fully understood, two types of molecules, CD81 and low density lipoprotein (LDL) receptors, have been recently implicated in mediating the entry of HCV into the target cells (216, 217). HCV E2 glycoprotein has been shown to bind to CD81 on the cell surface. There is no evidence, however, that the E2-CD81 interaction is sufficient to support viral entry and subsequent replication processes. It is likely that CD81 may be a co-receptor molecule for HCV, similar to the chemokine receptors for HIV. Blocking the interaction between HCV envelope and its receptor molecule(s) on the cell surface may serve as another valid approach for identification of novel anti-HCV therapeutic agents.

Cis-acting RNA elements. In addition to the enzymatic targets discussed above, cis-acting RNA elements in HCV genome can also serve as potential targets for intervention. In particular, the 5' and 3' untranslated regions (UTR) that contain conserved RNA structural motifs and are required for viral RNA translation and replication. The 5'-UTR contains an IRES (internal ribosomal entry site) element that is required for polyprotein translation (218). The IRES of HCV has been a target of drug discovery efforts because it is a virus-specific function and assays to screen for inhibitors are readily available. Several approaches have been applied to interfere with the IRES function, including antisense oligonucleotides, hammerhead ribozymes and "decoy" RNA. The major hurdles for nucleic acid-based approaches include the low specificity of such molecules and their delivery to the specific sites where HCV replicates. The ribozyme approach has so far produced the most promising results. Ribozymes are a new class of antiviral therapeutic agents that are capable of binding and cleaving viral RNA in a sequence-specific manner. Chemical modifications have created ribozyme molecules that are resistant to nuclease digestion while retaining biological activity *in vivo*. A ribozyme, which is highly specific for cleaving the 5' IRES of HCV genome, has been developed (219, 220). This ribozyme has demonstrated *in vitro* activity against HCV IRES function and prevention of viral replication in a surrogate replication system (HCV IRES-chimeric poliovirus) (221). More recently, a pharmacokinetics and liver distribution study showed that the ribozyme was able to accumulate in the liver of mice administered the ribozyme, with peak concentrations greater than the concentration necessary to inhibit HCV IRES function in cell culture (222).

In addition, several small molecules have been reported to show inhibitory activity of HCV IRES-dependent translation *in vitro* (Figure 7) (223, 224).

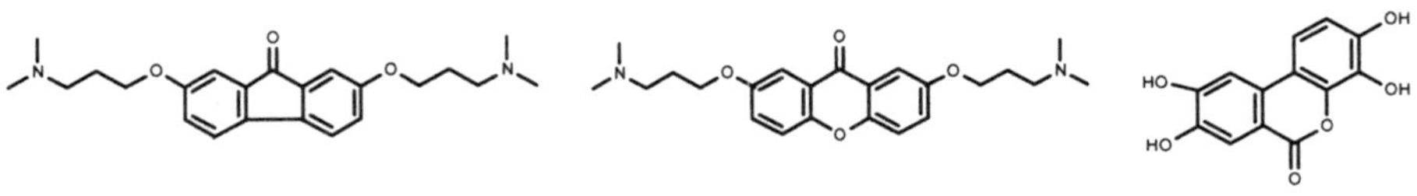

Figure 7. Inhibitors of HCV IRES-dependent translation

Mutagen approach. Despite the clinical efficacy shown by ribavirin in significantly improving the effectiveness of IFN treatment of chronic hepatitis C, its mechanism of action remains controversial. A recent study showed that ribavirin triphosphate can be used by an RNA virus (poliovirus) RdRp and is mis-incorporated into the progeny RNA (225). Since ribavirin is a purine analogue which can basepair with both cytidylate and uridylate, incorporation of ribavirin into viral replicating RNA will subsequently result in G-to-A or C-to-U mutations. Indeed, the presence of ribavirin resulted in a significant increase in the frequency of mutations in poliovirus RNA and a decrease in the production of infectious virus particles without affecting viral RNA synthesis (225). Therefore, as another possible mechanism, ribavirin acts as an RNA virus mutagen and causes decreased production of infectious progeny virus particles.

Based on this possible mechanism, it is conceivable that ribavirin analogues with increased mutagen capability will likely improve the effectiveness of the combination treatment. This can be achieved by modification of ribavirin, or similar nucleoside analogues, towards a more promiscuous/universal basepairing with the natural bases. Incorporation of such molecules into viral RNA by HCV RdRp is expected to significantly increase the frequency of mutations and eventually cause the "genetic meltdown" of this virus.

3.3 Conclusions

With the recent breakthroughs in establishing a robust, cell-based replication system for HCV, the anti-HCV drug discovery process is expected to accelerate significantly in the future years. A detailed understanding of the molecular targets involved in HCV replication will aid the design of more effective inhibitors that directly target viral functions.

Further immunomodulatory approaches will likely play a role in identifying the optimal treatment regimen against HCV infection. It is clearly within our reach to address the medical need of chronic hepatitis C with combinations of direct anti-HCV agents generated from the novel screening approaches, more effective immunomodulatory agent, and the improved, long-lasting pegylated-IFN.

4. THERAPY FOR HDV

HDV contains a small, approximately 1700 nucleotide, single-stranded circular RNA genome complexed with hepatitis delta antigen (HDAg). This ribonucleoprotein complex must associate with the hepatitis B virus (HBV) envelope proteins (HBsAg) to form infectious viral particles. HDV therefore relies on HBV as a helper virus for its transmission. Because of its dependence on HBV for infection, the treatment strategy for HDV is to use the therapies currently available for HBV infection. The current and future therapy possibilities for HBV infection have already been covered in a previous section. This section will focus on the limited studies that have looked specifically at treating HDV infection.

IFN-α treatment can result in loss of HDV RNA and normalization of serum ALT levels for up to 50% of patients but the sustained response rate is less than 20% (226-228). A sustained response is dependent upon the loss of HBsAg (229), which occurs infrequently after interferon treatment.

Lamivudine treatment for HBV is very effective in reducing serum levels of HBV DNA in HBV infected patients. A pilot study evaluating the efficacy of lamivudine for HDV reported that a 12-month course of lamivudine lowered HBV levels but had no effect on the levels of circulating HBsAg or HDV RNA (230). These results imply that the pool of hepatocytes containing HBV cccDNA (the template that drives HBsAg expression) and infected with HDV was not affected by lamivudine treatment. As is the case for IFN-α treatment, the HBsAg seroconversion rate with lamivudine is low. A study combining lamivudine and IFN-α also failed to show a sustained response after therapy (231). More potent therapies for treating HBV infection that will eradicate HBV cccDNA will be needed in order for HDV infection to be controlled.

An HDV specific target that has been explored recently as a potential therapy is the prenylation of HDV large antigen. Prenylation has been shown to be required for virion assembly by site-directed mutagenesis of the prenylation site (232). Cells treated with BZA-5B, a specific prenyltransferase inhibitor (233), were inhibited in the production of virus like particles suggesting that the prenylation of HDV large antigen could be targeted for inhibition of HDV assembly (234).

HDV infection in patients with chronic hepatitis B is associated with a more severe outcome than patients who are not co-infected with HDV (235). Fortunately, the outcome for HDV infected patients who have end-stage liver disease requiring liver transplants is good (236, 237). Immunoprophylaxis with anti-HBsAg is effective in preventing re-infection of the graft resulting in 5-year survival rates of 88%. In patients who were not re-infected with HBV, 88% had evidence of HDV re-infection (HDAg and HDV RNA) but there was no hepatitis and the HDV markers were no longer detectable in 95% of these patients after two years. Thus, HDV pathogenesis can be remedied by curing the patient of HBV infection.

In conclusion, current therapies for HDV infection are inadequate for effecting a sustained response. Future therapies for HBV infection that result in HBsAg seroconversion will be required in order to effectively treat HDV infections.

5. THERAPY FOR HEV

HEV is responsible for epidemic outbreaks in the tropical regions of Asia, Africa, and in Mexico (238). Sporadic acute hepatitis is also common in these endemic areas and has also been documented in South America (239, 240) and in Europe (241, 242). As for hepatitis A virus (HAV), HEV infections are self-limiting with no evidence of chronic infections. The severity of HEV infections is comparable to HAV except for the greatly increased fatality rate (up to 20%) among pregnant women (243). No therapy is currently available for treatment of HEV. Attempts to prevent HEV infection by the administration of globulin have been unsuccessful (244, 245).

HEV can be transmitted to primates and this allowed the viral genome to be cloned and sequenced (246, 247). The HEV genome consists of a positive sense RNA strand 7.5 kb in length. Despite this knowledge of the HEV genome sequence, the molecular biological analysis of HEV has been hampered by the lack of a straightforward cell culture system. However, the HEV clones have enabled the expression of recombinant HEV proteins that show promise in vaccine studies with primates (248, 249). Further development may one day yield effective vaccines to prevent hepatitis E epidemics and protect pregnant women and travelers in endemic areas.

REFERENCES

1 Lemon S.M. Type A viral hepatitis: new development in an old disease. N Engl J Med 1985; 313:1059-1067.

2 Lemon S.M., Robertson B.H. Current perspectives in the virology and molecular biology of hepatitis A virus. Semin Virol 1993; 4:285-295.

3 Ozsoyen S., Kocak N. Acute hepatic failure related to hepatitis A. Lancet 1989; 1:901.

4 Lemon S.M., Thomas D.L. Vaccines to prevent viral hepatitis. N Eng J Med 1997; 336:196-204.

5 Innis B.L., Snitbhan R., Kunasol P. Protection against hepatitis A by an inactivated vaccine. JAMA 1994; 271:1328-1334.

6 Werzberger A., Mensch B., Kuter B. A controlled trial of a formalin-inactivated hepatitis A vaccine in healthy children. N Engl J Med 1992; 327:453-457.

7 Sanchez F.S., Sosa I.R.G., Vargas G.M. Treatment of type A hepatitis with ribavirin. Clinical Applications of Ribavirin. Academic Press, Inc., 1984.

8 Hoofnagle J.H., Di Bisceglie A.M. Treatment of chronic viral hepatitis. N Engl J Med 1997; 336:347-356.

9 Wong D.K.H., Cheung A.M., O'Rourke K., Naylor C.D., Detsky A.S., Heathcote J. Effect of alpha-interferon treatment in patients with hepatitis B e antigen-positive chronic hepatitis B: a meta-analysis. Ann Intern Med 1993; 119:312-323.

10 Perrillo R.P., Rogenstein F.G., Peters M.G., DeSchryver-Kecskemeti K., Bodicky C.J., Campbell C.R., Kuhns M.C. Prednisone withdrawal followed by a recombinant alpha interferon in the treatment of chronic type B hepatitis. A randomized, controlled trial. Ann Intern Med 1988; 109:95-100.

11 Severini A., Liu X.Y., Wilson J.S., Tyrrell D.L.J. Mechanism of inhibition of duck hepatitis B virus polymerase by (-)-b-L-2',3'-dideoxy-3'-thiacytidine. Antimicrob Agents Chemother 1995; 39:1430-1435.

12 Lai C.L., Chien R.N., Leung N.W.Y., Chang T.T., Guan R., Tai D.I., Ng K.Y., Wu P.C., Dent J.C., Barber J., Stephenson S.L., Gray D.F., Group A.H.L.S. A one-year trial of lamivudine for chronic hepatitis B. N Eng J Med 1998; 339:61-68.

13 Dienstag J.L., Schiff E.R., Mitchell M., Casey D.E.J., Gitlin N., Lissoos T., Gleb L.D., Condreay L., Crowther L., Rubin M., Brown N. Extended lamivudine retreatment for chronic hepatitis B: maintenance of viral suppression after discontinuation of therapy. Hepatology 1999; 30:1082-1087.

14 Liaw Y.F., Leung N.W., Chang T.T., Guan R., Tai D.I., Ng K.Y., Chien R.N., Dent J., Roman L., Edmundson S., Lai C.L. Effects of extended lamivudine therapy in Asian patients with chronic hepatitis B. Asia Hepatitis Lamivudine Study Group. Gastroenterology 2000; 119:172-180.

15 Tassopoulos N.C., Volpes R., Pastore G., Heathcote J., Buti M., Goldin R.D., Hawley S., Barber J., Condreay L., Gray D.F. Efficacy of lamivudine in patients with hepatitis B e antigen-negative/hepatitis B virus DNA-positive (precore mutant) chronic hepatitis B. Lamivudine Precore Mutant Study Group. Hepatology 1999; 29:889-896.

16 Villeneuve J.-P., Condreay L.D., Willems B., Pomier-Layrargues G., Fenyves D., Bilodeau M., Leduc R., Peltekian K., Wong F., Margulies M., Heathcote E.J. Lamivudine treatment for decompensated cirrhosis resulting from chronic hepatitis B. Hepatology 2000; 31:207-210.

17 Dienstag J.L., Perrillo R.P., Schiff E.R., Bartholomew M., Vicary C., Rubin M. A preliminary trial of lamivudine for chronic hepatitis B infection. N Engl J Med 1995; 333:1657-1661.

18 Nevens F., Main J., Honkoop P., Tyrrell D.L., Barber J., Sullivan M.T., Fevery J., De Man R.A., Thomas H.C. Lamivudine therapy for chronic hepatitis B: a six-month randomized dose-ranging study. Gastroenterology 1997; 113:1258-1263.

19 Tuttleman J.S., Pourcel C., Summers J. Formation of the pool of covalently closed circular viral DNA in hepadnavirus-infected cells. Cell 1986; 47:451-460.

20 Civitico G.M., Locarnini S.A. The half-life of duck hepatitis B virus supercoiled DNA in congenitally infected primary hepatocyte cultures. Virology 1994; 203:81-89.

21 Moraleda G., Saputelli J., Aldrich C.E., Averett D., Condreay L., Mason W.S. Lack of effect of antiviral therapy in nondividing hepatocyte cultures on the closed circular DNA of woodchuck hepatitis virus. J Virol 1997; 71:9392-9399.

22 Dandri M., Burda M.R., Will H., Petersen J. Increased hepatocyte turnover and inhibition of woodchuck hepatitis B virus replication by adefovir in vitro do not lead to reduction of the closed circular DNA. Hepatology 2000; 32:139-146.

23 Zhu Y., Yamamoto T., Cullen J., Saputelli J., Aldrich C.E., Miller D.S., Litwin S., Furman P.A., Jilbert A.R., Mason W.S. Kinetics of hepadnavirus loss from the liver during inhibition of viral DNA synthesis. J Virol 2001; 75:311-322.

24 Fischer K.P., Tyrell D.L.J. Generation of duck hepatitis B virus polymerase mutants through site-directed mutagenesis which demonstrate resistance to lamivudine ((-)-beta-L-2',3'-dideoxy-3'-thiacytidine) *in vitro*. Antimicrob Agents Chemother 1996; 40:1957-1960.

25 Tipples G.A., Ma M.M., Fischer K.P., Bain V.G., Kneteman N.M., Tyrrell D.L.J. Mutation in HBV RNA-dependent DNA polymerase confers resistance to lamivudine *in vivo*. Hepatology 1996; 24:714-717.

26 Chang T.T., Lai C.L., Liaw Y.F., Guan R., Lim S.G., Lee C.M., Ng K.Y., et al. Incremental increases in HBeAg seroconversion and continued ALT normalization in Asian chronic HBV (CHB) patients treated with lamivudine for four years [abstract]. Antiviral Therapy 2000; 5(Suppl 1):44.

27 Liaw Y.F., Chien R.N., Yeh C.T., Tsai S.L., Chu C.M. Acute exacerbation and hepatitis B virus clearance after emergence of YMDD motif mutation during lamivudine therapy. Hepatology 1999; 30:567-572.

28 Bartenschlager R., Junker-Niepmann M., Schaller H. The P-gene product of hepatitis B virus is required as a structural component for genomic RNA encapsidation. J Virol 1990; 64:5324-5332.

29 Hirsch R.C., Lavine J.E., Chang L.J., Varmus H.E., Ganem D. Polymerase gene products of hepatitis B viruses are required for genomic RNA packaging as well as for reverse transcription. Nature (London) 1990; 344:552-555.

30 Nassal M., Rieger A. A bulged region of the hepatitis B virus RNA encapsidation signal contains the replication origin for discontinuous first-strand DNA synthesis. J Virol 1996; 70:2764-2773.

31 Tavis J.E., Perri S., Ganem D.E. Hepadnavirus reverse transcription initiates within the stem-loop of the RNA packaging signal and employs a novel strand transfer. J Virol 1994; 68:3536-3543.

32 Bartenschlager R., Junker-Niepmann M., Schaller H. The amino-terminal domain of the hepadnaviral P-gene encodes the terminal protein (genome-linked protein) believed to prime reverse transcription. EMBO J 1990; 11:4185-4192.

33 Wang G.H., Seeger C. The reverse transcriptase of hepatitis B virus acts as a protein primer for viral DNA synthesis. Cell 1992; 71:663-670.

34 Summers J., Mason W.S. Replication of the genome of a hepatitis B-like virus by reverse transcription of an RNA intermediate. Cell 1982;. 29:403-415.

35 Radziwill G., Tucker W., Schaller H. Mutational analysis of the hepatitis B virus P gene product: domain structure and Rnase H activity. J Virol 1990; 64:613-620.

36 Lanford R.E., Notvall L., Beames B. Nucleotide priming and reverse transcriptase activity of hepatitis B virus polymerase expressed in insect cells. J Virol 1995; 69:4431-4439.

37 Lee H.J., Kwon Y.T., Rho H.M., Jung G. Expression of hepatitis B virus polymerase gene in *E. coli*. Biotechnology Letters 1993; 15:821-826.

38 Seifer M., Hamatake R., Bifano M., Standring DN. Generation of replication-competent hepatitis B virus nucleocapsids in insect cells. J Virol 1998; 72:2765-2776.

39 Sells M.A., Chen M.L., Acs G. Production of hepatitis B virus particles in Hep G2 cells transfected with cloned hepatitis B virus DNA. Proc Nal Acad Sci USA 1987; 84:1005-1009.

40 Ladner S.K., Miller T.J., Otto M.J., King R.W. The M539V HBV polymerase variation responsible for 3TC-resistance also confers cross-resistance to other nucleoside analogs. Antivir Chem Chemother 1998; 9:65-72.

41 Fu L., Cheng Y.C. Characterization of novel human hepatoma cell lines with stable hepatitis B virus secretion for evaluating new compounds against lamivudine- and penciclovir-resistant virus. Antimicrob Agents Chemother 2000; 44:3402-3407.

42 Korba B.E., Milman G. A cell culture assay for compounds which inhibit hepatitis B virus replication. Antivir Res 1991; 15:217-228.

43 Ying C., De Clercq E., Nicholson W., Furman P., Neyts J. Inhibition of the replication of the DNA polymerase M550V mutation variant of human hepatitis B virus by adefovir, tenofovir, L-FMAU, DAPD, penciclovir and lobucavir. J Viral Hepat 2000; 7:161-165.

44 Perrillo R., Schiff E., Yoshida E., Statler A., Hirsch K., Wright T., Gutfreund K., Lamy P., Murray A. Adefovir dipivoxil for the treatment of lamivudine-resistant hepatitis B mutants. Hepatology 2000; 32:129-134.

45 Doong S.L., Tsai C.H., Schinazi R.F., Liotta D.C., Cheng Y.C. Inhibition of the replication of hepatitis B virus *in vitro* by 2',3'-dideoxy-3'-thiacytidine and related analogues. Proc Natl Acad Sci USA 1991; 88:8495-8499.

46 Korba B.E., Boyd M.R. Penciclovir is a selective inhibitor of hepatitis B virus replication in cultured human hepatoblastoma cells. Antimicrob Agents Chemother 1996; 40:1282-1284.

47 Innaimo S.F., Seifer M., Bisacchi G.S., Standring D.N., Zahler R., Colonno R.J. Identification of BMS-200475 as a potent and selective inhibitor of hepatitis B virus. Antimicrob Agents Chemother 1997; 41:1444-1448.

48 Furman P.A., Davis M., Liotta D.C., Paff M., Frick L.W., Nelson D.J., Dornsife R.E., Wurster J.A., Wilson L.J., Fyfe J.A. The anti-hepatitis B virus activities, cytotoxicities, and anabolic profiles of the (-) and (+) enantiomers of cis-5-fluoro-1-[2-(hydroxymethyl)-1,3-oxathiolan-5-yl]cytosine. Antimicrob Agents Chemother 1992; 36:2686-2692.

49 Heijtink R.A., De Wilde G.A., Kruining J., Berk L., Balzarini J., De Clerq E., Holy A., Schalm S.W. Inhibitory effect of 9-(2-phosphonylmethoxyethyl)-adenine (PMEA) on human and duck hepatitis B virus infection. Antiviral Res 1993; 21:141-153.

50 Chu C.K., Ma T., Shanmuganathan K., Wang C., Xiang Y., Pai S.B., Yao G.Q., Sommadossi J.P., Cheng Y.C. Use of 2'-fluoro-5-methyl-β-L-arabinofuranosyluracil as a novel antiviral agent for hepatitis B virus and Epstein-Barr virus. Antimicrob Agents Chemother 1995; 39:979-981.

51 Bryant M.L., Bridges E.G., Placidi L., Faraj A., Loi A.G., Pierra C., Dukhan D., Gosselin G., Imbach J.L., Hernandez B., Juodawlkis A., Tennant B., Korba B., Cote P., Marrion P., Cretton-Scott E., Schinazi R.F., Sommadossi J.P. Antiviral L-Nucleosides specific for hepatitis B virus infection. Antimicrob Agents Chemother 2001;. 45:229-235.

52 Xiong X., Flores C., Yang H., Toole J.J., Gibbs C.S. Mutations in hepatitis B polymerase associated with resistance to lamivudine do not confer resistance to adefovir *in vitro*. Hepatology 1999; 28:1669-1673.

53 Fu L., Liu S.H., Cheng Y.C. Sensitivity of L-(-)2',3'-dideoxythiacytidine resistant hepatitis B virus to other antiviral nucleoside analogues. Biochem Pharmacol 1999; 55:1351-1359.

54 Chang C., Enders G., Sprengel R., Peters N., Varmus H.E., Ganem D. Expression of the precore region of an avian hepatitis B virus is not required for viral replication. J Virol 1987; 61:3322-3325.

55 Chen H.S., Kew M.C., Hornbuckle W.E., Tennant B.C., Cote P.J., Gerin J.L., Purcell R.H., Miller R.H. The precore gene of the woodchuck hepatitis virus genome is not essential for viral replication in the natural host. J Virol 1992; 66:5682-5684.

56 Milich D.R., Jones J.E., Hughes J.L., Price J., Raney A.K., McLachlan A. Is a function of the secreted hepatitis B e antigen to induce immunologic tolerance *in utero*? Proc Natl Acad Sci USA 1990; 87:6599-6603.

57 Seifer M., Standring D.N. Assembly and antigenicity of hepatitis B virus core particles. Intervirology 1995; 38:47-62.

58 Nassal M. The arginine-rich domain of the hepatitis B virus core protein is required for pregenome encapsidation and productive viral positive-strand DNA synthesis but not for virus assembly. J Virol 1992; 66:4107-4116.

59 Kock J., Wieland S., Blum H.E., von Weizsacker F. Duck hepatitis B virus nucleocapsids formed by N-terminally extended or C-terminally truncated core proteins disintegrate during viral DNA maturation. J Virol 1998;. 72:9116-9120.

60 Hatton T., Zhou S., Standring D.N. RNA- and DNA-binding activities in hepatitis B virus capsid protein: a model for their roles in viral replication. J Virol 1992; 66:5232-5241.

61 Pogam S.L., Yuan T.T., Sahu G.K., Chatterjee S., Shih C. Low-level secretion of human hepatitis B virus virions caused by two independent, naturally occurring mutations (P5T and L60V) in the capsid protein. J Virol 2000; 74:9099-9105.

62 Yuan T.T., Sahu G.K., Whitehead W., Greenberg R., Shih C. The mechanism of an "immature secretion" phenotype of a highly frequent naturally occurring missense mutation at codon 97 of human hepatitis B virus core antigen. J Virol 1999; 73:5731-5740.

63 Liao W., Ou J.-H. Phosphorylation and nuclear localization of the hepatitis B virus core protein: significance of serine in the three repeated SPRRR motifs. J Virol 1995; 69:1025-1029.

64 Scaglioni P.P., Melegari M., Wands J.R. Characterization of hepatitis B virus core mutants that inhibit viral replication. Virology 1994; 205:112-120.

65 Dyson M.R., Murray K. Selection of peptide inhibitors of interactions involved in complex protein assemblies: Association of the core and surface antigens of hepatitis B virus. Proc Natl Acad Sci USA 1995; 92:2194-2198.

66 Bottcher B., Tsuji N., Takahashi H., Dyson M.R., Zhao S., Crowther R.A., Murray K. Peptides that block hepatitis B virus assembly: analysis by cryomicroscopy, mutagenesis and transfection. EMBO J. 1998;17:6839-6845.

67 King R.W., Ladner S.K., Miller T.J., Zaifert K., Perni R.B., Conway S.C., Otto J.J. Inhibition of human hepatitis B virus replication by AT-61, a phenylpropenamide

derivative, alone and in combination with (-)β-L-2',3'-dideoxy-3'-thiacytidine. Antimicrob Agents Chemother 1998; 42:3179-3186.

68 Heerman K.H., Gerlich W.H. "Surface proteins of hepatitis B viruses." In *Molecular Biology of the Hepatitis B virus*, A. McLachlan, ed., Boca Raton, FL: CRC, pp 109-143. 1991;

69 Kuroki K., Eng F., Ishikawa T., Turck C., Harada F., Ganem D. gp180, a host cell glycoprotein that binds duck hepatitis B virus particles, is encoded by a member of the carboxypeptidase gene family. J Biol Chem 1995; 270:15022-15028.

70 Tong S., Li, J., Wands J.R. Interaction between duck hepatitis B virus and a 170-kilodalton cellular protein is mediated through a neutralizing epitope of the pre-S region and occurs during viral infection. J Virol 1995; 69:7106-7112.

71 Tong S., Li J., Wands J.R. Carboxypeptidase D is an avian hepatitis B virus receptor. J Virol 1999; 73:8696-8702.

72 Urban S., Breiner K.M., Fehler F., Klingmuller U., Schaller H. Avian hepatitis B virus infection is initiated by the interaction of a distinct pre-S subdomain with the cellular receptor gp180. J Virol 1998; 72:8089-8097.

73 Urban S., Schwarz C., Marx U.C., Zentgraf H., Schaller H., Multhaup G. Receptor recognition by a hepatitis B virus reveals a novel mode of high affinity virus-receptor interaction. EMBO J 2000; 19:1217-1227.

74 Deng H., Liu R., Ellmeier W., Choe S., Unutmaz D., Burkhart M., Di M.P., Marmon S., Sutton R.E., Hill C.M., Davis C.B., Peiper S.C., Schall T.J., Littman D.R., Landau N.R. Identification of a major co-receptor for primary isolates of HIV-1. Nature (London) 1996; 381:661-666.

75 Feng Y., Broder C.C., Kennedy P.E., Berger E.A. HIV-1 entry cofactor: functional cDNA cloning of a seven-transmembrane, G protein-coupled receptor. Science 1996; 272:872-877.

76 Mehta A., Lu X., Block T.M., Blumberg B.S., Dwek R.A. Hepatitis B virus (HBV) envelope glycoproteins vary drastically in their sensitivity to glycan processing: Evidence that alteration of a single N-linked glycosylation site can regulate HBV secretion. Proc Natl Acad Sci USA 1997; 94:1822-1827.

77 Block T.M., Lu X., Platt F.M., Foster G.R., Gerlich W.H., Blumberg B.S., Dwek R.A. Secretion of human hepatitis B virus is inhibited by the imino sugar N-butyldeoxynojirimycin. Proc Natl Acad Sci USA 1994; 91:2235-2239.

78 Block T.M., Lu X., Mehta A.S., Blumberg B.S., Tennant B., Ebling M., Korba B., Lansky D.M., Jacob G.S., Dwek R.A. Treatment of chronic hepadnavirus infection in a woodchuck animal model with an inhibitor of protein folding and trafficking. Nature Med 1998; 4:610-614.

79 Blum H.E., Zhang Z.S., Galun E., von Weizsacker F., Garner B., Liang T.J., Wands J.R. Hepatitis B virus X protein is not central to the viral life cycle in vitro. J Virol 1992; 66:1223-1227.

80 Yaginuma K., Shirakata Y., Kobayashi M., Koike K. Hepatitis B virus (HBV) particles are produced in a cell culture system by transient expression of transfected HBV DNA. Proc Natl Acad Sci USA 1987; 84:2678-2682.

81 Zoulim F., Saputelli J., Seeger C. Woodchuck hepatitis virus X protein is required for viral infection in vivo. J Virol 1994; 68:2026-2030.

82 Chen H.S., Kaneko S., Girones R., Anderson R.W., Hornbuckle W.E., Tennant B.C., Cote P.J., Gerin J.L., Purcell R.H., Miller R.H. The woodchuck hepatitis virus X gene is important for establishment of virus infection in woodchucks. J Virol 1993; 67:1218-1226.

83 Chisari F.V., Ferrari C. Hepatitis B virus immunopathogenesis. Annu Rev Immunol 1995; 13:29-60.

84 Bertoletti A., Maini M.K. Protection or damage: a dual role for the virus-specific cytotoxic T lymphocyte response in hepatitis B and C infection? Curr Opin Microbiol 2000; 3:387-392.

85 Bertoletti A., Ferrari C., Fiaccadori F., Penna A., Margolskee R., Schlicht H.J., Fowler P., Guilhot S., Chisari F.V. HLA class I-restricted human cytotoxic T cells recognize endogenously synthesized hepatitis B virus nucleocapsid antigen. Proc Natl Acad Sci USA 1991; 88:10445-10449.

86 Nayersina R., Fowler P., Guilhot S., Missale G., Cerny A., Schlicht H.J., Vitiello A., Chesnut R., Person J.L., Redeker A.G., et al. HLA A2 restricted cytotoxic T lymphocyte responses to multiple hepatitis B surface antigen epitopes during hepatitis B virus infection. J Immunol 1993; 150:4659-71.

87 Maini M.K., Boni C., Ogg G.S., King A.S., Reignat S., Lee C.K., Larrubia J.R., Webster G.J., McMichael A.J., Ferrari C., Williams R., Vergani D., Bertoletti A. Direct ex vivo analysis of hepatitis B virus-specific CD8(+) T cells associated with the control of infection. Gastroenterology 1999; 117:1386-96.

88 Rehermann B., Fowler P., Sidney J., Person J., Redeker A., Brown M., Moss B., Sette A., Chisari F.V. The cytotoxic T lymphocyte response to multiple hepatitis B virus polymerase epitopes during and after acute viral hepatitis. J Exp Med 1995; 181:1047-1058.

89 Bertoletti A., Costanzo A., Chisari F.V., Levrero M., Artini M., Sette A., Penna A., Guiberti A., Fiaccadori F., Ferrari C. Cytotoxic T lymphocyte response to a wild-type hepatitis B virus carrying substitutions within the epitope. J Exp Med 1994; 180:933-943.

90 Guidotti L.G., Ishikawa T., Hobbs M.V., Matzke B., Schreiber R., Chisari F.V. Intracellular inactivation of the hepatitis B virus by cytotoxic T lymphocytes. Immunity 1996; 4:25-36.

91 Guidotti L.G., Rochford R., Chung J., Shapiro M., Purcell R., Chisari F.V. Viral clearance without destruction of infected cells during acute HBV infection. Science 1999; 284:825-9.

92 Webster G.J., Reignat S., Maini M.K., Whalley S.A., Ogg G.S., King A., Brown D., Amlot P.L., Williams R., Vergani D., Dusheiko G.M., Bertoletti A. Incubation phase of acute hepatitis B in man: dynamic of cellular immune mechanisms. Hepatology 2000; 32:1117-24.

93 Biron C.A., Nguyen K.B., Pien G.C., Cousens L.P., Salazar-Mather T.P. Natural killer cells in antiviral defense: function and regulation by innate cytokines. Ann Rev Immunol 1999; 17:189-220.

94 Guo J.T., Zhou H., Liu C., Aldrich C., Saputelli J., Whitaker T., Barrasa M.I., Mason W.S., Seeger C. Apoptosis and regeneration of hepatocytes during recovery from transient hepadnavirus infections. J Virol 2000; 74:1495-1505.

95 Kajino K., Jilbert A.R., Saputelli J., Aldrich C.E., Cullen J., Mason W.S. Woodchuck hepatitis virus infections: very rapid recovery after a prolonged viremia and infection of virtually every hepatocyte. J Virol 1994; 68:5792-5803.

96 Livingston B.D., Crimi C., Grey H., Ishioka G., Chisari F.V., Fikes J., Grey H., Chesnut R.W., Sette A. The hepatitis B virus-specific CTL responses induced in humans by lipopeptide vaccination are comparable to those elicited by acute viral infection. J Immunol 1997; 159:1383-139.

97 Heathcote J., McHutchison J., Lee S., Tong M., Benner K., Minuk G., Wright T., Fikes J., Livingston B., Sette A., Chestnut R. A pilot study of the CY-1899 T-cell vaccine in subjects chronically infected with hepatitis B virus. Hepatology 1999; 30:531-536.

98 Pol S., Michel M.L., Brechot C. Immune therapy of hepatitis B virus chronic infection. Hepatology 2000; 31:548-549.

99 Gurunathan S., Wu C.Y., Freidag B.L., Seder R.A. DNA vaccines: a key for inducing long-term cellular immunity. Curr Opin Immunol 2000; 12:442-447.

100 Rollier C., Sunyach C., Barraud L., Madani N., Jamard C., Trepo C., Cova L. Protective and therapeutic effect of DNA-based immunization against hepadnavirus large envelope protein. Gastroenterology 1999; 116:658-665.

101 Roy M.J., Wu M.S., Barr L.J., Fuller J.T., Tussey L.G., Speller S., Culp J., Burkholder J.K., Swain W.F., Dixon R.M., Widera G., Vessey R., King A., Ogg G., Gallimore A., Haynes J.R., Heydenburg F.D. Induction of antigen-specific CD8+ T cells, T helper cells, and protective levels of antibody in humans by particle-mediated administration of a hepatitis B virus DNA vaccine. Vaccine 2000; 19:764-778.

102 Musch E., Hogemann B., Gerritzen A., Fischer H.P., Wiese M., Kruis W., Malek M., Gugler R., Schmidt G., Huchzermeyer H., Gerlach U., Dengler H.J., Sauerbruch T. Phase II clinical trial of combined natural interferon-beta plus recombinant interferon-gamma treatment of chronic hepatitis B. Hepatogastroenterology 1998; 45:2282-2294.

103 Lau J.Y., Lai C.L., Wu P.C., Chung H.T., Lok A.S., Lin H.J. A randomised controlled trial of recombinant interferon-gamma in Chinese patients with chronic hepatitis B virus infection. J Med Virol 1991; 34:184-187.

104 Cavanaugh V.J., Guidotti L.G., Chisari F.V. Interleukin-12 inhibits hepatitis B virus replication in transgenic mice. J Virol 1997; 71:3236-3243.

105 Carreno V., Zeuzem S., Hopf U., Marcellin P., Cooksley W.G., Fevery J., Diago M., Reddy R., Peters M., Rittweger K., Rakhit A., Pardo M. A phase I/II study of recombinant human interleukin-12 in patients with chronic hepatitis B. J Hepatol 2000; 32:317-324.

106 Schalm S.W., Heathcote J., Cianciara J., Farrell G., Sherman M., Willems B., Dhillon A., Moorat A., Barber J., Gray D.F. Lamivudine and alpha interferon combination treatment of patients with chronic hepatitis B infection: a randomised trial. Gut 2000; 46:562-568.

107 Boni C., Bertoletti A., Penna A., Cavalli A., Pilli M., Urbani S., Scognamiglio P., Boehme R., Panebianco R., Fiaccadori F., Ferrari C. Lamivudine treatment can restore T cell responsiveness in chronic hepatitis B. J Clin Invest 1998; 102:968-975.

108 Sidwell R.W., Huffman J.H., Khare G.P., Allen L.B., Witknowski J.T., Robins R.K. Broad-spectrum antiviral activity of Virazole: 1-β-D-ribofuranosyl-1,2,4-triazole-3-carboxamide. Science 1972; 177:705-706.

109 Tam R.C., Pai B., Bard J., Lim C., Averett D.R., Phau U.T., Milovanovic T. Ribavirin polarizes human T cell responses towards a Type 1 cytokine profile. J Hepatol 1999; 30:376-382.

110 Tam R.C., Lim C., Bard J., Pai B. Contact hypersensitivity responses following ribavirin treatment in vivo are influenced by Type 1 cytokine polarization, regulation of IL-10 expression, and costimulatory signaling. J Immunol 1999; 163:3709-3717.

111 Hultgren C. Milich D.R., Weiland O., Sallberg M. The antiviral compound ribavirin modulates the T helper (Th) 1/Th2 subset balance in hepatitis B and C virus-specific immune responses. J Gen Virol 1998; 79:2381-2391.

112 Korba B.E., Cote P., Hornbuckle W., Tennant B.C., Gerin J.L. Treatment of chronic woodchuck hepatitis virus infection in the eastern woodchuck (*Marmota monax*) with nucleoside analogues is predictive of therapy for chronic hepatitis B virus infection in humans. Hepatology 2000;. 31:1165-1175.

113 Markland W., McQuaid T.J., Jain J., Kwong A.D. Broad-spectrum antiviral activity of the IMP dehydrogenase inhibitor VX-497: a comparison with ribavirin and

demonstration of antiviral additivity with alpha interferon. Antimicrob Agents Chemother 2000; 44:859-866.

114 Fried M.W., Fong T.L., Swain M.G., Park Y., Beames M.P., Banks S.M., Hoofnagle J.H., Di Bisceglie A.M. Therapy of chronic hepatitis B with a 6-month course of ribavirin. J Hepatol 1994; 21:145-150.

115 Kakumu S., Yoshioka K., Wakita T., Ishikawa T., Takayanagi M., Higashi Y. Pilot study of ribavirin and interferon-β for chronic hepatitis B. Hepatology 1993; 18:258-263.

116 Galban-Garcia E., Vega-Sanchez H., Gra-Oramas B., Jorge-Riano J.L., Soneiras-Perez M., Haedo-Castro D., Rolo-Gomez F., Lorenzo-Morejon I., Ramos-Sanchez V. Efficacy of ribavirin in patients with chronic hepatitis B. J Gastroenterol 2000; 35:347-352.

117 Cotonat T., Quiroga J.A., Lopez-Alcorocho J.M., Clouet R., Pardo M., Manzarbeitia F., Carreno V. Pilot study of combination therapy with ribavirin and interferon alfa for the retreatment of chronic hepatitis B e antibody-positive patients. Hepatology 2000; 31:502-506.

118 Tam R.C., Ramasamy K., Bard J., Pai B., Lim C., Averett D.R. The ribavirin analog ICN 17261 demonstrates reduced toxicity and antiviral effects with retention of both immunomodulatory activity and reduction of hepatitis-induced serum alanine aminotransferase levels. Antimicrob Agents Chemother 2000; 44:1276-1283.

119 Houghton M. "Hepatitis C viruses." In *Fields Virology*, Fields B.N., Knipe D.M., Howley P.M., eds., New York: Raven Press, 1996; pp 1035-1058.

120 Rice C.M. "Flaviviridae: the viruses and their replication." In *Fields Virology*, Fields B.N., Knipe D.M., Howley P.M., eds. New York: Raven Press, 1996; pp 931-960.

121 Seeff L.B. Natural history of hepatitis C. Am J Med 1999; 107:10S-15S.

122 Saito I., Miyamura T., Ohbayashi A., Harada H., Katayama T., Kikuchi S., Watanabe T.Y., Koi S., Onji M., Ohta Y., Choo Q.-L., Houghton M., Kuo G. Hepatitis C virus infection is associated with the development of hepatocellular carcinoma. Proc Natl Sci Acad USA 1990; 87:6547-6549.

123 World Health Organization. Hepatitis C. Seroprevalence of hepatitis C virus (HCV) in a population sample. Wkly Epidemiol Rec 1996; 71:346-349.

124 McHutchison J.G., Gordon S.C., Schiff E.R., Shiffman M.L., Lee W.M., Rustgi V.K., Goodman Z.D., Ling M.H., Cort S., Albrecht J.K. Interferon alfa-2b alone or in combination with ribavirin as initial treatment for chronic hepatitis C. Hepatitis Interventional Therapy Group. N Engl J Med 1998; 339:1485-1492.

125 Davis G.L., Esteban-Mur R., Rustgi V., Hoefs J., Gordon S.C., Trepo C., Shiffman M.L., Zeuzem S., Craxi A., Ling M.H., Albrecht J. Interferon alfa-2b alone or in combination with ribavirin for the treatment of relapse of chronic hepatitis C. N Engl J Med 1998; 339:1493-1499.

126 Poynard T., Marcellin P., Lee S.S., Niederau C., Minuk G.S., Ideo G., Bain V., Heathcote J., Zeuzem S., Trepo C., Albrecht J. Randomised trial of interferon alpha2b plus ribavirin for 48 weeks or for 24 weeks versus interferon alpha2b plus placebo for 48 weeks for treatment of chronic infection with hepatitis C virus. Lancet 1998; 352:1426-1432.

127 Reichard O., Norkrans G., Fryden A., Braconier J.H., Sonnerborg A., Weiland O., Group T.S.S. Randomised, double-blind, placebo-controlled trial of interferon alpha-2b with and without ribavirin for chronic hepatitis C. Lancet 1998; 351:83-87.

128 Goodbourn S., Didcock L., Randall R.E. Interferons: cell signalling, immune modulation, antiviral response and virus countermeasures. J Gen Virol 2000; 81:2341-2364.

129 Davis G.L., Balart L.A., Schiff E.R., Lindsay K., Bodenheimer H.C.J., Perrillo R.P., Carey W., Jacobson I.M., Payne J., Dienstag J.L.. Treatment of chronic hepatitis C

with recombinant interferon alfa. A multicenter randomized, controlled trial. N Engl J Med 1989; 321:1501-1506.

130 Bartenschlager R. Candidate targets for hepatitis C virus-specific antiviral therapy. Intervirology 1997; 40:378-393.

131 Davis G.L., Lau J.Y.N. Factors predictive of a beneficial response to therapy of hepatitis C. Hepatology 1997; 26:122S-127S.

132 Patterson J.L., Fernandez-Larsson R. Molecular mechanisms of action of ribavirin. Rev Infect Dis 1990; 12:1132-1146.

133 Dusheiko G., Main J., Thomas H., Reichard O., Lee C., Dhillon A., Rassam S., Fryden A., Reesink H., Bassendine M., Norkrans G., Cuypers T., Lelie N., Telfer P., Watson J., Weegink C., Sillikens P., Weiland O. Ribavirin treatment for patients with chronic hepatitis C: results of a placebo-controlled study. J Hepatol 1996; 25:591-598.

134 Bodenheimer H.C.J., Lindsay K.L., Davis G.L., Lewis J.H., Thung S.N., Seeff L.B. Tolerance and efficacy of oral ribavirin treatment of chronic hepatitis C: a multicenter trial. Hepatology 1997; 26:473-477.

135 Lee J.H., von Wagner M., Roth W.K., Teuber G., Sarrazin C., Zeuzem S. Effect of ribavirin on virus load and quasispecies distribution in patients infected with hepatitis C virus. J Hepatol 1998; 29:29-35.

136 Ramasamy K.S., Tam R.C., Bard J., Averett D.R. Monocyclic L-nucleosides with type 1 cytokine-inducing activity. J Med Chem 2000; 43:1019-1028.

137 Cramp M.E., Rossol S., Chokshi S., Carucci P., Williams R., Naoumov N.V. Hepatitis C virus-specific T-cell reactivity during interferon and ribavirin treatment in chronic hepatitis C. Gastroenterology 2000; 118:346-355.

138 Lohmann V., Overton H., Bartenschlager R. Selective stimulation of hepatitis C virus and pestivirus NS5B RNA polymerase activity by GTP. J Biol Chem 1999; 274:10807-10815.

139 Smith J.P. Treatment of chronic hepatitis C with amantadine. Dig Dis Sci 1997; 42:1681-1687.

140 Carlsson T., Lindahl K., Schvarcz R., Wejsta L.R., Uhnoo I., Shev S., Reichard O. HCV RNA levels during therapy with amantadine in addition to interferon and ribavirin in chronic hepatitis C patients with previous nonresponse or response/relapse to interferon and ribavirin. J Virol Hepat 2000; 7:409-413.

141 Zeuzem S., Teuber G., Naumann U., Berg T., Raedle J., Hartmann S., Hopf U.. Randomized, double-blind, placebo-controlled trial of interferon alfa2a with and without amantadine as initial treatment for chronic hepatitis C. Hepatology 2000; 32:835-841.

142 Fong T.L., Fried M.W., Clarke-Platt J. A pilot study of rimantadine for patients with chronic hepatitis C unresponsive to interferon therapy. Am J Gastroenterol 1999; 94:990-993.

143 Khalili M., Denham C., Perrillo R. Interferon and ribavirin versus interferon and amantadine in interferon nonresponders with chronic hepatitis C. Am J Gastroenterol 2000;95:1284-1289.

144 Jubin R., Murray M.G., Howe A.Y., Butkiewicz N., Hong Z., Lau J.Y. Amantadine and rimantadine have no direct inhibitory effects against hepatitis C viral protease, helicase, ATPase, polymerase, and internal ribosomal entry site-mediated translation. J Infect Dis 2000; 181:331-334.

145 O'Brien C.B., Moonka D.K., Henzel B.S. A randomized, double-blind, placebo-controlled study of multiple, ascending doses of recombinant human interleukin-12 (rIL-12) in patients with chronic hepatitis C (HCV) who previously failed therapy with interferon (IFN). Hepatology 1998; 28:574A (Abstract).

146 O'Brien C.B., Henzel B.S., Moonka D.K. Dosing kinetics of single ascending doses of recombinant human interleukin-12 (rIL-12) subcutaneously administered to adults with chronic hepatitis C infection previously treated with interferon-alpha. Hepatology 1998; 28:575A (Abstract).

147 Choo Q.-L., Kuo G., Weiner A.J., Overby L.R., Bradley D.W., Houghton M. Isolation of a cDNA clone derived from a blood-born Non-A, Non-B viral hepatitis genome. Science 1989; 244:359-364.

148 Bartenschlager R., Lohmann V. Replication of hepatitis C virus. J Gen Virol 2000; 81:1631-1648.

149 Hong Z., Beaudet-Miller M., Lanford R.E., Guerra B., Wright-Minogue J., Skelton A., Baroudy B.M., Reyes G.R., Lau J.Y.N. Generation of Transmissible Hepatitis C Virions from a Molecular Clone in Chimpanzees. Virolology 1999; 256:36-44.

150 Kolykhalov A.A., Agapov E.V., Blight K.J., Mihalik K., Feinstone S.M., Rice C.M. Transmission of hepatitis C by intrahepatic inoculation with transcribed RNA. Science 1997; 277:570-574.

151 Yanagi M., Purcell R.H., Emerson S.U., Bukh J. Transcripts from a single full-length cDNA clone of hepatitis C virus are infectious when directly transfected into the liver of a chimpanzee. Proc Natl Acad Sci USA 1997; 94:8738-8743.

152 Beard M.R., Abell G., Honda M., Carroll A., Gartland M., Clarke B., Suzuki K., Lanford R., Sangar D.V., Lemon S.M. An infectious molecular clone of a Japanese genotype 1b hepatitis C virus. Hepatology 1999; 30:316-624.

153 Yanagi M., Purcell R.H., Emerson S.U., Bukh J. Hepatitis C virus: an infectious molecular clone of a second major genotype (2a) and lack of viability of intertypic 1a and 2a chimeras. Virology 1999; 262:250-263.

154 Lohmann V., Korner F., Koch J.-O., Herian U., Theilmann L., Bartenschlager R. Replication of subgenomic hepatitis C virus RNAs in a hepatoma cell line. Science 1999; 285:110-113.

155 Pietschmann T., Lohmann V., Rutter G., Kurpanek K., Bartenschlager R. Characterization of cell lines carrying self-replicating hepatitis C virus RNAs. J Virol 2001; 75:1252-1264.

156 Blight K.J., Kolykhalov A.A., Rice C.M. Efficient initiation of HCV RNA replication in cell culture. Science 2000; 290:1972-1975.

157 Lohmann V., Korner F., Dobierzewska A., Bartenschlager R. Mutations in hepatitis C virus RNAs conferring cell culture adaptation. J Virol 2001; 73:1437-1449.

158 Beames B., Chavez D., Guerra B., Notvall L., Brasky K.M., Lanford R.E. Development of a primary tamarin hepatocyte culture system for GB virus-B: a surrogate model for hepatitis C virus. J. Virol. 2000; 74:11764-11772.

159 Muerhoff A.S., Leary T.P., Simons J.N., Pilot-Matias T.J., Dawson G.J., Erker J.C., Chalmers M.L., Schlauder G.G., Desai S.M., Mushahwar I.K. Genomic organization of GB viruses A and B: two new members of the Flaviviridae associated with GB agent hepatitis. J. Virol. 1995; 69:5621-5630.

160 Simons J.N., Pilot-Matias T.J., Leary T.P., Dawson G.J., Desai S.M., Schlauder G.G., Muerhoff A.S., Erker J.C., Buijk S.L., Chalmers M.L., Van Sant C.L., Mushahwar I.K. Identification of two flavivirus-like genomes in the GB hepatitis agent. Proc Natl Acad Sci USA 1995; 92:3401-3405.

161 Bukh J., Apgar C.L., Yanagi M. Toward a surrogate model for hepatitis C virus: an infectious molecular clone of the GB virus-B hepatitis agent. Virology 1999; 262:470-478.

162 Mendez E., Ruggli N., Collett M.S., Rice C.M. Infectious bovine viral diarrhea virus (strain NADL) RNA from stable cDNA clones: a cellular insert determines NS3 production and viral cytopathogenicity. J Virol 1998; 72:4737-4745.

163 Vassilev V.B., Collett M.S., Donis R.O. Authentic and chimeric full-length genomic cDNA clones of bovine viral diarrhea virus that yield infectious transcripts. J Virol 1997; 71:471-478.

164 Collett M.S., Larson R., Gold C., Strick D., Anderson D.K., Purchio A.F. Molecular cloning and nucleotide sequence of the pestivirus bovine viral diarrhea virus. Virology 1988; 165:191-199.

165 Failla C., Tomei L., De Francesco R. Both NS3 and NS4A are required for proteolytic processing of hepatitis C virus nonstructural proteins. J Virol 1994; 68:3753-3760.

166 Kim J.L., Morgenstern K.A., Lin C., Fox T., Dwyer M.D., Landro J.A., Chambers S.P., Markland W., Lepre C.A., O'Malley E.T., Harbeson S.L., Rice C.M., Murcko M.A., Caron P.R., Thomson J.A. Crystal structure of the hepatitis C virus NS3 protease domain complexed with a synthetic NS4A cofactor peptide. Cell 1996; 87:343-355.

167 Lin C., Thomson J.A., Rice C.M. A central region in the hepatitis C virus NS4A protein allows formation of an active NS3-NS4A serine proteinase complex *in vivo* and *in vitro*. J Virol 1995; 69:4373-4380.

168 Bartenschlager R., Ahlborn-Laake L., Mous J., Jacobsen H. Nonstructural protein 3 of the hepatitis C virus encodes a serine-type proteinase required for cleavage at the NS3/4 and NS4/5 junctions. J Virol 1993; 67:3835-3844.

169 Grakoui A., McCourt D.W., Wychowski C., Feinstone S.M., Rice C.M. Characterization of the hepatitis C virus-encoded serine proteinase: determination of the proteinase-dependent polyprotein cleavage sites. J Virol 1993; 67:2832-2843.

170 Hijikata M., Mizushima H., Akagi T., Mori S., Kakiuchi N., Kato N., Tanaka T., Kimura K., Shimotohno K. Two distinct proteinase activities required for the processing of a putative nonstructural precursor protein of hepatitis C virus. J Virol 1993; 67:4665-4675.

171 Steinkuhler C., Urbani A., Tomei L., Biasiol G., Sardana M., Bianchi E., Pessi A., De Francesco R. Activity of purified hepatitis C virus protease NS3 on peptide substrates. J Virol 1996; 70:6694-6700.

172 Kolykhalov A.A., Mihalik K., Feinstone S.M., Rice C.M. Hepatitis C virus-encoded enzymatic activities and conserved RNA elements in the 3' nontranslated region are essential for virus replication in vivo. J Virol 2000; 74:2046-2051.

173 Love R.A., Parge H.E., Wickersham J.A., Hostomsky Z., Habuka N., Moomaw E.W., Adachi T., Hostomska Z. The crystal structure of hepatitis C virus NS3 proteinase reveals a trypsin-like fold and a structural zinc binding site. Cell 1996; 87:331-342.

174 Yan Y., Li Y., Munshi S., Sardana V., Cole J.L., Sardana M., Steinkuehler C., Tomei L., De Francesco R., Kuo L.C., Chen Z. Complex of NS3 protease and NS4A peptide of BK strain hepatitis C virus: a 2.2 A resolution structure in a hexagonal crystal form. Protein Sci 1998; 7:837-847.

175 Yao N., Reichert P., Taremi S., Prosise W.W., Weber P.C. Molecular views of viral polyprotein processing revealed by the crystal structure of the hepatitis C virus bifunctional protease-helicase. Structure 1999; 7:1353-1363.

176 De Francesco R., Urbani A., Nardi M.C., Tomei L., Steinkuhler C., Tramontano A. A zinc binding site in viral serine proteinases. Biochem 1996; 35:13282-13287.

177 Sali D.L., Ingram R., Wendel M., Gupta D., McNemar C., Tsarbopoulos A., Chen J.W., Hong Z., Chase R., Risano C., Zhang R., Yao N., Kwong A.D., Ramanathan L., Le H.V., Weber P.C. Serine protease of hepatitis C virus expressed in insect cells as the NS3/4A complex. Biochem 1998; 37:3392-3401.

178 Narjes F., Brunetti M., Colarusso S., Gerlach B., Koch U., Biasiol G., Fattori D., De Francesco R., Matassa V.G., Steinkuhler C. Alpha-ketoacids are potent slow binding inhibitors of the hepatitis C virus NS3 protease. Biochemistry 2000; 39:1849-1861.

179 Ingallinella P., Altamura S., Bianchi E., Taliani M., Ingenito R., Cortese R., De Francesco R., Steinkuhler C., Pessi A. Potent peptide inhibitors of human hepatitis C virus NS3 protease are obtained by optimizing the cleavage products. Biochem 1998; 37:8906-8914.

180 Steinkuhler C., Biasiol G., Brunetti M., Urbani A., Koch U., Cortese R., Pessi A., De Francesco R. Product inhibition of the hepatitis C virus NS3 protease. Biochem 1998; 37:8899-8905.

181 Bianchi E., Orru S., Dal Piaz F., Ingenito R., Casbarra A., Biasiol G., Koch U., Pucci P., Pessi A. Conformational changes in human hepatitis C virus NS3 protease upon binding of product-based inhibitors. Biochemistry 1999; 38:13844-13852.

182 Sudo K., Matsumoto Y., Matsushima M., Konno K., Shimotohno K., Shigeta S., Yotoka T. Novel HCV protease inhibitors: 2,4,6-trihydroxy-3-nitro-benzamide derivatives. Antiviral Chemistry and Chemotherapy 1997; 8:541-544.

183 Sudo K., Matsumoto Y., Matsushima M., Fujiwara M., Konno K., Shimotohno K., Shigeta S., Yokota T. Novel hepatitis C virus protease inhibitors: thiazolidine derivatives. Biochem Biophys Res Commun 1997; 238:643-647.

184 Kakiuchi N., Komoda Y., Komoda K., Takeshita N., Okada S., Tani T., Shimotohno K. Non-peptide inhibitors of HCV serine proteinase. FEBS Letters 1998; 421:217-220.

185 Takeshita N., Kakiuchi N., Kanazawa T., Komoda Y., Nishizawa M., Tani T., Shimotohno K. An enzyme-linked immunosorbent assay for detecting proteolytic activity of hepatitis C virus proteinase. Anal Biochem 1997; 247:242-246.

186 Chu M., Mierzwa R., He L., King A., Patel M., Pichardo J., Hart A., Butkiewicz N., Puar M.S. Isolation and structure of Sch 351633: a novel (HCV) NS3 serine protease inhibitor from the fungus *Penicillium griseofulvum*. Bioorganic and Medicinal Chemistry Letters 1999; 9:1949-1952.

187 Chu M., Mierzwa R., Truumees I., King A., Patel M., Berrie R., Hart A., Butkiewicz N., Dasmahapatra B., Chan T., Puar M.S. Structure of Sch 68631: a new HCV proteinase inhibitor from Streptomyces sp. Tetrahedron Letters 1996; 37:7229-7232.

188 Kim D.W., Gwack Y., Han J.H., Choe J. C-terminal domain of the hepatitis C virus NS3 protein contains an RNA helicase activity. Biochem Biophys Res Commun 1995; 215:160-166.

189 Kim D.W., Kim J., Gwack Y., Han J.H., Choe J. Mutational analysis of the hepatitis C virus RNA helicase. J Virol 1997; 71:9400-9409.

190 Kim D.W., Gwack Y., Han J.H, Choe J. Towards defining a minimal functional domain for NTPase and RNA helicase activities of the hepatitis C virus NS3 protein. Virus Res 1997; 49:17-25.

191 Suzich J.A., Tamura J.K., Palmer-Hill F., Warrener P., Grakoui A., Rice C.M., Feinstone S.M., Collett M.S. Hepatitis C virus NS3 protein polynucleotide-stimulated nucleoside triphosphatase and comparison with the related pestivirus and flavivirus enzymes. J Virol 1993; 67:6152-6158.

192 Tai C.L., Chi W.K., Chen D.S., Hwang L.H. The helicase activity associated with hepatitis C virus nonstructural protein 3. J Virol 1996; 70:8477-8484.

193 Gwack Y., Kim D.W., Han J.H., Choe J. Characterization of RNA binding activity and RNA helicase activity of the hepatitis C virus NS3 protein. Biochem Biophys Res Commun 1996; 225:654-659.

194 Hong Z., Ferrari E., Wright-Minogue J., Chase R., Risano C., Seelig G., Lee C.-G., Kwong A.D. Enzymatic characterization of hepatitis C virus NS3/4A complexes

expressed in mammalian cells using the herpes simplex virus amplicon system. J Virol 1996; 70:4261-4268.

195 Gu B., Liu C., Lin-Goerke J., Maley D.R., Gutshall L.L., Feltenberger C.A., Del Vecchio A.M. The RNA helicase and nucleotide triphosphatase activities of the bovine viral diarrhea virus NS3 protein are essential for viral replication. J Virol 2000; 74:1794-1800.

196 Grassmann C.W., Isken O., Behrens S.-E. Assignment of the multifunctional NS3 protein of bovine viral diarrhea virus during RNA replication: an in vitro and in vivo study. J Virol 1999; 73:9196-9205.

197 Kim J.L., Morgenstern K.A., Griffith J.P., Dwyer M.D., Thomson J.A., Murcko M.A., Lin C., Caron P.R. Hepatitis C virus NS3 RNA helicase domain with a bound oligonucleotide: the crystal structure provides insights into the mode of unwinding. Structure 1998; 6:89-100.

198 Yao N., Hesson T., Cable M., Hong Z., Kwong A.D., Le H.V., Weber P.C. Structure of the hepatitis C virus RNA helicase domain. Nature Struct Biol 1997; 4:463-467.

199 Diana G.D., Bailey T.R., Nitz T.J. 1997; WO97/36554. .

200 Diana G.D., Bailey T.R. 1997; US Patent 5,633,388. .

201 Lesburg C.A., Cable M.B., Ferrari E., Hong Z., Mannarino A.F., Weber P.C. Crystal structure of the RNA-dependent RNA polymerase from hepatitis C virus reveals a fully encircled active site. Nature Struct Biol 1999; 6:937-943.

202 Bressanelli S., Tomei L., Roussel A., Incitti I., Vitale R.L., Mathieu M., De Francesco R., Rey F.A. Crystal structure of the RNA-dependent RNA polymerase of hepatitis C virus. Proc Natl Acad Sci USA 1999; 96:13034-13039.

203 Ago H., Adachi T., Yoshida A., Yamamoto M., Habuka N., Yatsunami K., Miyano M. Crystal structure of the RNA-dependent RNA polymerase of hepatitis C virus. Structure Fold Des 1999; 7:1417-26.

204 Ferrari E., Wright-Minogue J., Fang J.W.S., Baroudy B.M., Lau J.Y.N., Hong Z. Characterization of soluble hepatitis C virus RNA-dependent RNA polymerase expressed in *Escherichia coli*. J Virol 1999; 73:1649-1654.

205 Behrens S.-E., Tomei L., De Francesco R. Identification and properties of the RNA-dependent RNA polymerase of hepatitis C virus. EMBO J 1996; 15:12-22.

206 Ishii K., Tanaka Y., Yap C.C., Aizaki H., Matsuura Y., Miyamura T. Expression of hepatitis C virus NS5B protein: characterization of its RNA polymerase activity and RNA binding. Hepatology 1999; 29:1227-35.

207 Lohmann V., Korner F., Herian U., Bartenschlager R. Biochemical properties of hepatitis C virus NS5B RNA-dependent RNA polymerase and identification of amino acid sequence motifs essential for enzymatic activity. J Virol 1997; 71:8416-8428.

208 Lohmann V., Roos A., Korner F., Koch J.O., Bartenschlager R. Biochemical and kinetic analyses of NS5B RNA-dependent RNA polymerase of the hepatitis C virus. Virology 1998; 249:108-118.

209 Oh J.W., Ito T., Lai M.M. A recombinant hepatitis C virus RNA-dependent RNA polymerase capable of copying the full-length viral RNA. J Virol 1999; 73:7694-702.

210 Zhong W., Uss A.S., Ferrari E., Lau J.Y., Hong Z. De novo initiation of RNA synthesis by hepatitis C virus nonstructural protein 5B polymerase. J Virol 2000; 74:2017-22.

211 Zhong W., Ingravallo P., Wright-Minogue J., Uss A.S., Skelton A., Ferrari E., Lau J.Y., Hong Z. Template/primer requirements and single nucleotide incorporation by hepatitis C virus nonstructural protein 5B polymerase. J Virol 2000; 74:9134-9143.

212 Hong Z., Cameron C.E., Lau J.Y.N., Zhong W. Hepatitis C virus NS5B polymerase employs a novel mechanism to ensure terminal initiation during replication. 2001; Submitted .

213 Hijikata M., Mizushima H., Tanji Y., Komoda Y., Hirowatari Y., Akagi T., Kato N., Kimura K., Shimotohno K. Proteolytic processing and membrane association of putative nonstructural proteins of hepatitis C virus. Proc Natl Acad Sci USA 1993; 90:10773-10777.

214 Pieroni L., Santolini E., Fipaldini C., Pacini L., Migliaccio G., La Monica N. In vitro study of the NS2-3 protease of hepatitis C virus. J Virol 1997; 71:6373-6380.

215 Wu Z., Yao N., Le H., Weber P.C. Mechanism of autoproteolysis at the NS2-NS3 junction of the hepatitis C virus polyprotein. TIBS 1998; 23:92-94.

216 Agnello V., Abel G., Elfahal M., Knight G.B., Zhang Q.-X. Hepatitis C virus and other *Flaviviridae* viruses enter cells via low density lipoprotein receptor. Proc Natl Acad Sci USA 1999; 96:12766-12771.

217 Pileri P., Uematsu Y., Campagnoli S., Galli G., Falugi F., Petracca R., Weiner A.J., Houghton M., Rosa D., Grandi G., Abrignani S. Binding of Hepatitis C Virus to CD81. Science 1998; 282:938-941.

218 Tsukiyama-Kohara K., Iizuka N., Kohara M., Nomoto A. Internal ribosome entry site within hepatitis C virus RNA. J Virol 1992; 66:1476-1483.

219 Roberts E.C., Malmstrom T.A., Pavco P.A. Synthesis and testing of nuclease resistant hammerhead ribozymes directed against hepatitis C virus RNA. Hepatology 1998; 28:398A.

220 Macejak D.G., Jensen K.L., Bellon L., Pavco P.A., Blatt L.M. Inhibition of viral replication by nuclease resistant hammerhead ribozymes directed against hepatitis C virus RNA. Hepatology 1999; 30:abstract 995.

221 Macejak D.G., Jensen KL, F. J.S., Domenico K., Roberts E.C., Chaudhary N., von Carlowitz I., Bellon L., Tong M.J., Conrad A., Pavco P.A., Blatt L.M. Inhibition of hepatitis C virus (HCV)-RNA-dependent translation and replication of a chimeric HCV poliovirus using synthetic stabilized ribozymes. Hepatology 2000; 31:769-776.

222 Lee P.A., Blatt L.M., Blanchard K.S., Bouhana K.S., Pavco P.A., Bellon L., Sandberg J.A. Pharmacokinetics and tissue distribution of a ribozyme directed against hepatitis C virus RNA following subcutaneous or intravenous administration in mice. Hepatology 2000; 32:640-646.

223 Ikeda M., Sakai T., Tsuai S., Zuao I., Ryan H., Iyan S., Kai Y., Kako Y., Tsukada I., Yanagisawa M. 1996; JP-08268890. .

224 Kai Y., Sasho M., Kaku Y., Tsukada I., Yanagisawa M. 1996; JP-1010591. .

225 Crotty S., Maag D., Arnold J.J., Zhong W., Lau J.Y., Hong Z., Andino R., Cameron C.E. The broad-spectrum antiviral ribonucleoside ribavirin is an RNA virus mutagen. Nature Medicine 2000; 6:1375-1379.

226 Lau J.Y., King R., Tibbs C.J., Catterall A.P., Smith H.M., Portmann B.C., Alexander G.J., Williams R. Loss of HBsAg with interferon-alpha therapy in chronic hepatitis D virus infection. J Med Virol 1993; 39:292-296.

227 Hoofnagle J.H. Therapy of acute and chronic viral hepatitis. Adv Intern Med 1994; 39:241-275.

228 Saracco G., Rizzetto M. A practical guide to the use of interferons in the management of hepatitis virus infections. Drugs 1997; 53:74-85.

229 Battegay M., Simpson L.H., Hoofnagle J.H., Sallie R., Di Bisceglie A.M. Elimination of hepatitis delta virus infection after loss of hepatitis B surface antigen in patients with chronic delta hepatitis. J Med Virol 1994; 44:389-92.

230 Lau D.T., Doo E., Park Y., Kleiner D.E., Schmid P., Kuhns M.C., Hoofnagle J.H. Lamivudine for chronic delta hepatitis. Hepatology 1999; 30:546-9.

231 Wolters L.M., van Nunen A.B., Honkoop P., Vossen A.C., Niesters H.G., Zondervan P.E., de Man R.A. Lamivudine-high dose interferon combination therapy for chronic hepatitis B patients co-infected with the hepatitis D virus. J Viral Hepat 2000; 7:428-34.

232 Glenn J.S., Watson J.A., Havel C.M., White J.M. Identification of a prenylation site in delta virus large antigen. Science 1992; 256:1331-1333.

233 Marsters J.C., Jr., McDowell R.S., Reynolds M.E., Oare D.A., Somers T.C., Stanley M.S., Rawson T.E., Struble M.E., Burdick D.J., Chan K.S., et al. 1994; Benzodiazepine peptidomimetic inhibitors of farnesyl-transferase. Bioorg Med Chem 2:949-957.

234 Glenn J.S., Marsters J.C., Jr., Greenberg H.B. Use of a prenylation inhibitor as a novel antiviral agent. J Virol 1998; 72:9303-9306.

235 Fattovich G., Boscaro S., Noventa F., Pornaro E., Stenico D., Alberti A., Ruol A., Realdi G. Influence of hepatitis delta virus infection on progression to cirrhosis in chronic hepatitis type B. J Infect Dis 1997; 155:931-935.

236 Samuel D., Zignego A.L., Reynes M., Feray C., Arulnaden J.L., David M.F., Gigou M., Bismuth A., Mathieu D., Gentilini P., Benhamou J., Brechot C., Bismuth H. Long-term clinical and virological outcome after liver transplantation for cirrhosis caused by chronic delta hepatitis. Hepatology 1995; 21:333-9.

237 Marzano A., Salizzoni M., Rizzetto M. Liver transplantation in viral hepatitis. New insights. Acta Gastroenterol Belg 1999; 62:342-347.

238 Bradley D.W. Hepatitis E: Epidemiology, aetiology and molecular biology. Med Virol 1992; 2:19-28.

239 Hyams K.C., Yarbough P.O., Gray S., Callahan J., Gotuzzo E., Gutierrez J., Vasquez P.B., Hayes C.G., Watts D.M. Hepatitis E virus infection in Peru. Clin Infect Dis 1996; 22.

240 Pujol F.H., Favarov M.O., Marcaano T., Este J.A., Magris M., Liprandi F., Khudyakov Y.E., Khydyakova N.S., Fields H.A. Prevalence of antibodies against Hepatitis E virus among urban and rural populations in Venezuela. J Med Virol 1994; 42.

241 Quiroga J.A., Cotonat T., Castillo I., Carreno V. Hepatitis E virus seroprevalence in acute viral hepatitis in a developed country confirmed by a supplemental assay. J Med Virol 1996; 50:16-19.

242 Zaaijer J.L., Kok M., Lelie P.N., Timmerman R.J., Chau K., van der Pal H.J.H. Hepatitis E in the Netherlands: Imported and endemic. Lancet 1993; 341:826.

243 Purcell R.H., Ticehurst J.R. "Enterically transmitted non-A, non-B hepatitis: epidemiology and clinical characteristics.". In *Viral Hepatitis and Liver Disease*, Zuckerman A., ed., New York: Alan R. Liss, 1988; pp 131-137.

244 Joshi Y.K., Babu S., Sarin S., Tandon B.N., Gandhi B.M., Chaturvedi V.C. Immunoprophylaxis of epidemic non-A non-B hepatitis. Indian J Med Res 1985; 81:18-19.

245 Khuroo M.S., Dar M.Y. Hepatitis E: evidence for person-to-person transmission and inability of low dose immune serum globulin from an Indian source to prevent it. Indian J Gastroenterol 1992; 11:113-116.

246 Reyes G.R., Purdy M.A., Kim J., Luk K.C., Young L.M., Fry K.M., Bradley D.W. Isolation of cDNA from the virus responsible for enterically transmitted non-A, non-B hepatitis. Science 1990; 247:1335-1339.

247 Tam A.W., Smith M.M., Guerra M.E., Huang C.C., Bradley D.W., Fry K.E., Reyes G.R. Hepatitis E Virus (HEV): Molecular cloning and sequencing of the full-length viral genome. Virology 1991; 185:120-131.

248 Purdy M.A., McCaustland K.A., Krawezynski K., Spelbring J., Reyes G.R., Bradley D.W. Preliminary evidence that a trpE-HEV fusion protein protects cynomolgus

macaques against challenge with wild-type hepatitis E virus (HEV). J Med Virol 1993; 41:90-94.

249 Yarbough P.O., Krawezynski K., Tam A.W., McAtee C.P., McCaustland K.A., Zhang Y., Garcon N., Spellbring J., Carson D., Myriam F., Lifson J.D., Slaoui M., Prieels J.P., Margolis H., Fuerst T.R. "Prevention of hepatitis E using r62K subunit vaccine: Full protection against heterologous HEV challenge in cynomolgus macaques." In *Viral Hepatitis and Liver Disease*, Rizzetto M., Purcell R.H., Gerin J.L., Verme G., eds.., Turin: Turin Edizioni Minerva Medica, 1997; pp 650-655.

Index

A

C

D

E

F

G

H

I

J

L

M

N

P

R

S

T

V

W